Facies interpretation and the stratigraphic record

Facies interpretation and the stratigraphic record

A. HALLAM
University of Birmingham

W. H. Freeman and Company
Oxford and San Francisco

W. H. Freeman and Company Limited
20 Beaumont Street, Oxford, OX1 2NQ
660 Market Street, San Francisco, California 94104

Library of Congress Cataloging in Publication Data

Hallam, Anthony.
Facies interpretation and the stratigraphic record.

Bibliography: p.
Includes indexes.
1. Facies (Geology) 2. Geology, Stratigraphic.
I. Title.
QE651.H18 551.7 80–24276
ISBN 0-7167-1291-1

Phototypeset in V.I.P. Times by
Western Printing Services Ltd, Bristol
Printed in the United States of America

Preface

The record of sedimentary strata and their contained fossils is a veritable treasure house for those who wish to investigate Earth history but considerable disagreement exists about how best to teach the subject of stratigraphy in university courses. Some see the need for a rigorous system-by-system treatment with emphasis on correlation and nomenclature. This, however, can all too easily degenerate into a dry recitation of facts unrelieved by any serious reference to concepts; a dull catalogue of rock formation and fossil names, together with legalistic definitions, tends both to bore and bemuse the great majority of students. The alternative is to lay emphasis on environmental interpretation—what has been called dynamic stratigraphy or 'stratigraphy without tears'. Unfortunately much of the choicest material is often siphoned off, as it were, for specialist sedimentology or palaeoecology courses, with only broad and sometimes rather nebulous generalization left for the stratigraphy course.

There is a quite commonly held view, often expressed in a slightly pejorative way, that stratigraphy is one of the 'classical' divisions of geology. In earlier times it has occupied a central role in university courses. This dominant position has been gradually eroded over the years as other more specialized subjects have been introduced. Modern advances in Earth science are often thought to relate largely to the development of new research fields. Another widely held view is that a subject is somehow more scientific if more or less complex apparatus is required for its study. Stratigraphy is not usually associated with much instrumentation, though in reality instrumentation has played an increasingly important role through the last few decades, with gamma-ray and electric logging equipment, magnetometers and mass spectrometers (for magnetic and isotope stratigraphy) and advanced equipment for drilling and seismic reflection studies. Nevertheless it remains true that much useful work can still be done with long-established simple tools such as the hammer and petrographic microscope. In the words of Samuel Butler, 'there's many a good tune played on an old fiddle'. In actual fact the complexity of the instrumentation is quite irrelevant to how scientific a given subject is, because the essence of science is the critical evaluation and testing of conceptual models.

I feel strongly that at this stage in our science the bugbear of subject fragmentation should be more actively resisted. The pioneering work in sedimentology and palaeoecology has frequently demanded a narrowing of focus for effective pursuit of the goals. There are, however, limits to what can be done further for such research to be both original and significant. In these days of rapid communication research results quickly become assimilated into the body of knowledge and readily absorbed into undergraduate teaching. To make further advances we must 'play back' information into the stratigraphic mainstream in order to tackle both classical and newly posed questions of major geological importance. There is no longer much justification for palaeontologists ignoring sediments or sedimentologists disregarding both fossils and stratigraphy. Both groups must also pay due attention to new advances in geotectonics and oceanography.

What in effect I am calling for is a greater acknowledgement of the lessons that should have been learned as a consequence of the Earth sciences revolution of the late 1960s and early 1970s. We saw clearly then the value of interdisciplinary research, when such disparate groups as palaeontologists, petrologists and geophysicists came closer together in the pursuit of common goals than they had ever been before.

Lyell's *Principles of Geology* opens eloquently with the following statement:

> Geology is the science which investigates the successive changes that have taken place in the organic and inorganic kingdoms of Nature; it enquires into the causes of these changes, and the influence which they have exerted in modifying the surface and external structure of our planet.

This statement remains just as true today and points up the central role of stratigraphy, which in its broader sense is synonymous with historical geology, and is by its very nature an integrative discipline. A capacity to bring together pertinent information from different research fields, to focus on questions of Earth history, is the hallmark of a good geologist, requiring judgement as well as skill, besides the special outlook acquired from studying subjects in a stratigraphic context. This is something frequently denied those who enter geology after advanced training in physics, chemistry or biology, however accomplished they may be in their own subjects.

This book is intended as a contribution towards restoring stratigraphy to its rightful place as the core discipline of geology. Rather than endeavouring in the conventional way to slog through the stratigraphic column, I have preferred to select a number of important themes for examination, drawing upon relevant examples from a wide range of areas and ages. The book is designed primarily for advanced undergraduate and graduate courses, though I hope it may also be of some value to professional geologists who are too busy to read much literature outside their own specialist field. A basic knowledge is assumed of sedimentology, palaeontology and the principles of stratigraphic correlation. Stage terms are the *lingua franca* of stratigraphers and are frequently cited in this book, but as many are not especially familiar to students I have included in an appendix a list of the more widely used terms.

After an introductory chapter on the principles and techniques of facies analysis and two more giving a concise account of modern sedimentary environments and their ancient facies equivalents, seven major topics are dealt with chapter by chapter. Each chapter can be treated as a discrete entity but they are deliberately arranged in a certain order and there is a good deal of cross referencing. I hope in consequence that an overall coherence emerges, culminating in one of the grandest themes in the whole of Earth science—the relationship of organic evolution and distributions to environmental change. Although the term *facies analysis* is appropriate to the detailed studies on restricted topics undertaken by most individual researchers, I prefer *facies interpretation* in the book's title because the topics of chapters 4 to 10 also involve a considerable measure of synthesis.

All teachers should endeavour to strike a balance between instruction and stimulation. Since this book is not a treatise I have tried to avoid overwhelming the reader with data and have preferred to concentrate on particular examples, but ample references are provided for those who wish to pursue given topics in more detail. Stimulation is harder to guarantee, and one has to acknowledge that it often involves provoking to disagreement, but while I have a definite point of view on some subjects I have sought to highlight areas of uncertainty.

It should go without saying that I could not have attempted such a work without the encouragement and assistance of a large number of friends and colleagues, who have either stimulated me with their ideas, made constructive criticism of my own, provided me with information or directed me to relevant literature. I should like to single out the following for special mention: Nick Badham, Martin Bradshaw, Russell Coope, Al Fischer, Ken Hsü, John Hudson, John Imbrie, Hugh Jenkyns, Erle Kauffman, Gerry Middleton, Bruce Sellwood, Finn Surlyk, Peter Vail, Jerry Van Andel and Jim Valentine. I ought also to point out that it was Perce Allen who originally persuaded me to write a book on facies interpretation in a stratigraphic context. Naturally I bear the sole responsibility for what is written and can only hope that those cited may approve of the product. Finally, I am especially indebted to Christine Sturch for typing the manuscript.

A. HALLAM

January 1980

Acknowledgements

Thanks are due to the following societies, publishers and journals and to the authors concerned (who are acknowledged in the underlines) for permission to reproduce the illustrations listed below.

The Royal Society, Figs. 4.1, 7.12; The Greenland Geological Survey, Fig. 4.2; The Geological Society of London, Figs. 5.3, 5.10, 6.1, 6.9, 6.10, 8.7; Cambridge University Press, Figs 6.5, 7.6; *Nature*, Figs. 5.5, 10.11; Scottish Academic Press, Fig. 7.1; The Geologists Association, Figs. 4.14, 4.15; The Royal Dutch Geological and Mining Society, Fig. 8.6; The Systematics Association, Fig. 10.3; Academic Press Inc. Figs. 2.1, 2.3, 2.4; *Paleobiology*, Figs. 10.1, 10.8; University of Chicago Press, Figs. 6.14, 6.15, 10.5; *Scientific American*, Figs. 8.1, 9.1; *American Journal of Science*, Fig. 7.3; The American Geophysical Union, Fig. 7.10; The Society of Economic Paleontologists and Mineralogists, Figs. 1.1, 1.2, 3.13, 4.3, 4.10, 8.8, 8.9, 9.2; The American Association of Petroleum Geologists, Figs. 5.2, 5.12, 6.3, 6.7, 6.8, 6.11; Blackwell Scientific Publications Ltd, Figs. 2.7, 2.8, 2.9, 2.10, 2.14, 3.1, 3.2, 3.5, 3.9, 3.10, 3.11, 3.12, 4.5, 4.7, 4.13, 5.13, 5.14, 8.2, 8.3, 10.10, 10.12; The Geological Society of America, Figs. 2.12, 4.8, 4.9, 4.11, 4.12, 6.13, 6.16, 10.4; Elsevier Scientific Publishing Company, Figs. 1.3, 2.5, 3.6, 5.1, 5.4, 6.6, 7.2, 7.4, 7.5, 7.14, 8.4, 8.10, 10.6; Springer Verlag, Figs. 2.2, 2.6, 3.3; Prentice-Hall Inc., Fig. 3.4.

Contents

1. Principles and techniques of facies analysis

The concept of facies

Facies is a latin word meaning face, figure, appearance, aspect, look or condition, and thus signifies an abstract idea. It has been used by geologists in a variety of contexts, but we are only concerned here with its usage in the study of sedimentary rocks (Teichert 1958).

Its presently accepted meaning as the sum total of lithological and faunal characteristics of a given stratigraphical unit derives from the Swiss geologist Gressly (1838), who introduced the term in his description of Upper Jurassic strata in the region of Solothurn in the Jura Mountains. Prévost in France arrived at a similar concept at about the same time, but used the term *formation* as an equivalent of Gressly's facies. Subsequently the Austrian geologist Mojsisovics proposed the terms *isopic*, for rocks of the same, and *heteropic* for rocks of different facies but, useful as they appear to be, they have not been widely adopted.

The most penetrating discussion of facies relationships by a nineteenth century geologist was by the German Johannes Walther. Walther is celebrated today for his 'law' of correlation of facies (actually I prefer the term 'rule'—physics has *laws*, geology has *rules*; sorting out the exceptions is part of the fun). This so-called law, or rule, has been variously interpreted, and Middleton (1973) has endeavoured to sort out possible confusion by going back to Walther's original statement on the subject. The relevant passage (Walther 1894, p. 979) is translated by Middleton as follows:

> The various deposits of the same facies-area and similarly the sum of the rocks of different facies areas are formed beside each other in space, though in a cross section we see them lying on top of each other . . . this is a basic statement of far-reaching significance that only those facies and facies-areas can be superimposed primarily which can be observed beside each other at the present time.

This statement has been generally interpreted to indicate that facies occurring in a conformable vertical sequence of strata were formed in laterally adjacent environments. Although actualistic comparisons are demanded, Walther indicated elsewhere his awareness that there may be no close modern equivalents of some facies. He also stressed that his 'law' only applies to successions without major disconformities indicated by erosional surfaces and/or stratigraphic gaps.

Until the last few decades the facies concept was taken up more actively in continental Europe than in Great Britain and North America. For instance, it was applied with great success by Heim and others to unravelling the structural complexities of the Alps, by distinguishing different facies units, now juxtaposed, that must have been originally deposited in widely separated regions. In contrast the influential stratigraphic teaching early this century of Buckman in England and Ulrich in the United States very largely disregarded the significance of facies differences; it was almost axiomatic that rocks of different facies in a given region were to be treated as being of different age. The remarkable parallels between the thinking of these independently working palaeontologists are brought out by comparing the critical accounts of Buckman by Arkell (1933) and Ulrich by Dunbar and Rodgers (1957), which make salutary reading.

At present, of course, virtually everyone appreciates the significance of facies differences, but the term is still often used somewhat indiscriminately for inferred environments, e.g. geosynclinal, deltaic, bathyal facies. In the appropriate context this is perhaps not too objectionable but it nevertheless involves a degree of subjective judgement. It seems desirable therefore to make the usage more descriptive, e.g. bituminous shale, coral limestone facies, with a subsequent interpretation of environment. A distinction of *lithofacies* and *biofacies* is useful in referring respectively to the petrological and faunal or floral characteristics of stratigraphic units, though strictly speaking lithofacies is a synonym of facies, because fossils constitute part of the rock.

Techniques

Facies analysis is of necessity an integrative discipline, in which evidence from a variety of fields is brought to bear on the interpretation of ancient environments. No single research field is paramount, and each one has its particular strengths and weaknesses but obviously in particular cases some may be more important or relevant than others.

INORGANIC SEDIMENTARY STRUCTURES

The most striking features of many sedimentary rocks observed in the field are such structures as cross bedding, graded bedding, ripple marks and load and flute casts. It is therefore rather odd that their intensive study did not commence until about two decades ago, despite the notable pioneer work of Sorby in England and Gilbert in the United States, dating back into the last century. The present rise to dominance of this field of research, coincident with the widespread acknowledgement of *sedimentology* as a separate subdiscipline, is reflected in the large number of textbooks and monographs partly or entirely devoted to describing and interpreting sedimentary structures, of which the following are to be particularly recommended: Allen (1970), Blatt, Middleton and Murray (1979), Potter and Pettijohn (1977), Reineck and Singh (1973) and Selley (1976a).

The great value of studying such structures is that we can learn a great deal about the nature of the water or air movements responsible for transport and deposition of the sediment, about the energy level, current character and direction; also in some instances about the rate of sedimentation and sediment supply (thus sets of climbing ripples are generally held to signify a high rate). The distinctive structures and general lithological characteristics of tillites are the best indicators of glacial conditions. The major limitation of sedimentary structures is that they tell us little about the general character of the depositional environment, such as the depth, temperature and salinity of water. There are also uncertainties in distinguishing between the effects of traction and turbidity currents, wind-induced as opposed to tidal currents and aeolian as opposed to some shallow marine dunes.

GRAIN-SIZE DISTRIBUTION AND TEXTURES

Analysis of these features is a laboratory exercise based on sieving and microscopy. The grain-size distribution of sandy sediments has been studied for many years and it has become well established, for instance, that fluvial sands are more poorly sorted than beach sands. Over the years progressively more refined statistical techniques have been applied involving, in particular, the analysis of skewness and kurtosis (Friedman 1961, 1967, 1979; Visher 1969). Electron microscopic study of the surface textures of quartz sand grains can yield information on whether the environment of deposition was glacial, littoral or aeolian (Krinsley and Donahue 1968; Krinsley and Doornkamp 1973).

Unfortunately the great resistance to destruction of quartz sand grains allows them to be readily transported from one depositional regime to another and recycled from an older to a younger group of sediments. Diagenetic corrosion and crystalline overgrowths can considerably modify the original grain size and texture, and bioturbation, which is extremely common in marine environments, can play havoc with grain-size distributions. For reasons such as these, and because of the widespread recognition that sedimentary structures are usually more informative, the subject has tended to go into eclipse in recent years.

On the other hand, textural studies are of paramount importance in the study of carbonate deposits, because limestones and dolomites rarely reveal their secrets until examined under the microscope. These rocks are invariably the product of diagenesis, and analysis of diagenetic fabrics must precede any attempt to elucidate the original character of the sediment. The outstanding textbook on the subject is by Bathurst (1975).

MINERALOGY

Mineralogical analysis under the microscope is obviously necessary to distinguish between the various types of sandy rocks such as orthoquartzites, arkoses and greywackes and can give important information on sediment provenance and hinterland weathering as well as depositional environment (Greensmith 1978). Between the two world wars there was a vogue for the analysis of heavy minerals in sandstones, such as tourmaline, zircon, rutile, staurolite and kyanite. In fact heavy-mineral analysis became almost synonymous with sedimentary petrology. Whereas such study can in some favourable instances give useful information on sediment provenance, there are a number of serious snags. The more stable heavy minerals, like zircon, are strongly resistant to destruction and hence can survive several cycles of sedimentation. The less stable ones can be destroyed by diagenesis. Moreover the analysis is unusually laborious for the (often) modest results that are obtained. Thus today heavy-mineral analysis is very much a minority interest.

Whereas carbonates have a rather dull mineralogy—calcite and dolomite are the only minerals of importance—this is not true of chemically deposited sediments such as evaporites (which give reliable evidence of arid climates), ironstones and phosphorites, and accordingly mineralogical investigation assumes great importance for these rocks. The danger to guard against is assuming that the minerals examined under the microscope necessarily provide information about the depositional environment. Evaporite minerals are notoriously susceptible to diagenetic replacement (Stewart 1963) and precise inference of original depositional conditions of ironstone-forming environments on the basis of siderite, pyrite and magnetite distributions (e.g. Garrels 1960; James 1966) is unjustified because these minerals formed during diagenesis.

It is a commonplace that sedimentologists tend to fall clearly into two camps—students of sandstones and students of carbonates, which have little communion. Both groups tend to ignore the volumetrically most important sediments of all, the argillaceous rocks. Yet clay mineralogy by X-ray diffractometry is not an especially esoteric technique, and is well within the grasp of most sedimentologists, with such textbooks as Grim (1968) and Millot (1970) to provide a suitable background.

One advantage is that there are few clay minerals of geological importance. The commonest, illite, is not especially informative, but a high proportion of kaolinite is likely to signify the proximity of, or derivation of sediment from, a warm, humid landmass (Griffin *et al.* 1968; Hallam 1975). In contrast, clays deposited in saline, lacustrine and paralic environments of arid regimes are often characterized by such magnesium-enriched species as sepiolite and palygorskite (Jeans 1978). Clays largely or entirely composed of smectite and accessory zeolites are likely to signify a volcanic source (Hallam 1975; Jeans *et al.* 1977). An abundance of chlorite (not a true clay mineral) may indicate a comparatively unweathered source of metamorphic rocks, as at the present day in high latitudes on the North Atlantic margins (Griffin *et al.* 1968).

A few other clay minerals or minerals most suitably analysed by X-ray diffractometry deserve a mention. Glauconite appears, with a few doubtful exceptions, to be a good indicator of marine conditions and, with collophane (calcium apatite), an indicator of slow depositional rate. The aluminium oxides boehmite and diaspore are, with haematite, the principal constituents of the climatically significant family of bauxites and laterites.

GEOCHEMISTRY

In the 1960s considerable work was done on certain trace elements in argillaceous rocks with a view to assessing their value as palaeosalinity indicators. Potter *et al.* (1963) undertook a study of many samples of sediments in North America which exhibited a wide variety in terms of age, source area, tectonic situation, sedimentation rate and climate. Their results for modern and ancient sediments were broadly similar, suggesting little post-depositional alteration of trace element content. Very generally, boron, chromium, copper, gallium, nickel and vanadium were significantly more enriched in marine than in non-marine deposits.

Boron has attracted far more attention than any other trace element (Harder 1970). Although it seems to have value as a palaeosalinity index a number of limitations have become apparent. Its uptake varies with the mineralogy and grain size of the clay and with temperature. Furthermore a considerable time must elapse before the element is absorbed to capacity. Samples should be standardized with respect to mineralogy, grain size and diagenetic history. Boron content is evidently not sensitive to rapid salinity fluctuations, which are much better assessed by analysing fauna. The greatest value of analysing the content of boron and other trace elements such as those cited above is probably for rocks lacking fauna of any kind, such as Proterozoic sediments.

Oxygen-isotope ratios of fossil shells were first studied in the 1950s as a potential palaeotemperature tool. However, the ratio also varies with salinity, and it has since become increasingly evident that, also because of post-depositional alteration by diagenesis and the circulation through rocks of meteoric waters, the $^{18}O/^{16}O$ ratio is an unreliable guide to palaeotemperature in rocks now exposed on the continents (e.g. Hallam 1975). On the other hand foraminiferal shells found in deep-sea cores extending back to the late Cretaceous have yielded interesting and apparently reliable results, as will be outlined in chapter 7. Taken in conjunction, however, with $^{13}C/^{12}C$ ratios, oxygen isotopes can yield valuable information about varying salinities in marginal marine and non-marine environments, provided that the effects of diagenesis can be shown to be minimal (Keith *et al.* 1964; Tan and Hudson 1974). For pre-Tertiary limestones more generally, combined $^{18}O/^{16}O$ and $^{13}C/^{12}C$ analysis is probably most useful as a tool in the study of carbonate diagenesis (Hudson 1977a). Oxygen-isotope ratios of cherts, however, appear to offer promise as palaeotemperature indicators, even as far back as the Precambrian (Knauth and Epstein 1976).

FAUNA AND FLORA

Environmental interpretation is enormously facilitated for rocks in which fossils are present. Thus the occurrence of such stenohaline invertebrate groups as corals, brachiopods, echinoderms, cephalopods or bryozoans is sufficient to indicate marine conditions. More generally, fossils are by far the best salinity indicators we have (Fig. 1.1). Less reliably, they are also among the best indicators of seawater depth (Fig. 1.2). Inshore, marginal marine environments are characterized by reduced diversity combined usually with increased individual abundance of the constituent species, as compared with fully marine environments.

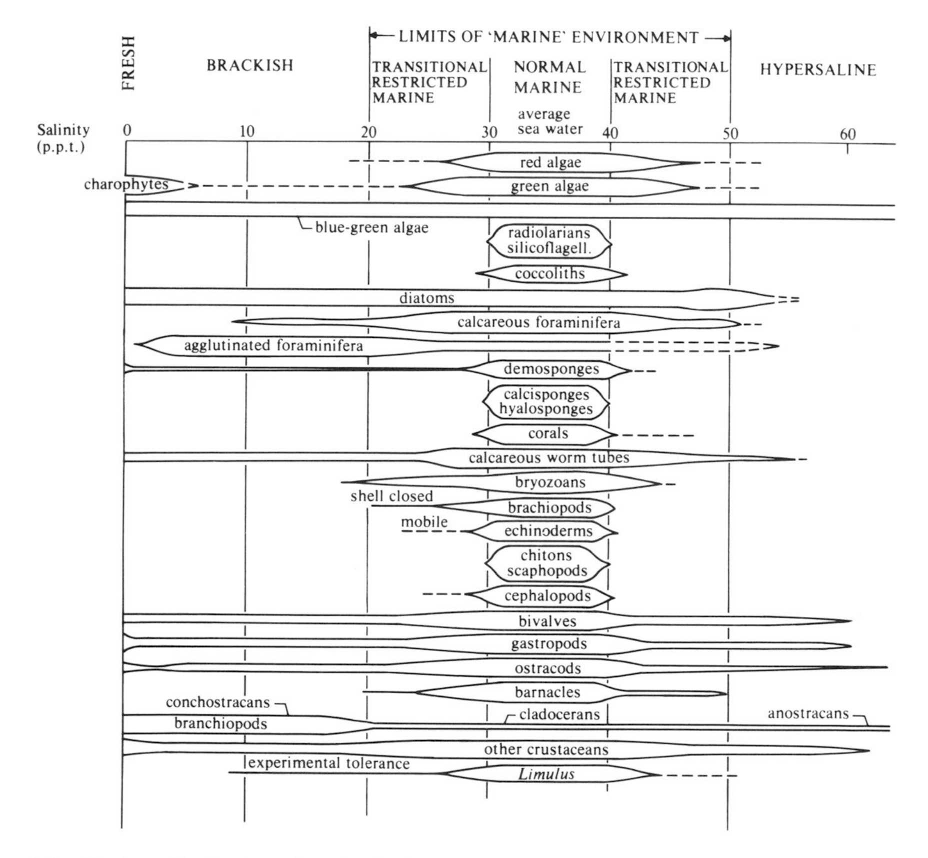

1.1 Modern distribution of major fossilizable invertebrate and algal groups in relation to salinity. After Heckel (1972).

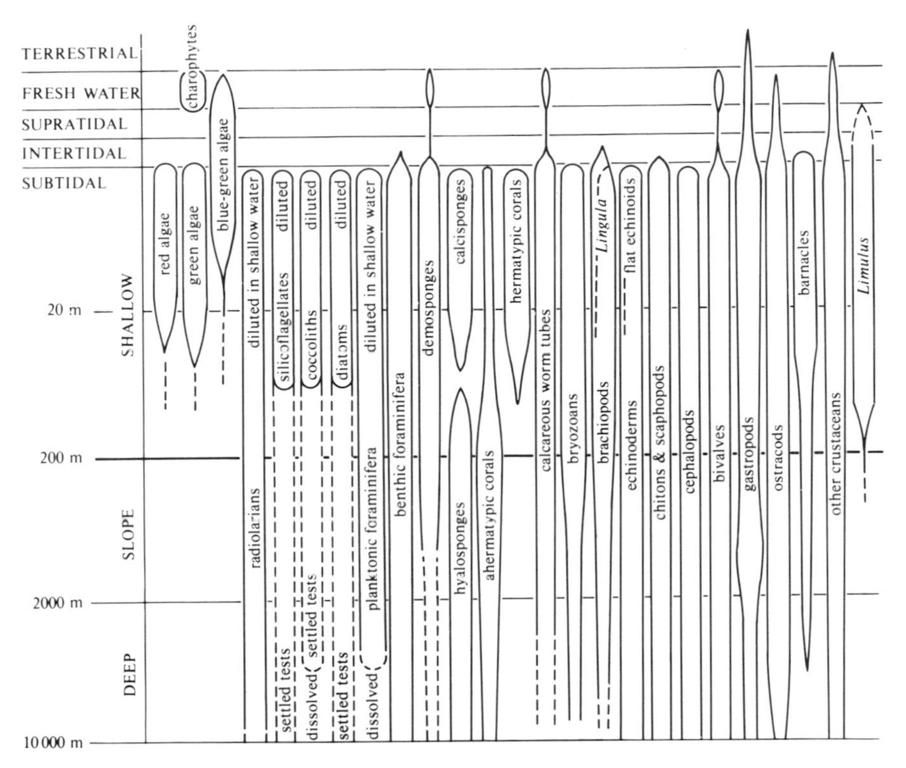

1.2 Modern distribution of major fossilizable invertebrate and algal groups in relation to depth of sea. After Heckel (1972).

Terrestrial plant fossils are the best available indicators of climate. If the floras are sufficiently young to have close modern relatives, temperatures can be established with remarkable precision (Wolfe 1978). Amongst marine invertebrates hermatypic corals are probably the best indicators of tropical conditions.

Biostratonomic studies of shell orientation, disarticulation and fragmentation provide valuable information on the movements of bottom waters in the depositional environment and particular types of sedimentary substrate are characterized by distinctive types of organisms, some of which are readily fossilizable (Rhoads and Young 1970; Schäfer 1972). Particularly valuable in these respects are the infilled dwelling and feeding burrows, surface trails and footprints known as *trace fossils*. The study of these sedimentary structures has expanded so much in recent years that some elevate it to a separate subdiscipline, *palaeoichnology* (Frey 1975). Trace fossils can inform on such elusive characteristics as firmness of substrate, rate of sedimentation, water movements and food supply, as well as more general environmental parameters such as salinity and depth of sea. Organically produced sedimentary structures which do not qualify as trace fossils, such as stromatolites, are among the best indicators of supratidal, intertidal or very shallow subtidal conditions.

Inevitably there are some limitations to the use of fossils as environmental indicators. Most obviously, they are absent from or extremely rare in many sediments. As we go further back in time we find progressively fewer fossils with living relatives whose ecological tolerances are known, and environmental inferences become correspondingly less precise. Similarly, the older the fossil the more doubtful becomes the assumption that ecological tolerance has persisted unchanged through time. Thus sessile benthic crinoids and the bivalve genus *Pholadomya* are present-day inhabitants of the deep sea and the bivalves *Astarte* and *Thracia* are now restricted to cold waters, whereas the general evidence of facies associations clearly indicates that back in the Jurassic they flourished in a warm, shallow neritic environment.

Hermatypic corals are probably good indicators of tropical, shallow waters within the photic zone back at least as far as the early Mesozoic, although hermatypic forms do not have any significant morphological characteristics in common to distinguish them from the bathymetrically more tolerant ahermatypic forms. Palaeozoic corals belong to quite different taxonomic groups, rendering of dubious validity any close ecological comparison with living forms, and the symbiotic algae (zooxanthellae) which restrict corals to the photic zone might not yet have evolved. Nevertheless, by taking into account the physical and chemical requirements for building substantial reef structures, together with the general facies association, we can infer with reasonable confidence that most Palaeozoic reef-building corals probably inhabited shallow waters.

In the same way no one seriously doubts that the long-extinct trilobites were marine organisms, because of their frequent association with representatives of phyla such as brachiopods and echinoderms that are exclusively marine at the present day.

Bathymetric estimations are notoriously difficult because depth of water is not an environmental variable *per se*, and we have to think in terms of depth-related factors such as carbonate solubility, incidence of light, food supply and so on (Hallam 1967a). The fact, for example, that the ratio of planktonic to benthic foraminifera in bottom sediment tends to increase with water depth correlates significantly with the more pelagic distribution of the planktonics and the greater rarity of benthics beyond the neritic zone (Funnell 1967). Not only are there exceptions to the rule, such as oceanic island margins, but it is impossible to acquire precise quantitative estimates of water depth in the past by this or any other technique using fossils.

Trace fossils provide an instructive case. They have been held to be among the best bathymetric indicators, and Seilacher (1967) has proposed several depth zones characterized by distinctive ichnogenera (Fig. 1.3). Whereas Seilacher's qualitative scheme appears to hold in many instances in the stratigraphic record there are notable exceptions. Thus *Zoophycos* is common in some English Carboniferous deposits such as the Yoredale beds which were undoubtedly deposited in very shallow water, because they include algal limestones and coals. Osgood (1970) gives further examples of *Zoophycos* occurring in shallow-water facies and also questions the existence of a sharp distinction between Seilacher's *Cruziana* and *Nereites* facies. Furthermore, Seilacher's deep-water *Nereites* facies is characterized by complex surface tracks including spirals and meanders, suggesting to him an efficient foraging behaviour adapted to a condition of minimal food supply. Photography at abyssal depths supports Seilacher's hypothesis for the Antarctic but not the Arctic Basin (Kitchell *et al.* 1978). Evidently other factors than simply depth and food supply are involved in controlling the distribution of the different trace-fossil facies.

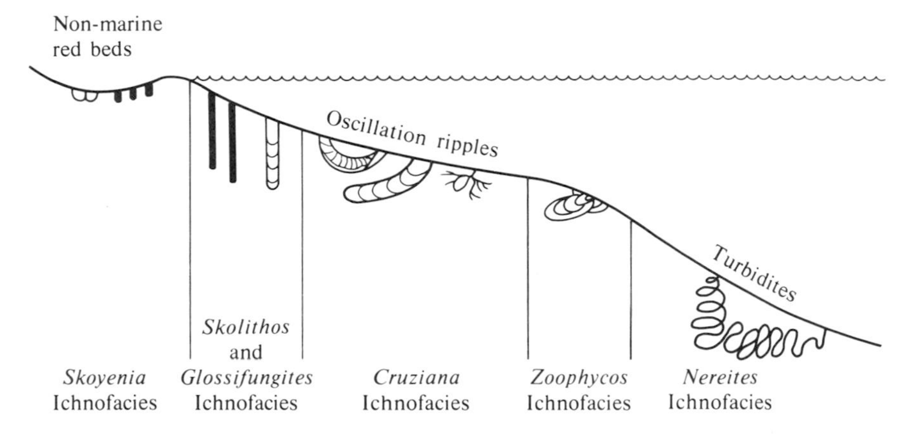

1.3 Bathymetric zonation of trace fossil assemblages according to Seilacher (1967).

STRATIGRAPHIC RELATIONSHIPS

This type of study is absolutely central to facies analysis if it is to be of any major significance for geology. A sedimentologist will not achieve successful interpretation of such interesting rocks as bituminous shales or oolitic ironstones without taking into consideration evidence from fossils and stratigraphy. No amount of intensive study of sedimentary structures in a sandstone formation and clay-mineral analysis of the overlying marine shale will resolve the really interesting question, which is whether the implied phase of water deepening was the result of local tectonic subsidence or eustatic rise of sea level. A major question to be asked of any major body of sandstone or reef limestone is: what are its spatial and temporal relationships? Is the sand body a lens, a sheet or an elongate 'stringer'? Is the reef mass elongate and hence a possible barrier? Are these lithological entities diachronous? For a geologist to divorce sedimentology from stratigraphy is about as sensible as trying to play football with one leg!

Nowadays there is increasing appreciation of the fact that sedimentation rates in given regions have varied enormously through time. Deposits laid down during marine transgressions are characteristically more condensed than those laid down during the intervening regressions. The breaks represented by modest-looking disconformities, marine hardgrounds or non-marine calcrete horizons may signal the passage of considerable time intervals. For particular rock sequences, such breaks may represent significantly more time than the intervening sediments (Ager 1973). The term *paraconformity* has been applied to widespread, palaeontologically significant breaks that leave little trace in the rock record (Newell 1967a). On a more detailed scale, many sedimentologists have come to recognize that the occasional violent hurricane or storm may have more important erosional and depositional consequences than more normal humdrum events observable most of the time.

There is no need to emphasize further what should be obvious, that a prime requirement of facies analysis is a sound stratigraphic framework based on a zonal subdivision of time units which is as refined as possible. Unfortunately it is rarely possible to achieve a high degree of stratigraphic precision in continental sequences because fossils are generally more scarce and of lower diversity than in marine sequences. They are also usually longer-ranging in time, though mammal remains in Tertiary deposits are a notable exception. Within a limited area, however, volcanic ash bands can be treated as precise time markers.

The newly developed technique of *magnetic polarity statigraphy* offers some hope for deposits laid down at times of comparatively frequent geomagnetic reversals. Thus it has been demonstrated for some Plio-Pleistocene terrestrial sequences that discontinuous lithological units can record remarkably uniform rates of sedimentation (Johnson *et al.* 1975). The new technique can also add precision to correlation between continental and marine deposits and it has been claimed on this basis that the mass extinction of dinosaurs and planktonic foraminifera at the end of the Cretaceous can be tied down to a time interval as brief (for geologists) as 10^5 years at most (Lerbekmo *et al.* 1979).

Facies models

The primary goal of facies analysis is to produce a *facies model* which is in effect a hypothesis about the environments signified by the rocks and fossils under study. While lengthy verbal presentation cannot be dispensed with it is very desirable to supplement it with diagrammatic illustrations, because one good picture can on occasion be more informative than a thousand words. Such diagrams take several forms. Purely spatial variations can be represented by palaeogeographic maps and temporal variations by stratigraphic columns with appropriate symbols for the rocks and contained fossils. Sometimes it is possible to combine the two and produce a block diagram illustrating variations in three dimensions. Examples of facies models are given by Blatt *et al.* (1979), Selley (1978) and Walker (1979) and more will be presented in the succeeding chapters. The recent book edited by Reading (1978) is an excellent and authoritative source of information on Recent sedimentary environments and their ancient facies analogues.

Actualistic comparisons are a necessary and vital part of facies analysis but, in attempting a more general interpretation of past environments we need to be clearer in our minds about what the notion of uniformitarianism really signifies. Gould (1965) has distinguished two separate strands of thought in Lyell's original concept:

1. A *methodological uniformitarianism* comprising the assumptions that natural laws are constant in space and time, and that no hypothetical unknown processes be invoked if observed historical results can be explained by presently observable processes.

2. A *substantive uniformitarianism*, which postulates a uniformity of material conditions or of rates of processes.

Now clearly what Gould calls methodological uniformitarianism is vital for interpreting the past. We have to assume, for example, that the ripples on a Cambrian sandstone bed formed under conditions that we can observe today. There is, however, no logical connection with substantive uniformitarianism, and no good reason to assume uniformity of rates of processes. Thus there may well have been times when rates of erosion on the continents were greater than now. Biological evolution is decidedly non-uniformitarian and must have effected considerable changes of environment. Geology is a historical science concerned with past configurations of the Earth, dealing with successions of unique, strictly unrepeatable events through time. In this it differs fundamentally from more theoretical sciences like physics, which deal with the finding and testing of universal laws.

Not only is the term uniformitarianism cumbersome, it is also misleading, and *actualism* is to be preferred. Actualistic comparisons require judgement as well as skill. Whereas the interpretation of a sandstone with trough cross bedding as the product of migrating sets of dunes may be essentially uncontroversial, a straightforward comparison between modern and ancient epicontinental seas might be seriously in error if the full implications of varying configurations of land and sea across the globe have not been adequately appreciated. Rather than uncritically repeating the old adage that 'the present is the key to the past' it might be more profitable to ask: to what extent is the present the key to the past?

A note on statistical techniques

Good science depends on careful observation, which demands precision and estimations of confidence in the results. Therefore simple quantification, backed up where relevant by measurement of standard deviations, has become routine, and is as pertinent to facies analysis as to any other field of geology. One school of geologists has, however, gone much further by applying ever-more sophisticated statistical models to the interpretation of stratigraphic data. The danger with this sort of research is that it may become too technical for the great majority to understand and one sometimes tends to get the impression that these 'mathematical geologists' communicate only with each other and get seduced away from significant geological problems by the beauty of the techniques they work with.

One also has to admit that nothing particularly momentous seems to have come from nearly two decades of such research. A common feeling is that a sledgehammer is being used to crack a nut, that a mountain of data processing and analysis has brought forth a mere molehill of a result, or at least a conclusion utterly unsurprising to someone well versed in the field. Nevertheless it is only fair to acknowledge that, provided the project is carefully thought out, based on a good prior understanding of the rocks under study, valuable results may emerge which could not have been obtained by less mathematically rigorous methods.

Thus sedimentary beds which appear to vary cyclically in sequence can be subjected to *Markov chain analysis* to determine departures from randomness or factors intrinsic to the sedimentational regime. Significant periodicities may be sought in varved sequences by *power spectrum analysis*, while in the study of spatial and temporal lithological trends it is more objective to smooth out randomly generated 'noise' by statistical means (Schwarzacher 1975).

2. Continental and marginal marine environments and facies

Because the deposits of continental and marginal marine environments are more accessible to study than those of fully marine environments, and because of the paucity of burrowing organisms that would destroy or disturb inorganic sedimentary structures, their conditions of formation are comparatively well understood. In consequence there is a wealth of information from which ancient facies equivalents can be interpreted with a generally high degree of confidence. On the other hand the general rarity of stratigraphically useful fossils, apart from pollen and spores, often makes precise correlation difficult if not impossible.

A conventional classification will be adopted, into alluvial, aeolian, lacustrine, glacial, and deltaic and coastal plain deposits.

Alluvial deposits

The predominantly siliciclastic deposits of alluvial regimes range considerably in coarseness, from the conglomerates of alluvial fans to the progressively sandier and muddier sediments laid down in braided and meandering river systems. There is a corresponding change in the pattern of sediment transport and deposition as base level is approached (Fig. 2.1).

Alluvial fans are localized deposits whose shape approximates the segments of a cone, which are laid down in areas of high relief where there is an abundant supply of sediment. They commonly coalesce to form a bajada and are often found along active fault scarps. Mostly they develop in regions of intense ephemeral flood discharge.

Collinson (1978a) has distinguished four types of deposit in alluvial fans:

1. *Debris flow deposits.* The region near the fan apex is subjected to high-density, high-viscosity mudflows that are strong enough to transport clasts up to boulder size and hence give rise to deposits that can superficially be confused with glacial boulder clay.

2. *Sheetflood deposits.* The moderately well-sorted sands and gravels deposited by sheet floods do not extend far from the downstream end of channels. Scouring is common but cross bedding not ubiquitous.

3. *Stream channel deposits.* These are lenticular-bedded, poorly sorted sands and gravels laid down predominantly in the upper reaches of fans under conditions of low-viscosity flow. The coarser layers may be imbricated and the sandier beds cross bedded. There is normally a channelled contact with the underlying sediment.

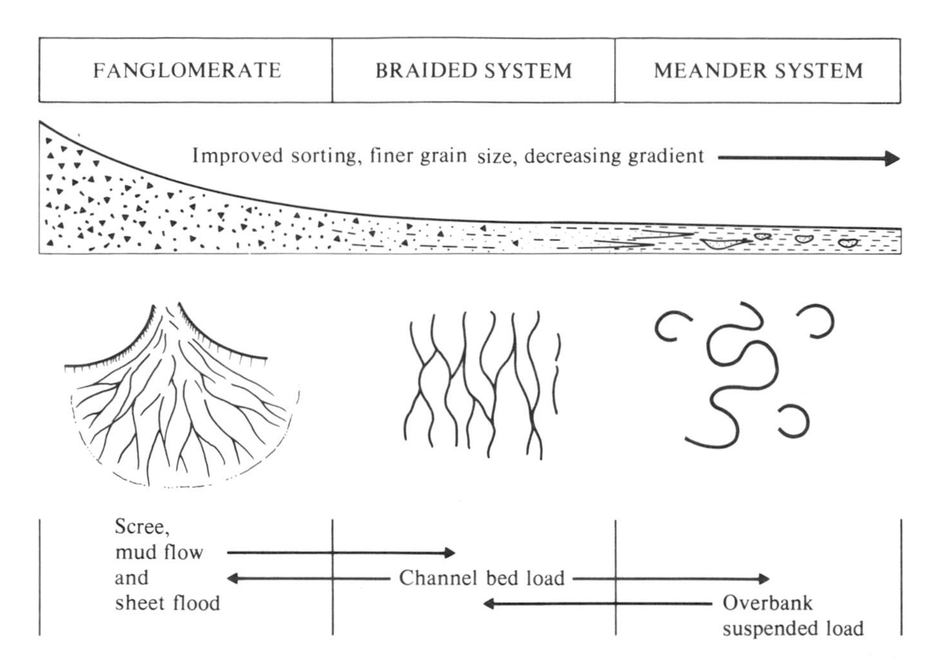

2.1 Changes in sedimentary processes and deposits from alluvial fans to braided and meandering channel systems. After Selley (1978).

4. *Sieve deposits.* A highly permeable older deposit causes flow to diminish rapidly as infiltration of water occurs, giving rise to a clast-supported gravel lobe. The resulting deposits are rather well sorted and poorly imbricated. In the course of burial the clast interstices are gradually filled with finer sediment.

Recognition of ancient alluvial fan deposits in the stratigraphic record is a comparatively straightforward matter, because there are few other types which have such an abundance of conglomerate (Bull 1972). Stratified conglomerates with little matrix probably signify the activity of stream channels and sheet floods while unstratified conglomerates with dispersed clasts in a muddy matrix signify mudflows (Fig. 2.2). Fining-upward sequences may have resulted from waning sediment supply as the source area was worn down. Coarsening-upward sequences are less common and probably resulted either from uplift of the source area and/or progradation of fan lobes. Good examples have been described from the Torridonian (late Precambrian) of north-west Scotland (Selley 1965; Williams 1969), the Permian of East Greenland (Collinson 1972) and the Triassic of north-west Scotland (Steel 1974).

Within the lower reaches of alluvial regimes, sand, silt and mud are the

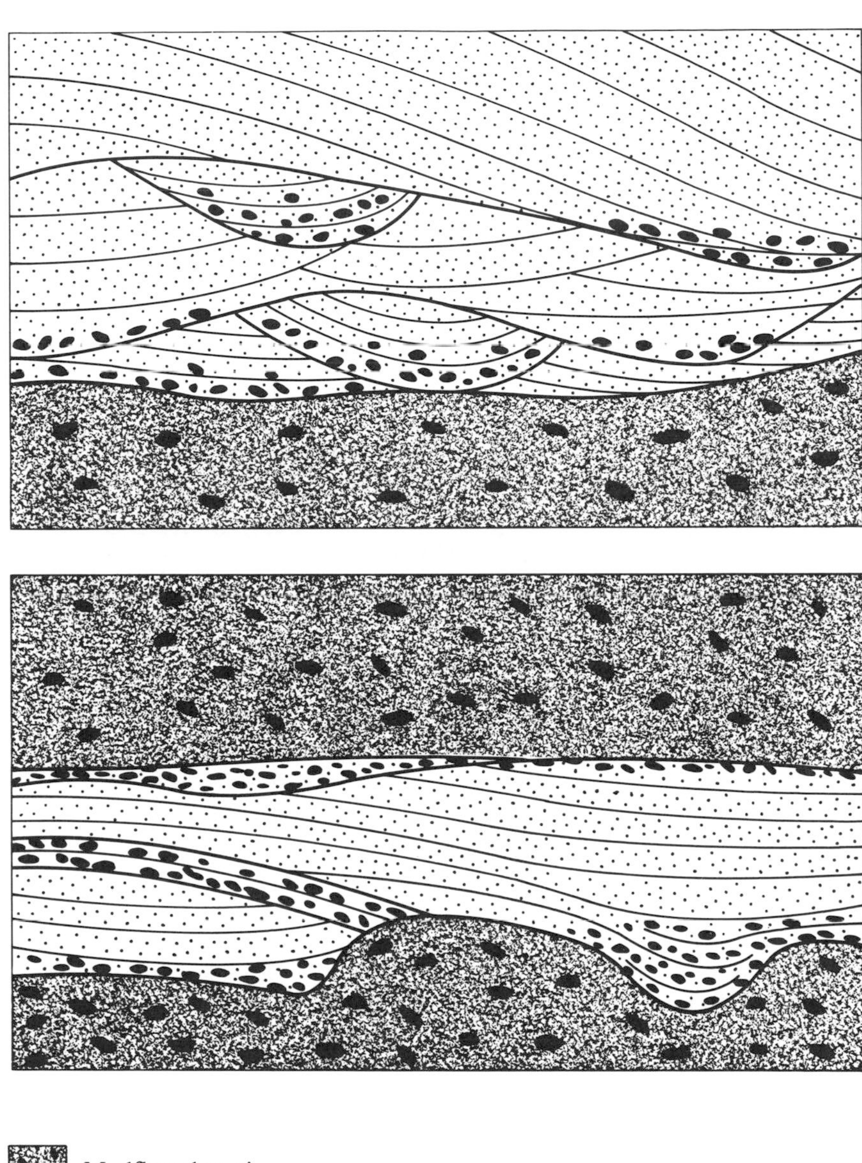

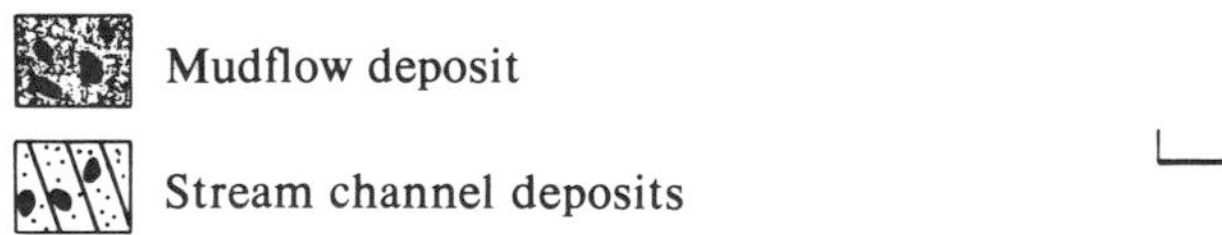

2.2 Bedding structures of alluvial fan deposits. After Reineck and Singh (1973).

predominant sediments, and a distinction can be made between the deposits of *low-sinuosity* and *high-sinuosity* rivers, although gradations between the two naturally occur (Collinson 1978a).

Low-sinuosity rivers are characteristically braided, and the deposits tend to be sandier than those of high-sinuosity rivers. Sedimentation occurs almost entirely in the rapidly shifting channel complex and silts are rarely deposited in abandoned channels; normally there is no flood plain (Fig. 2.3). There is a variety of bed features in the channels. Transverse *sand waves*, marked by a high ratio of length to height, continuous crests and minimal leeside scour, give rise to sets of tabular cross bedding, often alternating with ripple laminated beds. *Dunes*, characterized by a low length to height ratio, discontinuous crests and deep, leeside scour troughs give rise to trough cross bedding. Water stage fluctuations give rise to *scour-and-fill* structures and low-angle erosion surfaces known as *reactivation surfaces* may intersect the cross bedding.

High-sinuosity, or meandering, rivers occur in physiographic regimes of relatively low slopes, with a high ratio of suspended to bed load and cohesive bank materials, and a relatively steady discharge. There is a clearer separation of channel and overbank environments (Fig. 2.4). Periodically the channel bank is breached during times of flood, a process known as *avulsion*.

Whereas the river channel deposits are characteristically sandy, those of the flood plain are finer grained. The levées are usually composed of varying proportions of sand and silt, while silt and mud are laid down intermittently and slowly at times of flood over a more extensive region beyond the levées.

In ideal conditions of bankfull discharge and fully developed helicoidal flow, a characteristic fining-up sequence is produced by lateral migration of the channels (Fig. 2.4). Erosion takes place on the concave banks of meanders and deposition on the convex bank, at the *point bar*. As the channel migrates laterally a tabular sand unit above an erosion surface, perhaps with a lag conglomerate, exhibits an upward succession trough or planar cross bedding followed by ripple cross-laminated or parallel-laminated deposits. The sand is replaced eventually by more muddy deposits of the flood plain. Periodically these *overbank* deposits are interrupted by fans or tongues of sand that thin distally. These are the result of crevassing, and are known as *crevasse splays*.

In humid regimes the overbank deposits may pass locally into swamps whose vegetation may successively be converted after burial into peat, lignite and coal. Alternatively layers with plant rootlets may be the only surviving evidence of a former vegetation cover. In semi-arid and arid regimes thin layers of *calcrete* may form as a result of downward or upward percolation, evaporation and precipitation from calcareous groundwater. Isolated calcareous nodules may coalesce into more continuous layers

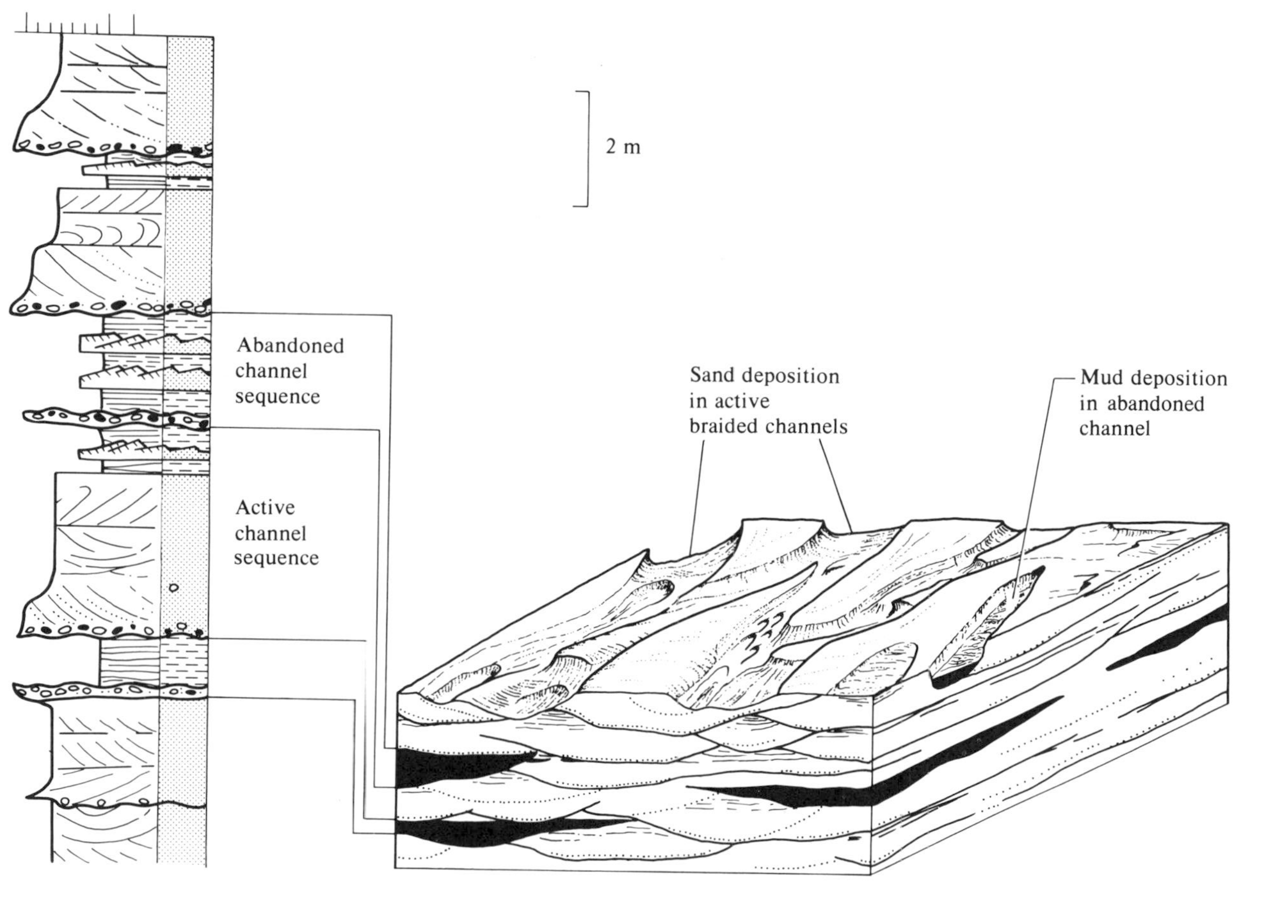

2.3 Physiography and sediments of a braided alluvial channel system. After Selley (1978).

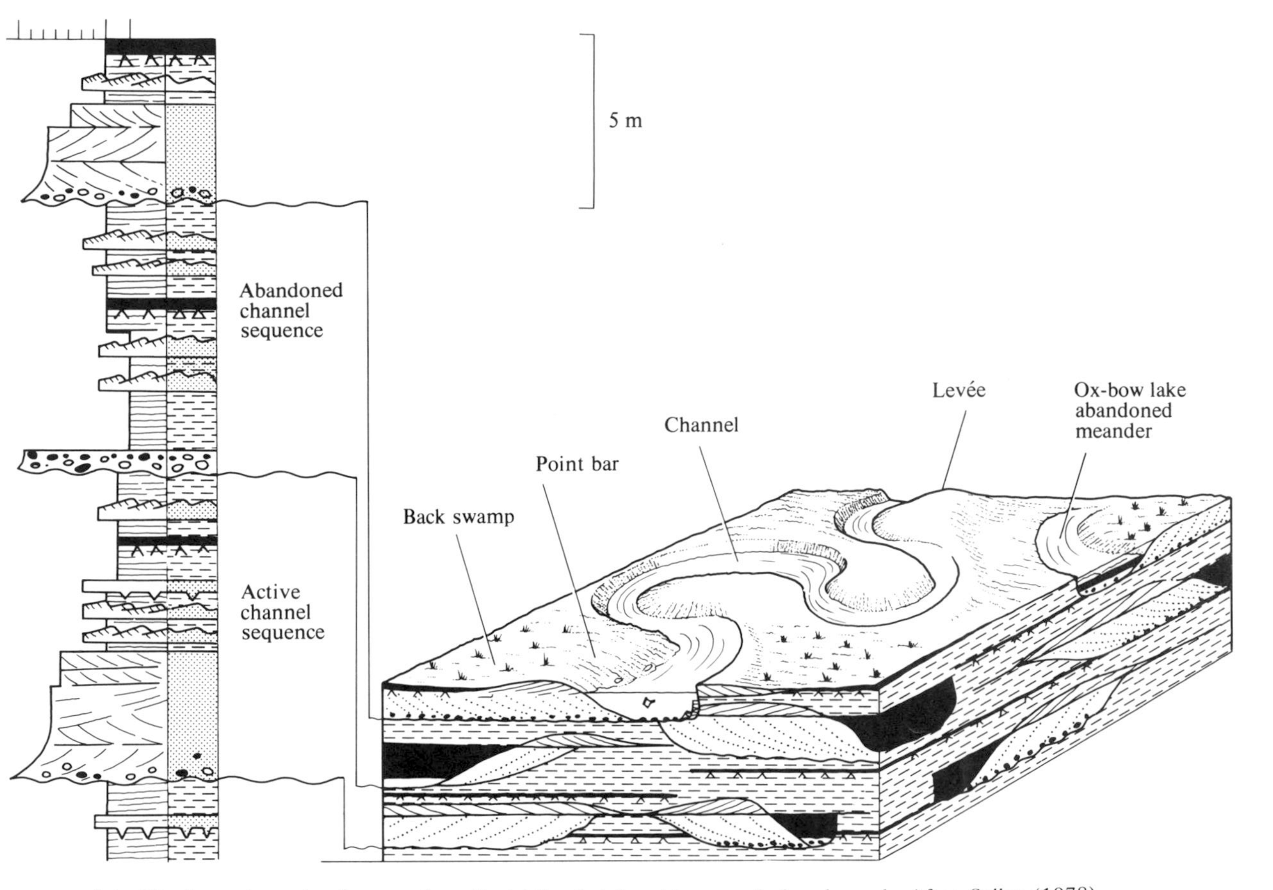

2.4 Physiography and sediments of an alluvial flood plain cut by meandering channels. After Selley (1978).

exhibiting characteristic structures such as buckling, brecciation and honeycomb patterns. There is still a great deal to learn about the formation of calcretes but it is evident that they form very slowly, of the order of thousands of years (Goudie 1973).

Examples of ancient deposits of braided and meandering flood plains are now widely recognized in the stratigraphic record. Apart from the differences of grain size and sedimentary structures, palaeocurrent analysis may help to distinguish the two types of regime. A wide distribution of palaeocurrent directions determined from the analysis of cross bedding should normally signify high-sinuosity, and a narrow distribution low-sinuosity streams, though the situation is probably more complex than this because the current directions may indicate patterns of bar movement rather than channel type. Allen (1964) was the first to establish that the now-classic fining-upward sequences of the Old Red Sandstone in the Anglo Welsh Basin are probably the deposits of rivers meandering in a flood plain, and Selley (1965) and Hubert (1978) give good examples of probably braided stream deposits, respectively in the Torridonian of north-west Scotland and the Triassic Newark Group of the north-eastern United States.

The Old Red Sandstone also contains well described calcrete horizons (Allen 1974), as does the Triassic of the Scottish Hebrides (Steel 1974) and the north-eastern United States (Hubert 1978). Hubert lists several criteria whereby the calcite nodules of calcrete horizons can be distinguished from normal diagenetic concretions. The Triassic examples he describes invariably contain casts of plant roots (thereby indicating that the climate was semi-arid rather than arid). Distinctive microstructures such as pedotubules and crystallaria are closely comparable to Quaternary examples, and the upper surfaces are often cut by erosional surfaces. The occurrence of reworked limestone clasts in overlying fluvial sandstones prove that the calcretes formed before subsequent stream erosion.

Many alluvial deposits of the past have a distinctive red coloration due to the presence of haematite. It is now well established that the coloration is post-depositional in origin, either the result of diagenetic reactions in which minerals like biotite and hornblende are broken down, and other grains coated by one of the breakdown products, haematite, or of a later groundwater circulation (Schluger and Robertson 1975; Van Houten 1973; Walker, T. R. 1967). The most favourable depositional environment is probably in a semi-arid climate with the sediment having free access to oxidizing groundwaters.

Aeolian deposits

Hot deserts have a variety of depositional environments including aeolian sand seas, alluvial fans, ephemeral streams and playas or inland sabkhas (Fig. 2.5).

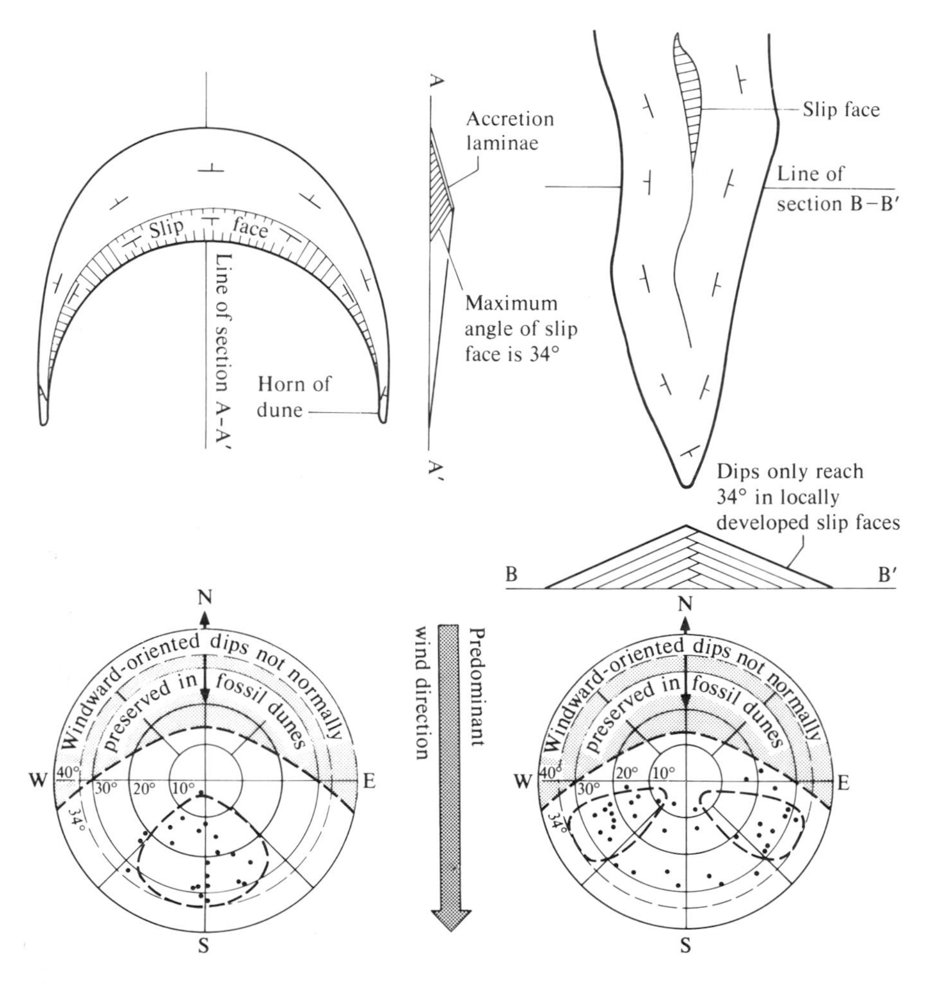

2.5 Polar nets of the distribution of dip attitudes in ideal barchan and seif dunes. After Glennie (1970).

Wind-deposited sands may exhibit either horizontal laminae or fine- and large-scale foresets with either a constant orientation, in the case of seif dunes, or multiple orientation in the case of barchan dunes (Glennie 1970 and Fig. 2.6), although the detailed palaeocurrent pattern may be more complicated because of the migration of superimposed dunes on the flanks (Collinson 1978b). The individual laminae tend to be well sorted, and sharp differences in grain size between adjacent laminae are common. Grains larger than 5 mm diameter are rare and clay drapes very rare. The sands are free of clay and mica is usually absent; quartz grains are well rounded and often have a frosted or polished surface.

McKee (1966) lists several criteria considered to be characteristic of aeolian cross stratification.

1. There are many sets of medium- to large-scale size, with a maximum height of about 7 m; foreset dips as often as high as 34°.
2. The bounding surfaces separating individual sets are commonly sub-horizontal and dip downwind at a low angle.
3. Vertical sections often show an upward tendency towards progressively thinner cross-bed sets.
4. Planar cross bedding is dominant.

Interdune areas are characterized by stony desert with faceted pebbles with a surface of desert varnish produced as a result of the action of dew together with polishing. Poorly sorted deposits with these *ventifacts*, with roughly horizontal bedding, probably signify sheet floods (Collinson 1978b).

Flat-bedded playa deposits consist of sand, silt and clay often exhibiting desiccation polygons and sometimes footprints and rainprints. Halite and

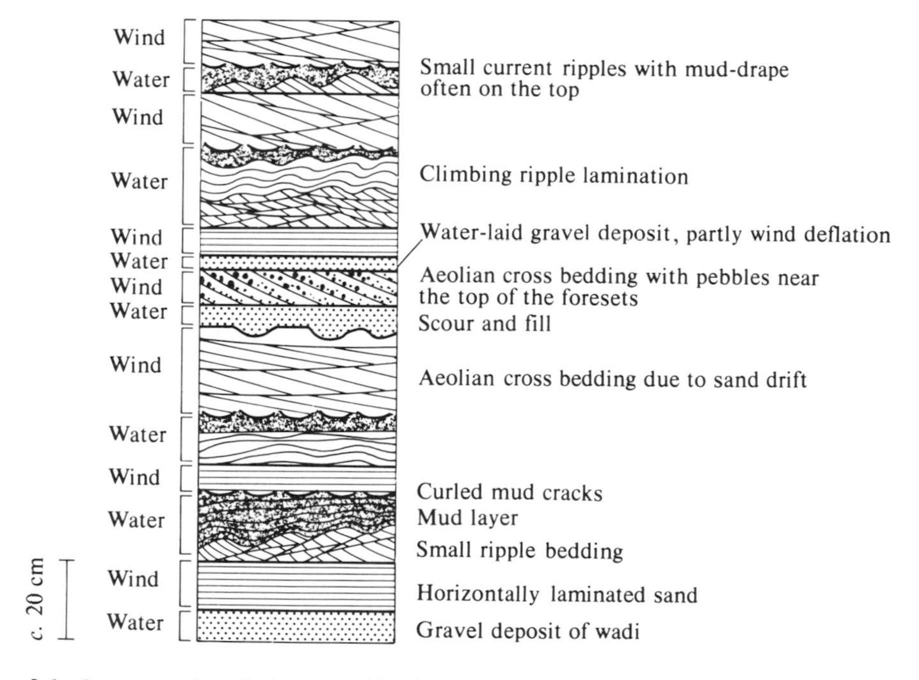

2.6 Sequence of wadi deposits with alternating wind- and water-laid sediments. After Reineck and Singh (1973).

gypsum occur as crusts or nodules and crystals that may disrupt the sedimentary layers; calcrete may also be present.

Turning to the stratigraphic record, the Permian Rotliegendes of north-west Europe, with equivalents extending into Great Britain, appears to be an outstanding example of a group of clastic deposits laid down in a hot desert. The unit occupied a basin up to 2000 by 500 km in extent bounded in the south by the Hercynian mountain belt, and locally exceeds 1500 m in thickness. Aeolian, alluvial and inland sabkha environments have been identified, with a predominantly easterly wind direction determined from dune cross-bedding sets (Glennie 1972). Clemmensen (1978) cites the association of large-scale cross bedding with gypsiferous beds of sabkha type as a good indication that the Triassic Gipsdalen Formation of East Greenland was laid down in an aeolian environment. Other examples of aeolian dune deposits include the Permian Lyons Sandstone of Colorado (Walker and Harms 1972) and part of the Upper Jurassic of New Mexico (Tanner 1965).

Pryor (1971) has challenged the conventional aeolian interpretation of dune sands in the Yellow Sands Formation of north-east England, which is equivalent to part of the Rotliegendes. He states that there is no unequivocal criterion for aeolian cross bedding, and large dunes and sand waves are known to form in some shallow-marine environments. Frosting of the component sand grains is attributed entirely to diagenetic corrosion and it is pointed out that these grains are not especially well rounded, and the sediments in general are only moderately well sorted, with a certain amount of silt and clay in the matrix. On the other hand he disregards the total absence of marine fossils (even burrows), the lack of obvious alluvial criteria and the close association in the sedimentary sequence with substantial Zechstein evaporites.

In a similar way, a simple aeolian origin for the celebrated dune-bedded Lower Jurassic Navajo Sandstone of the United States Western Interior has been put into question by Stanley *et al.* (1971). Ripple-marked and wavy-bedded horizons, shale seams associated with widespread truncation planes and dolomitic carbonate lenses point to a subaqueous origin for at least part of the unit. Evidently a definite boundary between subaerial and subaqueous conditions cannot be clearly defined.

Lacustrine deposits

Although lakes occupy only about 1% of the continental surface at the present day they might have been considerably more extensive in the past, so that due attention should be paid to their deposits.

Insofar as these deposits are laid down in bodies of standing water they should be easy to distinguish from alluvial and aeolian deposits but there is no difference discernible from many marine deposits if no palaeontological or geochemical criteria are available. Thus deltas are formed where rivers enter lakes as in the sea, turbidites may be deposited in the appropriate hydrographic regime and limestones may form where there is no siliciclastic influx. Wave base tends to be shallow because of the low fetch and deltas are of fluvial-dominated type because of the absence of tides and weakness of winds. Thus they tend to be elongate, with well-developed mouth bars (Collinson 1978c). Stratification of water layers is more usual than in the sea and may result in various types of annual layered or varved deposits.

Fauna and flora, which provide the most diagnostic criteria, are very restricted in diversity compared with the marine environment (Picard and High 1972a). Blue-green algae, which may form stromatolites or oncolites, are normally abundant. Other algae include diatoms and charophytes but calcified green algae like the Codiaceae and Dasycladaceae are exclusively, and red algae almost exclusively marine. Bivalves such as *Unio, Anodonta* and *Corbicula* and gastropods including *Viviparus, Valvata, Physa, Ancylus, Planorbis* and *Lymnaea* are characteristic, as are some ostracods. Bioturbation of sedimentary layers is restricted by the low diversity and abundance of burrowing fauna.

In warm, arid regions the lakes are normally saline rather than fresh water. Gypsum and aragonite are now being precipitated in the Dead Sea and in the early Holocene large quantities of halite were precipitated. The composition of the precipitated salts may differ appreciably from those resulting from the evaporation of sea water. In the highly alkaline soda lakes of East Africa, e.g. Lake Magadi, large quantities of sodium carbonate (trona) are precipitated together with zeolite minerals. Sodium sulphate is precipitated from Searles Lake, California.

Perhaps the classic example of ancient lake deposits is that of the Eocene Green River Formation of Wyoming and adjacent states. A variety of sediment types occur, including sandstones, shales, algal and oolitic limestones and dolomites indicating varying degrees of fluvial influence and fluctuating lake level (Bradley 1964; Picard and High 1972b). An especially interesting facies is the strongly bituminous oil shale, the fine laminae of which were interpreted long ago as annual varves in a classic paper by Bradley (1929). Fossils in the oil shales, whose bituminous matter was derived from the decay of planktonic algae, include superbly preserved fish.

The presumed varves are the primary basis for the conventionally accepted stratified lake model, but this has been rejected by Eugster and Surdam (1973) in favour of a model of a vast alkaline earth playa fringing an alkaline lake. This is based essentially on sedimentary structures indicative of frequent subaerial exposure and current transport, on the occurrence of trona deposits and the abnormally high proportion of magnesium in the carbonates. No close modern analogue is known.

Another well-described lacustrine deposit is the Triassic Lockatong

Formation of the Newark Group in the north-eastern United States (Van Houten 1964, 1965). A feature of particular interest is the abundance of the sodium-rich zeolite analcime. Small-scale detrital and chemical cycles are related to periodic fluctuations in rainfall. As with Bradley's interpretation of the Green River oil shales, carbonate–organic laminites in the Devonian Caithness Flags of north-east Scotland have been interpreted as seasonal layers in a lake. Further examples of ancient lake deposits are given in Matter and Tucker (1978).

Glacial deposits

About 10% of the Earth's surface is presently covered by ice and, primarily because of interest in the Pleistocene glaciations when the ice cover reached a maximum of some 30%, a great deal of work has been done on glacial deposits. Nevertheless there is still much to be learned, especially about the deposits that form beneath ice sheets.

The most characteristic deposit is boulder clay or *till* but it is to be noted that poorly sorted, non-glacial 'mixtites' formed by debris flow occur both in alluvial fan and deep-sea fan environments. The specifically glacial criteria that have been proposed for *tillites* (the ancient equivalent of tills) include exotic, far-travelled clasts, striations on the clasts, striations and crescentic gouges on the underlying pavement and great lateral extent. Glacial influence in marine sediments is signified by *dropstones*, with isolated large clasts, released by melting icebergs, occurring in otherwise well-bedded layers (Edwards 1978; Frakes 1979; Harland *et al.* 1966). It should be recognized that scattered pebbles from a distant source, in finer-grained sediments, can also be the result of organic rafting, if the deposits are sufficiently young. This is almost certainly the case with the Cenomanian Cambridge Greensand of southern England.

Glaciofluvial deposits are those laid down by meltwaters. Meltwater streams are characteristically braided, and give rise to interstratified sand and gravel. A drift apron may form around lobes of the ice margin to produce *kames.* Subglacial or englacial streams deposit the winding shoestring sand bodies known as *eskers.* Beyond the limit of glaciofluvial deposits, finer-grained particles of silt and clay may be transported by wind and deposited as *loess.* In glacial lakes seasonal meltwater runoff gives thin silt layers which alternate with clay deposited through the rest of the year, to produce the classic type of varve.

The weathering of tills at times of warmer climate can effect significant changes such as solution of carbonates, partial solution of silicates and disappearance of unstable minerals. The *gumbotils* so produced are important features for the Pleistocene stratigrapher. Alternating freeze and thaw phenomena can produce infilled frost wedges and involutions in glacial strata.

Confident identification of ancient glacial deposits is more than usually dependent on the study of whole facies associations (Fig. 2.7). Subglacially deposited massive and banded tills should exhibit a wide range of clast types and should be traceable laterally for at least several kilometres. The long axes of clasts often show a preferred azimuthal orientation related to ice flow, and there may be an erosive base. Isolated bodies of stratified sediment indicate the deposits of subglacial and englacial streams. The underlying strata may also be deformed into folds as a result of ice movement. Associated stratified conglomerate and sandstone probably signify the outwash of braided streams and alluvial fans, while laminites with isolated dropstones in a random orientation are characteristic of glaciomarine conditions (Edwards 1978).

Back beyond the Pleistocene, the best authenticated glacial deposits are those of the late Carboniferous and early Permian of the southern continents and India (Crowell and Frakes 1975) and the late Ordovician of the Sahara (Beuf *et al.* 1971). A large number of features have been recognized, including striated pavements, massive tillites, laminites, the deposits of outwash fans, eskers, kettleholes and freeze-and-thaw structures. There are also well-documented late Precambrian glacial deposits (Edwards 1975; Frakes 1979; Reading and Walker 1966; Spencer 1971).

Deltaic and coastal plain deposits

It is not convenient to separate these two categories too rigidly because deltas form part of the coastal plain. Whereas deltas, however, form distinct sedimentary entities with a significant subtidal component consideration of other parts of coastal plain will be largely restricted to those deposits formed in the intertidal and supratidal zone. This corresponds broadly to what is sometimes called the *paralic* zone.

DELTAS

The overall shape and sedimentary characteristics of deltas depend upon a variety of factors, such as the nature of the hinterland, climate, rate of subsidence, topography and the relative importance of river, wave and tidal influence. Of these the hydrographic characteristics are probably the most important and form the basis of modern classifications. Thus Fisher *et al.* (1969; Fig. 2.8) distinguish *constructive* deltas, dominated by fluvial processes, from *destructive* deltas, which are wave- and tide-dominated. Galloway (1975) proposes a three-fold subdivision, with the inevitable intermediates: *fluvial-dominated*, e.g. Mississippi, Po, Ebro, Danube; *tide-dominated*, e.g. Ganges, Brahmaputra, Mekong; *wave-dominated*, e.g. Senegal, Rhône, Saõ Francisco. Coleman and Wright (1975) undertook a multivariate analysis of 34 modern deltas, which they were able to group into six discrete delta models primarily on the basis of hydrographic conditions, topography and sediment load.

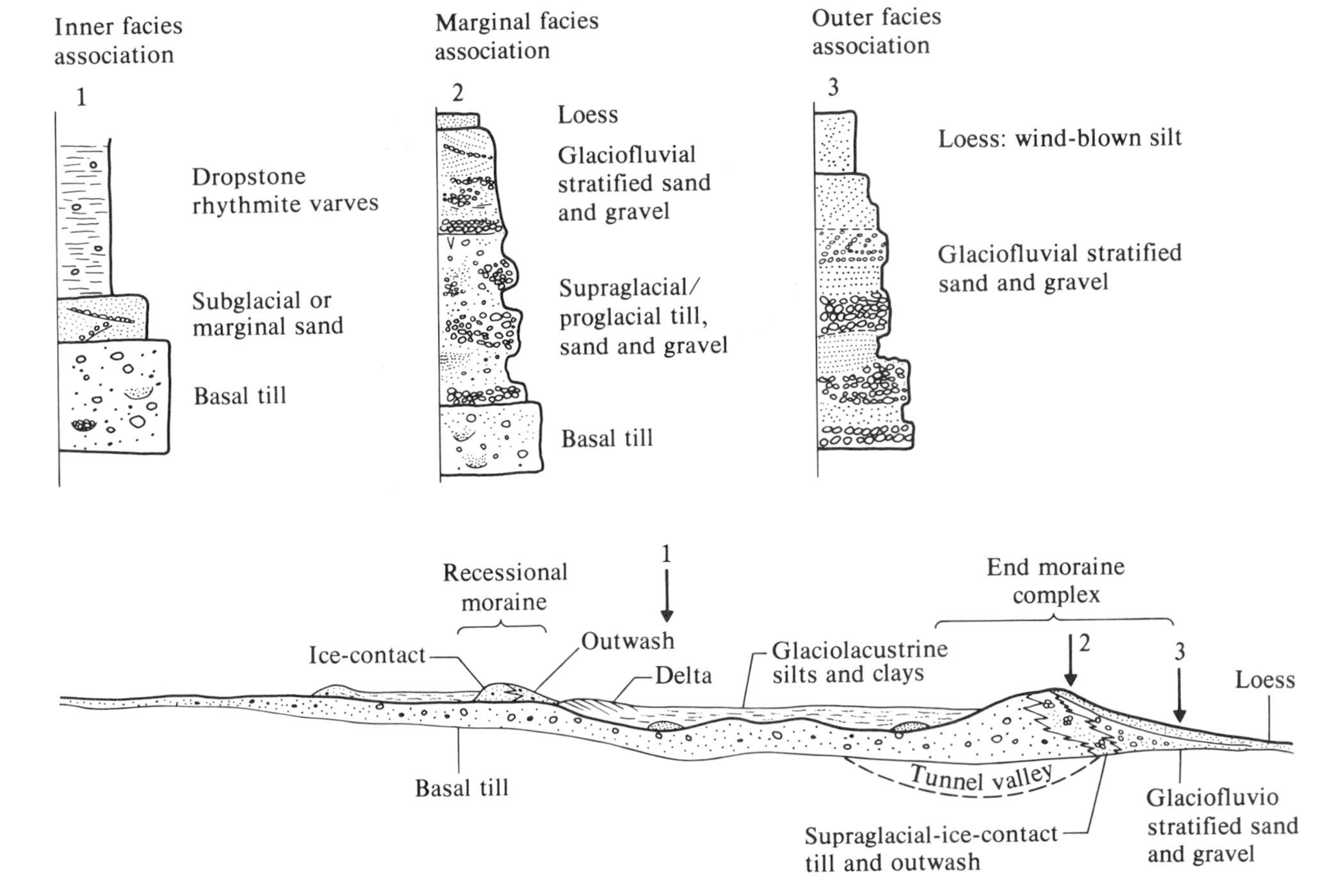

2.7 Simplified version of the facies associations and sequences that develop from the advance and retreat of a terrestrial ice sheet. Adapted from Edwards (1978).

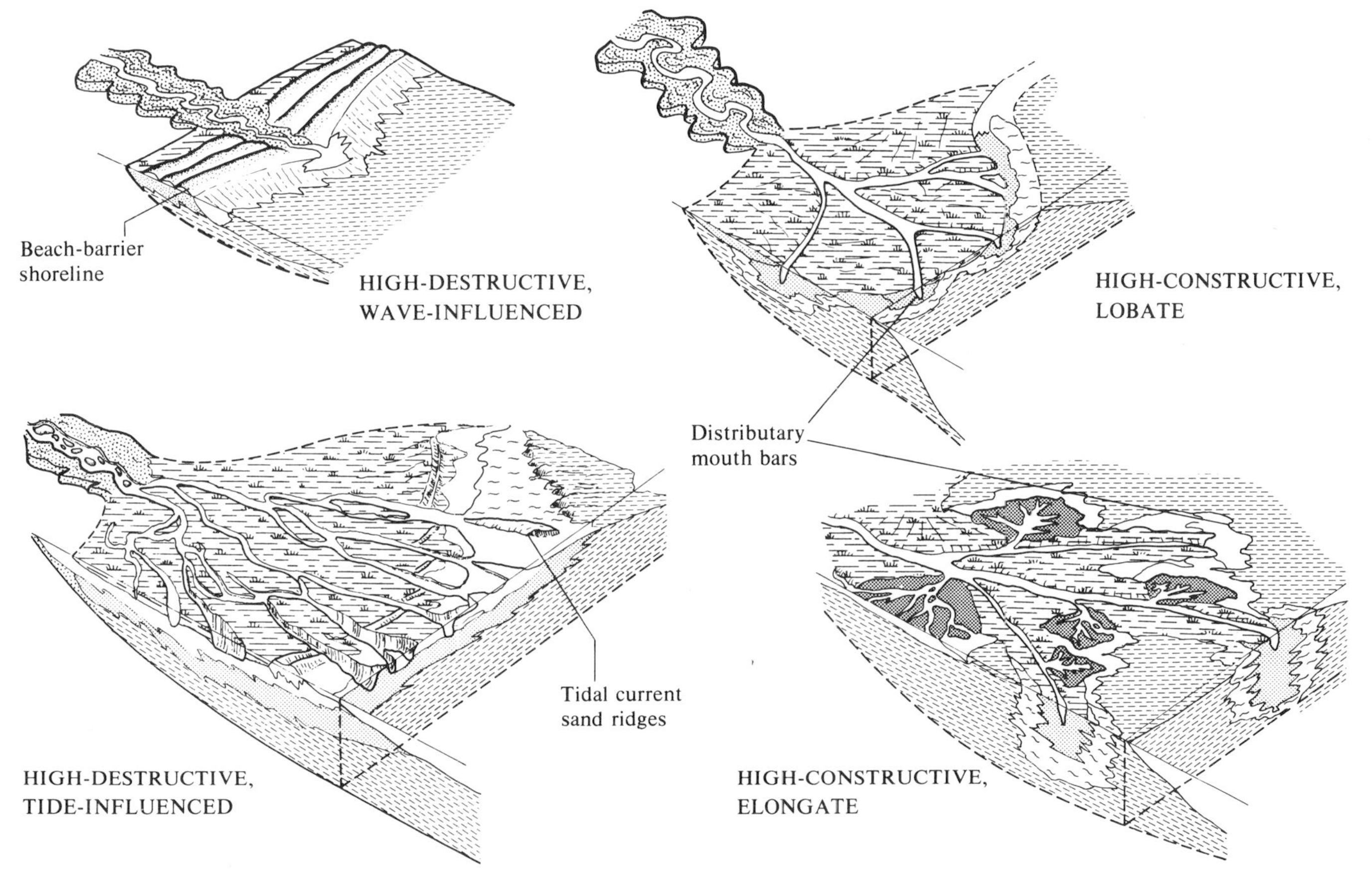

2.8 High-constructive and high-destructive delta types. After Elliott (1978a).

Elliott (1978a) stresses that it is important to distinguish the *delta plain* and *delta front* environments approximately above and below mean sea level.

The delta plain contains active and abandoned distributary channels, characteristically containing sandy sediments, in the midst of interdistributary areas with finer-grained sediments deposited at times of overbank flooding. These latter may be swampy, with water of lowered salinity compared with the sea, in regions of humid climate, but in semi-arid regions calcrete may form (e.g. the Ebro Delta) or gypsum and halite may precipitate from salinas (e.g. the Nile Delta).

On fluvial-dominated delta plains distributary channels experience more frequent avulsion than alluvial channels and hence abandoned courses are abundant. The sedimentary facies can resemble that of some alluvial environments in exhibiting basal lag deposits overlain by fining-upward sequences, but distributary sand bodies have a low ratio of width to depth compared with alluvial channels because of frequent switching of course. Levées are characteristically composed of rippled- or parallel-laminated sand. As in flood plains crevasse splays may give rise to lobes of sand in the interdistributary areas. The sequence of deposits laid down in these areas may be very complex in detail, as indicated in Fig. 2.9.

Tide-dominated delta plains have distributary channels with higher width to depth ratios and tidal flat sequences, the characteristics of which will be described later; fluvial processes may not be much in evidence.

On the delta front the sedimentary load is deposited as the river water outflow competence is decreased. With a low-density outflow, as off the Mississippi and Po deltas, a surface jet or plume may extend for some distance offshore, whereas with a high-density outflow density currents may bypass the shoreline and delta front development is restricted. Water bodies of equal density mix freely and rapid sedimentation ensues. There is a general tendency for coarse sediment to be deposited at the distributory mouth and for the sediments to become progressively finer in deeper water offshore, as the *prodelta* region is approached.

Progradation gives rise to a coarsening-up sequence (Fig. 2.10), which is often taken to be characteristic of deltas, but many complications may disturb this simple picture, notably because the sediment supply varies considerably along the delta front and because the rate of subsidence and/or sea level may change through time.

The Mississippi provides a good example of a fluvial-dominated delta front, where deposition at distributary mouths gives rise to a series of discrete lunate mouth bars which protrude into the Gulf of Mexico as radiating *bar finger sands*. Wave-dominated delta fronts, in contrast, have a relatively straight beach with only a slight deflection at the distributary mouth, and a relatively steep slope. Progradation involves the whole delta

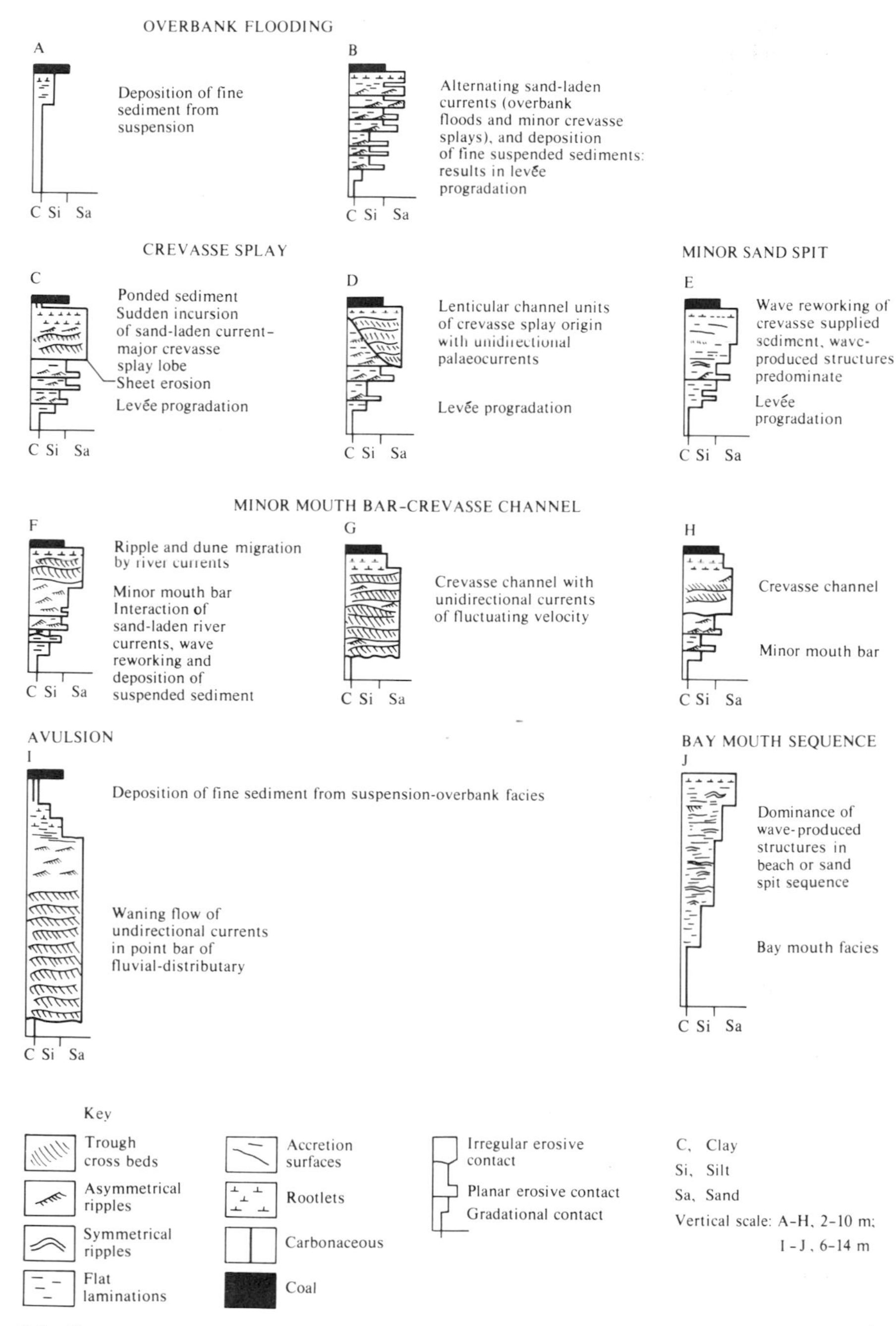

2.9 Sequences produced in fluvial-dominated interdistributary areas of deltas. After Elliott (1978a).

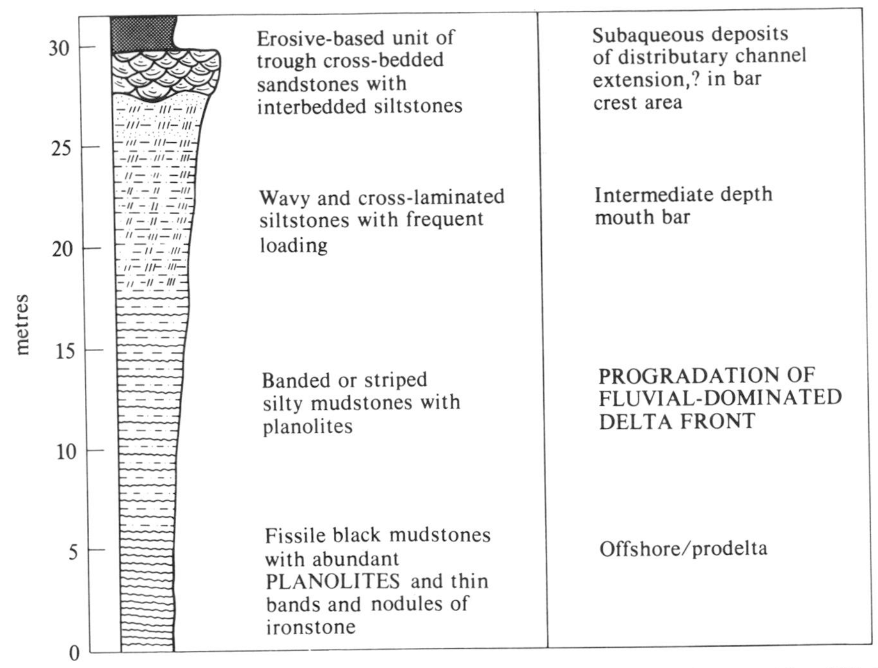

2.10 Idealized representation of a fluvial-dominated delta front sequence. After Elliot (1978a).

front and characteristically will give rise to a sequence of bioturbated marine mud passing up eventually into the well-sorted sand with low angle or parallel laminations typical of a beach. Tide-dominated delta fronts have tidal-current ridges radiating from the distributary mouth, and the sediments may show such characteristic features as bidirectional trough cross bedding and clay drapes.

Delta abandonment is initiated by avulsion resulting from over extension of the channel system as the delta progrades into the marine basin. Subsidence and sediment compaction continue and thin, slowly deposited but laterally persistent deposits are produced. Peats accumulated onshore may pass seawards via muds and silts with a very restricted marine fauna into an offshore, more faunally rich and bioturbated facies. The development of delta lobes and channel switching appears to be more characteristic of fluvial-dominated than other types of deltas.

With regard to the stratigraphic record, Elliott (1978a) claims that only fluvial-dominated deltaic sequences have been recognized, though Hubert *et al.* (1972) make a case for a tide-dominated delta in the Cretaceous of the United States Western Interior. Certainly the well-studied examples in the Tertiary of the United States Gulf Coast (Fisher *et al*. 1969) and the

Carboniferous of Europe and North America (Elliott 1978a) exhibit many characteristics of fluvial dominance, including bar finger sands, crevasse channel fills and channel abandonment features. Laterally persistent delta abandonment deposits such as coal seams may serve as marker horizons and thereby prove of considerable value in reconstructing deltaic facies associations (e.g. Ferm 1970 and Fig. 2.11).

Care must be taken in distinguishing the deposits of distributary and fluvial channels. Thus some of the well-known cross-bedded sandstones of the Millstone Grit (Namurian) of northern England, generally assumed to be deltaic in origin, have been plausibly reinterpreted by McCabe (1977) as alluvial channel deposits, and a comparable change of view is expressed for the almost equally celebrated Moor Grit of the Bathonian of the Yorkshire coast (Leeder and Nami 1979).

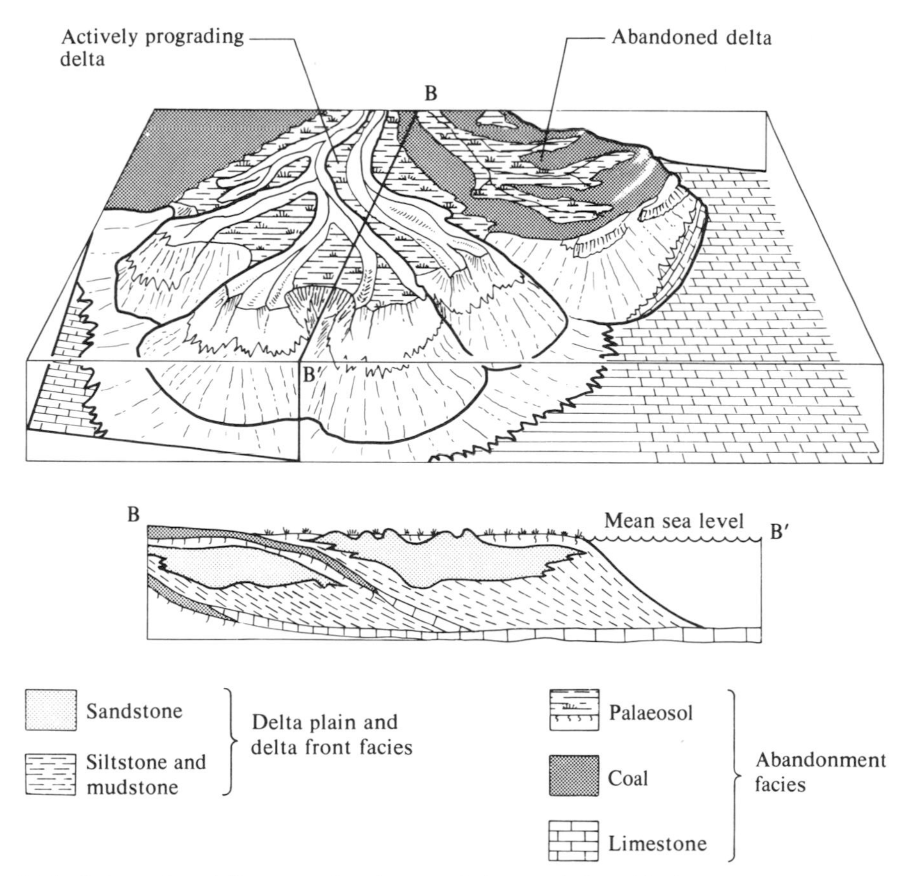

2.11 Palaeosol–coal–limestone abandonment facies association of Carboniferous delta fronts and delta plains in the United States. After Ferm (1970).

OTHER PARTS OF THE COASTAL PLAIN

The deposits of coastal plains beyond the influence of deltas are influenced most notably by the degree of exposure of the coast, tidal range and climate. Exposed coasts are characterized by sandy beaches whereas silt and mud deposition occurs on more protected coasts. Some degree of tidal influence is almost ubiquitous on the margins of the present ocean system. On most coastlines the tidal range is between 2 and 4 m (mesotidal) but a limited number of regions have ranges of less than 2 m(microtidal: most of the Arctic margins, Mediterranean, Baltic, Caribbean, southern Australia) or more than 4 m (macrotidal: north-west Europe, Bay of Fundy, East Africa, Argentina, north-west Australia). Tidal sand ridges are confined to the macrotidal regime, and tidal flats, salt marshes and estuaries are dominant in both the macrotidal and mesotidal regimes.

Sandy beaches are built by wave processes and the sand grains are characteristically well sorted and rounded. It is customary to distinguish several categories related to position with respect to sea level (Elliott 1978b). The *backshore* is the zone above high-tide mark and is only inundated during storms. It may be backed on its landward side by wind-moulded coastal dunes. The *foreshore* is the intertidal zone which is dominated by swash zone processes. A high rate of water flow generates current-lineated plane beds while a low rate favours the formation of ripples. Typically beach sands exhibit parallel laminations dipping gently 2–3° seaward; sometimes subtle planes of truncation can be discerned. Occasional layers of heavy minerals are attributed to grain segregation in the repeated swash-and-backwash flow. The *shoreface* is the subtidal part of the beach face and is more appropriately considered as part of the more fully marine regime.

Landward of the beach the coastal plain of humid regimes consists of salt marsh with halophytic plants that may leave tell-tale rootlets in muddy and silty sediment. The name *chenier* has been given to isolated sand ridges in a marshy coastal plain and are classically developed in Louisiana. They are apparently formed during reworking episodes by waves and currents if the sediment supply diminishes. Such features may range up to 3 m in height, 1000 m in width and 50 km long. They should be preserved as sand stringers in clay and silt and must be carefully distinguished from the bar finger sands of deltas.

Because of their restricted areal development and the high probability of reworking during subsequent transgressions or regressions, it is not to be expected that many beach deposits will be recognizable in the stratigraphic record. Consideration must be paid to the whole facies context, for instance the sandstone body in question should lie parallel to the inferred shoreline. A likely example has been described from the Cretaceous of the United States Western Interior, where a coarsening-up sequence has been interpreted as signifying a change from shoreface to foreshore to aeolian dunes (Davies *et al.* 1971).

Lagoons are extremely variable shallow-water environments isolated to a greater or lesser degree from the open sea by offshore barrier islands. In microtidal regimes abnormal salinity and hence restricted fauna is usual because of the paucity of tidal inlets; brackish water usually characterizes humid regions and hypersaline water arid regions, but salinity can also vary considerably in a given lagoon from season to season. In mesotidal lagoons salinity values are normal.

Because of the variability further generalizations are difficult to make but it is worth making brief mention of the deposits of one of the best-studied examples, the Laguna Madre of Texas (Rusnak 1960). On the wave-exposed shorelines carbonate coquinas and oolites occur together with rippled sand. Both horizontal and cross stratification occur. In the quieter water regions mudflats are found in association with algal mats, and gypsum crystals form in laminae below the sediment surface. Abnormal salinities are characteristic of this and other Texas coastal lagoons. In consequence the fauna is very restricted in diversity but reefs of *Crassostrea* may be common. Hudson (1962, 1963, 1970) has been able to make convincing comparisons between the sediments and faunas of the Texas coast bays and the deposits of the Middle Jurassic (Bathonian) Great Estuarine Series of north-west Scotland.

Where *tidal flats* contain both mud and sand, as in the well-studied examples of the North Sea margins, the muds may occur at a higher level than the sands (Fig. 2.12) and the areas where mud and sand are intermixed characteristically exhibit *flaser* and *lenticular* bedding. Flaser bedding is composed of mud-draped ripples and lenticular bedding of thin sand lenses and isolated ripple forms set in mud and silt. Neither feature is unique to tidal flat environments, however, and more restrictive criteria should be sought, such as the lag concentrates and other erosional effects produced by laterally migrating tidal channels, bimodal ('herringbone') cross bedding in the sands and abundant reactivation surfaces.

Interpretation is made more difficult if there is little to no sand, as in the northern part of the Gulf of California (Thompson 1975). The Colorado River brings in copious quantities of mud and silt and, because of insignificant barriers, tidal currents sweep across the flats with little restriction as a broad uniform flow and hence there is little tendency to develop channel systems. Salt is precipitated in the supratidal zone and diagenesis could produce a red-bed sequence which could easily be misinterpreted as having been formed in an arid, inland basin rather than in the paralic zone.

Some recent examples and presumed ancient analogues of tidal flat deposits are described in Ginsburg (1975). Klein (1971) has proposed a sedimentary model for determining palaeotidal range on the basis of the

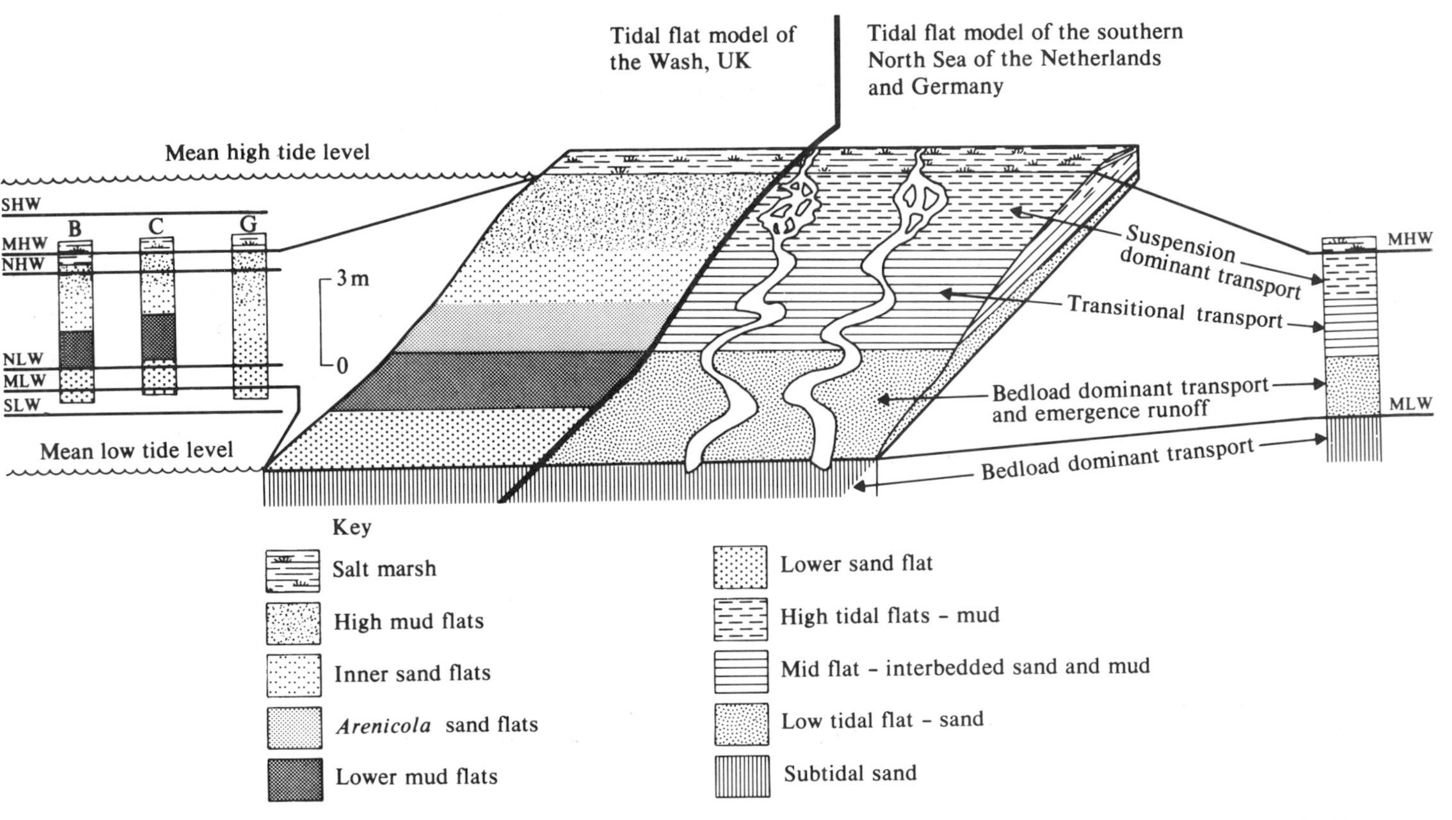

2.12 Intertidal zone facies models for the classic North Sea region. Numbers in key are sand-mud ratios. SHW, Spring high water; MHW, mean high water; NHW, neap high water; NLW, neap low water; MLW, mean low water; SLW, spring low water. After Klein (1971).

North Sea examples. Progradation should give rise to a fining-upward sequence from sand to mud, the thickness of which should give an indication of the mean tidal range. However, it is difficult to see how subsidence, compactional and erosional effects can be discounted. If a given fining-upward sequence can be traced laterally it is almost certain that it will be seen to vary in thickness. Which locality is then to be chosen as giving the best approximation to true range? Furthermore, the model assumes a still-stand of sea level and without an independent time control there is no way of determining how long the prograding sequence took to form.

Siliciclastic sediments are now predominant in coastal regions but carbonates occur in regions where no major rivers enter the sea, as in the Bahamas, southern Florida and the Yucatan, the Trucial Coast, Shark Bay, Western Australia and part of the west coast of Scotland and Ireland. These areas of comparatively limited extent have a disproportionately large interest because of the much more extensive deposits of similar type in the past.

The features of greatest environmental significance are the sedimentary laminae known as *stromatolites*, produced by periodic entrapment of particles by the gelatinous mats of blue-green algae. Numerous examples have been recognized in the stratigraphic record back into the early Precambrian (Walter 1976). Present-day examples seem to be largely confined to low latitudes.

Stromatolites have been widely held to be good indicators of intertidal conditions and Logan *et al*. (1964) were among the first to demonstrate how growth form varies with environment. The more common type of laterally linked hemispheres of the *Collenia* type is held to signify protected intertidal mudflats where wave action is slight, whereas the more sharply demarcated vertically stacked *Cryptozoon* structure is produced in more exposed mudflats, where the scouring action of waves prevents the growth of algal mats between the sharply projecting stromatolite heads (Fig. 2.13).

It is now known, however, that stromatolites can also form in shallow subtidal environments a few metres deep (Gebelein 1969; Monty 1967), so that additional criteria are required to establish periodic emergence above sea level. The presence of desiccation pores is very informative in this respect. According to Shinn (1968) such features are most abundant in the supratidal zone and are never found in the subtidal zone. After burial they become filled with sparry calcite to give rise to the distinctive rocks known as *birdseye limestones*. Contemporary breakup of rapidly consolidated laminae to produce horizons of intraformational breccia is another feature that seems to demand exposure, as is the occurrence of scattered gypsum crystals, which will normally be preserved as calcite pseudomorphs in limestones (Hudson 1970). A further emergence criterion is the occurrence of calcrete in the form of vadose pisolites (Bernoulli and Wagner 1971; Dunham 1969).

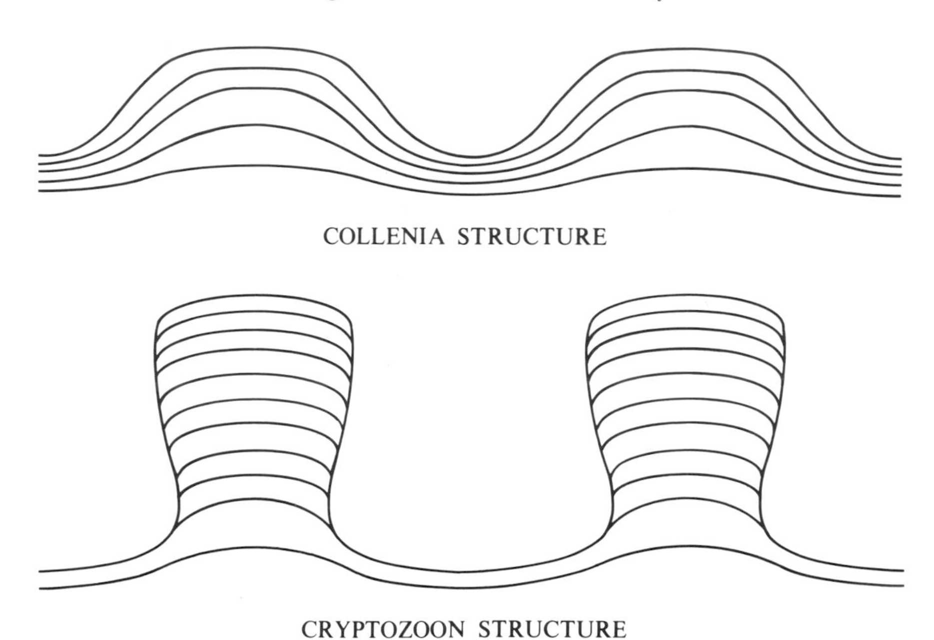

2.13 *Collenia* and *Cryptozoon* structures of stromatolites.

The supratidal zone in carbonate regimes also locally contains scattered crystals or crusts of dolomite, secondary after aragonite (Bathurst 1975).

In warm arid regions, evaporite minerals often occur, as in the well-described coastal sabkhas of the Trucial Coast on the southern side of the Persian Gulf (Till 1978). Pore waters in the carbonate sediments of these supratidal flats are highly concentrated and drawn towards the surface by evaporation. Anhydrite replaces gypsum diagenetically to form nodules, which may coalesce into a 'chicken-wire' fabric. Halite occurs as an ephemeral surface crust, while gypsum in the form of lozenge-shaped crystals is found within algal mat sediments of the upper intertidal zone. Sections through the sabkha show a regressive sequence passing down into aragonitic muds like those being deposited subtidally at present (Fig. 2.14).

In the transient salty pools on supratidal flats known as salinas halite may be precipitated. Petrographic examination of halite crystals precipitated from salinas in Baja California reveals a distinctive chevron structure (Shearman 1970).

Shearman (1966) was the first to make a convincing comparison of the Trucial Coast sabkha sequence with a rock sequence, namely limestone, containing stromatolites and nodular anhydrite, of the Lower Purbeck Beds (uppermost Jurassic) of southern England. The anhydrite was recognized in a borehole core. Where the beds outcrop at the surface, as in Dorset, the

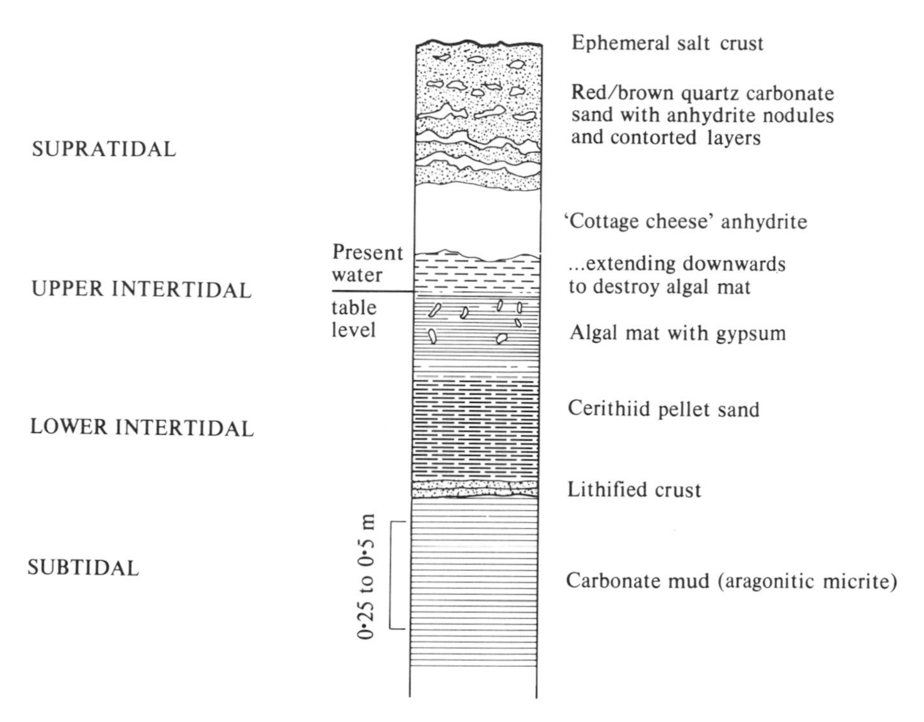

2.14 The characteristic regressive sabkha sequence in Abu Dhabi, Trucial Coast. After Till (1978).

calcium sulphate minerals have been replaced by calcite pseudomorphs (West 1975) and solution breccias occur. Shearman and others claim to have recognized other ancient sabkha sequences in much more significant evaporite deposits, such as the Devonian of the Elk Point Basin in Western Canada. This matter will be discussed further in chapter 5.

To conclude the chapter, let us attempt to indicate the potential hazards of facies analysis by reference to the work of Professor Perce Allen, who has devoted the whole of his professional life to wrestling with the environmental interpretation of the Wealden Beds of southern England—a thick Lower Cretaceous group of non-marine sandstones and clays. After many years of intensive work he arrived at what seemed a very plausible model involving a series of eustatically controlled transgressions and regressions, the latter being marked by progradation of deltaic sands from a northerly source into lake muds.

As time passed, Allen became increasingly vexed by a number of awkward facts which he could not fit neatly into the model. There were indications of marine as well as fluviatile invasions from the north. The purported offshore, 'prodelta' clays contained horsetails and dinosaur footprints!

Pebble beds on the presumed delta tops seemed to imply hinterland episodes of considerable emergence and erosion. A visit to look at modern river sediments in India helped to provide the key to a completely new interpretation (Allen 1975).

The depth of deposition in all environments is now considered never to have exceeded 10 m and was usually less than 2 m, with the deeper water deposits probably sand rather than mud. The Wealden regime is seen as a coastal-alluvial mud trap of freshwater pools and brackish lagoons traversed by muddy and sandy channels. Periodic tectonic uplift to the north led to an abrupt increase of river gradients and vigorous loaded streams built out sandy aprons from the nearby uplands.

Allen's changed views provoked the following 'discussion' published with his paper, and presented here with apologies to Lewis Carroll.

> *Dr B. W. Sellwood and Dr L. B. Halstead asked:*
>
> You are old, Father Allen, the young man said,
> And the Chalk has become very white,
> Yet the Wealden you've turned once more on its head,
> Do you think that this time you are right?
>
> *The author subsequently replied*:
>
> In my youth, Bruce and Bev, I sought panaceas,
> But no one answered my calls,
> So filling the vacuum with my own daft ideas,
> I've really got caught by the b***s.
>
> In my youth, Bruce and Bev, those deltas were right
> If under eustatic control,
> But soil beds and footprints have put them to flight,
> And I'm now in a hell of a hole.
>
> In my youth, Bruce and Bev, I fought the old men;
> You think I'm at it again.
> But this time, my friends, the battle's between
> Myself and my dreamland terrain.

The moral of this cautionary tale is surely that humility and scientific honesty should be retained in the face of rocks we think we know well, even if it means proving ourselves wrong on occasion.

3. Marine environments and facies

Study of marine sediments and their contained organisms is more difficult than that of the continents and continental margins for several reasons. They are less accessible because of complete submergence by water and hence have to be investigated by dredge samples and shallow cores backed up by underwater photography and study of larger bed forms by acoustic methods. In limited areas of clear, shallow water these methods may be supplemented by investigations by divers. Large areas of the continental shelf are covered by Pleistocene relict sands, which complicates any simple attempt to relate sediment type to the present hydrographic regime. Furthermore, the hydrographic processes, involving tides, normal waves and storms in shallow waters, and turbidity currents in deeper waters, are complex, and physically produced sedimentary structures are frequently destroyed in shallow waters by intensive bioturbation by burrowing organisms. Nevertheless sufficient has been learned in recent years about marine environments to justify some soundly based generalizations pertinent to the interpretation of ancient marine deposits.

It is convenient to distinguish shallow marine or *neritic* deposits on the continental shelf from more offshore, usually deeper-water deposits mainly of the bathyal and abyssal zones, and siliciclastic from carbonate deposits.

Shallow marine regimes

NORMAL SHOREFACE AND OFFSHORE REGIONS WITH SILICICLASTIC DEPOSITS

Most of the area of the present-day continental shelf is covered by a blanket of siliciclastic sediments. Very generally there is an outer-shelf, relict sand blanket of pre-Holocene deposits not in equilibrium with processes operating now, and an inner-shelf, nearshore, sand prism of beaches, shoreface and barrier islands. A modern-shelf mud blanket often occurs inshore of the relict sands but may cover the whole shelf or just the outer shelf. Direct sediment supply from rivers is negligible except at the mouths of large rivers such as the Amazon, Mississippi and Ganges-Brahmaputra, where mud blankets frequently extend across the full width of the shelf. Some estuaries actually receive sediment from offshore (Johnson 1978).

The benthic biomass reaches a maximum on the continental shelf, 150–500 g m^{-2} as compared with a mere 1 g m^{-2} on the abyssal plains (Menzies *et al.* 1973). Consequently benthic organisms are of major importance in modifying the sediment. While there is little benthic life in and on mobile substrates where active transport of bed load takes place, finer-grained deposits exhibit greater stability and contain a rich fauna. Deposit feeders are abundant and continuous reworking and flocculation can destabilize the top few centimetres of sediment to a thixotropic condition (Rhoads and Young 1970). On the other hand, sea grass, algal baffles and organic surface films can have a stabilizing influence.

Tidal current velocity broadly correlates with tidal range and reaches its highest values where resonant amplification occurs, as in the Bay of Fundy. This bay has the world's highest range of 15 m, with currents of 1–2 m s^{-1}. Velocities of 60–100 cm s^{-1} are reached where the range exceeds 3–4 m. The importance of occasional storms in affecting sedimentation has become increasingly appreciated, and has led to the storm versus fairweather concept (Swift 1969). It proves useful to distinguish between *tide-dominated* and *storm-dominated* shelf sedimentation (Johnson 1978).

Tide-dominated shelf sedimentation is characterized by a high tidal range exceeding 3–4 m, as in partly enclosed seas and gulfs such as in north-west Europe, the Persian Gulf and the Gulf of California. Distinctive bed forms arise from the reworking of Holocene sediments, notably *sand waves* and *sand ridges*. Sand waves are straight-crested and transverse to the current. They have been studied most intensively in the southern North Sea, where they are mostly 3–15 m high and 150–500 m wide. *Sand ridges*, in contrast, are parallel to the current and much larger; in the southern North Sea they are 10–40 m high, 1–2 km wide and up to 60 km long. Both types of structure are cross bedded.

Wind-driven waves provide the major source of energy on the more extensive storm-dominated shelves where the tidal range rarely exceeds 2–3 m and the tidal current velocity is less than 30 cm s^{-1} compared with wind-driven currents of up to 70 cm s^{-1}. Sand ridges occur on more exposed shelves such as the Atlantic Shelf of the United States but not in areas of more restricted fetch and less intense wave activity such as the Gulf of Mexico. Such ridges may have a similar morphology and origin to those on tide-dominated shelves, but dunes and sandwaves are common on ridge flanks only in strongly tidal seas, and sandwaves are of subordinate importance in storm-dominated seas. Johnson (1978) considers that significant bed-load movement of sand and development of dunes, sand waves and ridges, giving rise to extensively cross-bedded deposits, are mainly restricted to tide-dominated seas, but I have seen large hurricane-generated submarine barchans on the Great Bahama Bank, where the tidal range is negligible.

Where sand passes into quieter, usually deeper-water mud environments there may be discrete bodies of sand in a muddy matrix, either as storm-generated sheets or as 'starved ripples' if the sand supply is limited, giving rise to flaser and lenticular ('linsen') bedding. Frequently, however, the sand and mud are mixed as a result of bioturbation. Indeed this facies of contrasted sedimentary types is ideal for the preservation of trace fossils. The various types of inorganic sedimentary structures likely to be preserved in the stratigraphic record are shown in Fig. 3.1. Wave-ripple cross lamination is a distinctive type of structure characteristic of sandy deposits in shallow seas (Fig. 3.2). Wave-generated ripples may not be easy to distinguish from current-generated ripples when, as is often the case, the two processes interact.

One of the few studies that pays equally close attention to both sediments and fauna of a shallow sea with siliciclastic deposits is that of Howard *et al.* (1972) on the coastal region of Georgia. The downshore profile is divided into several zones, each with characteristic features of the sort that should be preservable in the stratigraphic record.

Backshore beach. This has ripple-laminated sand with little bioturbation because of low diversity and density of the macrobenthos, but burrows of the ghost crab *Ocypode* occur.

Foreshore beach. The sediments are similarly composed of clean, fine sand typically with low angle, seaward-dipping sets of parallel or subparallel laminae. In addition there are minor ridge and runnel systems with variably dipping stratification, ripples, wavy bedding and shell concentrations, thought to have only low-preservation potential. There is a fauna of low diversity and high density but only minor bioturbation, with burrows of the blind shrimp *Callianassa.*

Shoreface. This also has clean, fine sand in parallel to subparallel laminated sets from 0 to −1 m mean low water (MLW) and ripple lamination at −1 to −2 MLW. This is a high-energy regime with much reworking by waves. The fauna is of low diversity and density and there is no bioturbation.

Upper offshore. Between −2 and −10 m MLW the sediment changes to muddy fine sand. There is interbedded sand and mud in the upper part but below −5 m bioturbation has totally destroyed any lamination. *Callianassa* burrows are conspicuous and the fauna is of both high diversity and density.

Lower offshore. Below 10 m depth occurs clean, medium- to coarse-grained sand with megarippled cross bedding. The low diversity, low density fauna includes the heart urchin *Moira*, which is responsible for most of the modest amount of bioturbation. This is a Pleistocene relict sand.

Without the relict sand to complicate matters one would have expected the upper offshore zone to pass down gradually into one with faunally-rich bioturbated muds.

The use of box cores to study marine ‘lebenspuren’ was pioneered by

FACIES	SUBFACIES	TYPICAL LOG	INTERNAL STRUCTURE	SAND CONTENT	BED OR SET THICKNESS	INFERRED PROCESSES AND NOTES
SANDSTONE FACIES	Sa Cross bedded	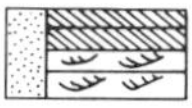	Tabular, Trough } Cross bedding	90–100%	*c.* 10–200 cm	Cross beds variable in type and set thickness. Represents dunes/megaripples (trough sets) and sand waves (tabular sets)
	Sb Flat bedded	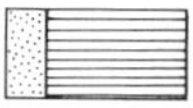	Parallel and low-angle lamination		variable	Wave- or current-formed lamination associated with high-energy conditions
	Sc Cross laminated	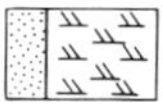	Cross lamination		1–5 cm	Cross lamination. Varies in relation to ripple type, notably current, combined-flow and wave ripples
HETEROLITHIC FACIES	Ha Sand-dominated	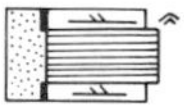	Parallel lamination	75–90%	5–20 cm (max 200 cm)	Alternations of parallel and cross-laminated sheet sandstones. Thicker sheet sandstones may form 20–90% of this subfacies. Amalgamation may be common. Sand deposited from suspension and as bedload. Variable reworking by current and wave ripples. Sheet sandstones commonly inferred to be the product of intense storm conditions. May contain transported shell debris. Bioturbation increases in the finer-grained intercalations
		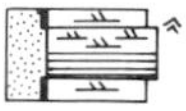	Parallel to cross lamination		5–20 cm (max 200 cm)	
		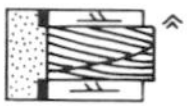	Low-angle and trough lamination		5–20 cm (max 50 cm)	
		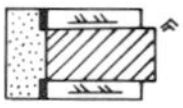	Isolated tabular cross bedding		5–20 cm (max 50 cm)	
		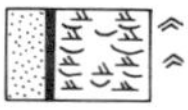	Sandy flaser bedding		1–5 cm	
	Hb Mixed	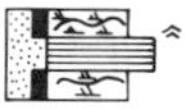	Parallel lamination	50–75%	1–10 cm	Mainly ripple laminated sandstones and mudstones with subordinate parallel-laminated sheet sandstones (10–50%). Variable types of cross lamination in response to current, combined-flow and wave ripples. Storm and fair weather increments may be recognized as above. Upper part of sheet sandstones bioturbated
		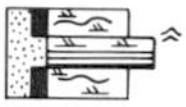	Parallel to cross lamination		1–10 cm	
			Low-angle lamination		1–10 cm	
			Flaser-wavy bedding		1–3 cm	
	Hc Mud-dominated		Parallel lamination	10–50%	1–5 cm	Mainly linsen bedding with rare sheet sandstones (5–10%). Sand lenses formed by current or wave processes. Sandstone interbeds deposited from suspension during storms. Suspension deposition of muds predominant fair weather process. Latter commonly intensively bioturbated
			Parallel to cross lamination		1–5 cm	
			Linsen bedding		1–3 cm	
MUD FACIES	Ma	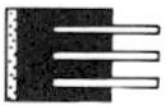	Graded sand and/or shell-rich layers	0–10%	0.1–2 cm	Mainly muds with thin sand interbeds and sand silt streaks. Deposition entirely from suspension. Wave and current activity only accompany rare storms. Intensive bioturbation, *in situ* or slightly transported benthic faunas
	Mb		Mud		< 0·5 cm	

3.1 Facies scheme for sublittoral siliciclastic sediments. After Johnson (1978).

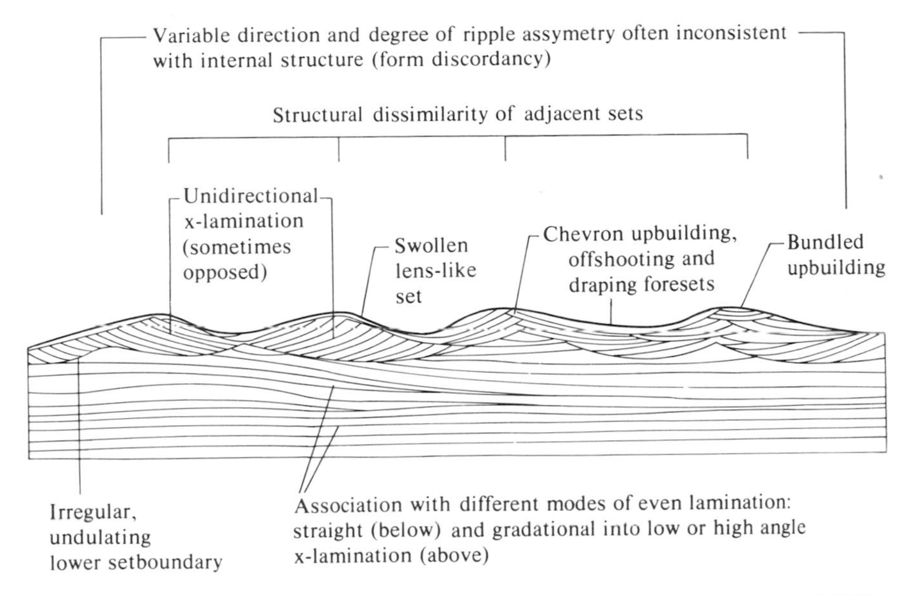

3.2 General features diagnostic of wave-ripple cross lamination. After de Raaf *et al.* (1977).

German workers at the Senckenberg Institute at Wilmhelmshaven. The results of detailed studies in several widely separated regions suggests that vertical burrows are almost the only forms in the shore zone between high tide level and wave base, a phenomenon evidently related to the intensity of physical disturbance. Below wave base there is a much greater diversity of forms including horizontal burrows (Fig. 3.3).

Turning to the stratigraphic record, bioturbated mudstones and clays with a rich benthic fauna, signifying a quiet but fairly shallow marine depositional environment, are so common that they do not call for any special comment or citation here. Attention will instead be devoted to considering to what extent the structures within sandstones can be used to distinguish between tide- and wave-dominated environments.

Tidally-created sand bodies should ideally exhibit herringbone cross bedding indicating two opposite directions of movement, reactivation surfaces and clay drapes (in deposits with a significant clay content). Unfortunately none of these criteria is foolproof (Johnson 1978). Reactivation surfaces and clay drapes are not unique to this type of environment, and the pattern of cross bedding is highly variable. The flood tide may be much stronger than the ebb, or vice versa, so that unidirectionality may predominate. Moreover oppositely dipping sets may represent much longer-term fluctuations than diurnal. Clay drapes are often up to 1 cm thick, which seems too much for diurnal fluctuations at a reasonable rate of sedimentation.

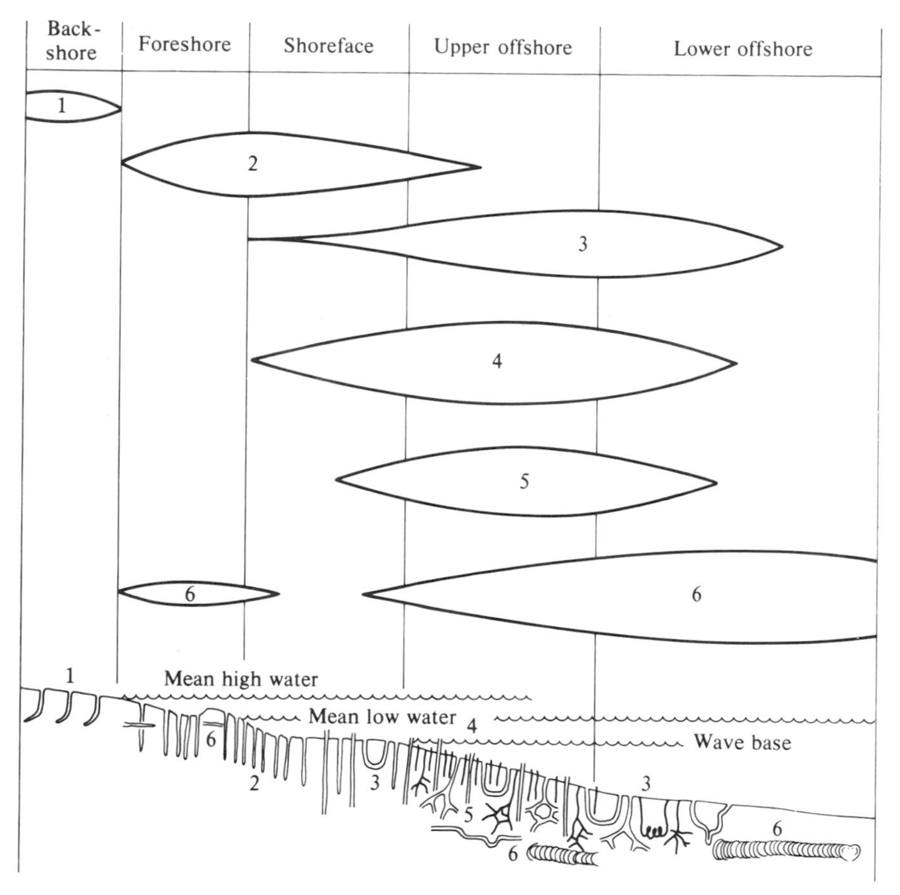

3.3 Distribution of organic traces in a generalized beach-offshore profile. 1, Burrows of terrestrial crustaceans; 2, straight vertical burrows of low-level suspension feeders; 3, U-shaped burrows of low-level suspension feeders or collectors; 4, tubes of high-level feeders; 5, dwelling burrows of intrasedimentary feeding animals; 6, crawling traces of intrasedimentary feeding animals. After Dörjes and Hertweck (1975).

One of the best-documented cases for a tidal sand body, the Cambrian Eriboll Sandstone of north-west Scotland, was made by Swett *et al.* (1971). They produced rose diagrams purporting to indicate crudely bimodal cross-bed trends suggesting NE–SW flowing tidal currents. Actually an examination of their figure 8 shows that the current patterns differ in each of the seven localities sampled, and only a minority approximate to a simple NE–SW direction. The supporting evidence cited includes trace fossils showing escape structures and stromatolites, birdseyes and desiccation cracks in the overlying Durness carbonates, but these features merely indicate very shallow-water conditions and tell us nothing about tidal range.

If sand waves indeed signify tidal activity, it is worth noting that several ancient examples have been claimed, for instance in Eocene sandstones of the Pyrenees (Nio 1976).

Wave- or storm-dominated environments should be signified by extensive sheet sandstone beds a few centimetres thick, by flaser bedding or by wave-ripple cross lamination, most distinctive if chevron structures are present. A probable example is described by de Raaf *et al.* (1977) from the Lower Carboniferous of Eire.

BARRIER ISLANDS

Apart from offshore coral reefs, the large coastal sand bodies, elongate parallel to the shoreline—barrier islands—are the only marine deposits that emerge above sea level. The best-studied examples occur a short distance offshore of the Atlantic and Gulf coasts of the United States. They are bounded on the shoreward side by shallow lagoons with only limited connections to the open sea by tidal channels (Fig. 3.4). The sand body, which normally coarsens upwards and culminates in aeolian dunes, may prograde seaward, as in the case of Galveston Island, or landward as a result of shoreface erosion and washover by marine storms. In the latter case marine transgression might result in the sort of sequence portrayed in Fig. 3.5.

Ancient barrier island sequences will only be recognizable by paying full attention to the broad facies context. Large lensoid bodies showing coarsening-upward sequences of well-sorted sandstone with large-scale cross bedding should lie a short distance offshore of the inferred shoreline and ideally should pass proximally into lagoonal and distally into finer-grained marine deposits. Possible examples which fulfil some of these conditions have been described from the Cretaceous of the United States Western Interior by Shelton (1965) and Davies *et al.* (1971). Insofar as modern transgressive barriers are related to the special conditions of the Holocene 'Flandrian' transgression, involving the reworking of Pleistocene offshore relict sands, it is not surprising that very few ancient examples have been recognized, but Bridges (1976) makes a case for one in the Silurian of South Wales.

CARBONATE PLATFORMS AND REEFS

Carbonate sediments are restricted to areas of high organic productivity and low siliciclastic input. Most carbonate shelves are found in the tropical–subtropical zone but smaller areas occur in higher latitudes, as off the west coast of Ireland and Scotland and the south coast of Australia (Sellwood 1978). Lees and Buller (1972) distinguish two types of association in waters of different latitudes, based on the characteristic calcite- and aragonite-secreting organisms. Their temperate water *foramol* association consists predominantly of benthic foraminifera, molluscs, bryozoans, calcareous red

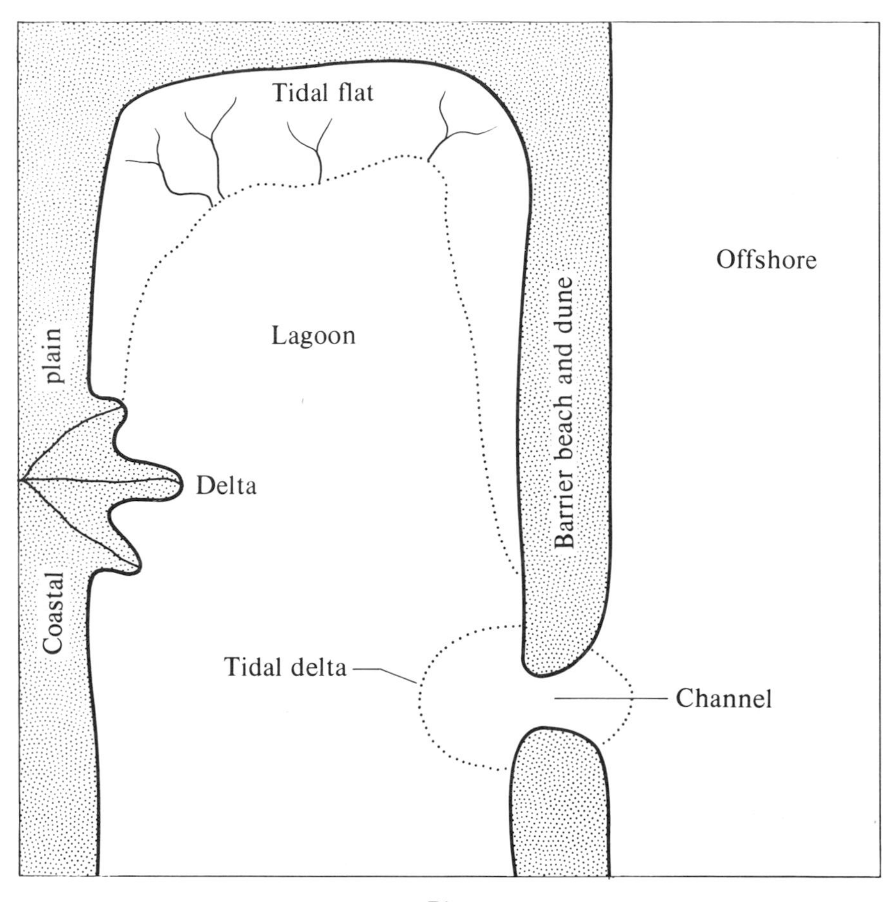

Plan

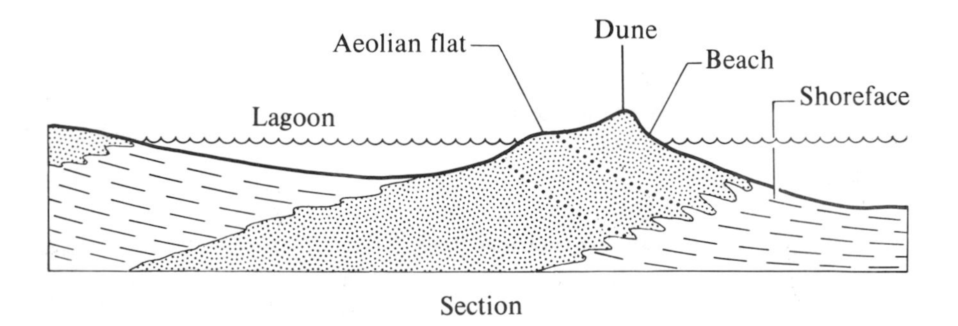

Section

3.4 Barrier island facies model based on Galveston Island, Texas. After Blatt *et al.* (1972).

	Description	
Dune	Fine-medium, well sorted Aeolian cross bedding	Barrier-beach
Beach-berm washover	Medium-very coarse, pebbly Low-angle lamination	
Back-barrier marsh	Peat-clayey sand	Outer lagoon margin
Tidal delta	Medium-pebbly, poorly sorted Abundant cross bedding	
Back-barrier lagoonal sand	Fine-medium, clean, well sorted Rare silt streaks	
Lagoon	Soft dark-grey clay-silt Abundant borings	Central lagoon
Beach		
Brackish salt marsh	Soft grey-brown organic mud Peat	Inner lagoon margin
Marsh fringe	Dark-brown muddy sand, roots	
Channel gravels (rare)		
Pleistocene		

3.5 Transgressive sequence of barrier island association deposited prior to shoreface erosion. After Elliott (1978b).

algae and barnacles, while their tropical water *chlorazoan* association contains in addition calcareous green algae and hermatypic corals. According to Lees (1975) inorganically-produced particle aggregates including ooids are restricted to the chlorazoan association, a phenomenon believed to be bound up with the higher salinities achieved in the tropics. Indeed, inorganic precipitation of aragonite aggregates is thought likely to occur only in warm waters of comparatively high salinity.

The two best-known areas of extensive carbonate sedimentation, on which facies interpretation depends heavily, are the Great Bahama Bank and the Persian Gulf.

The Great Bahama Bank occupies a microtidal area of 103 600 km^2 north, west and south of Andros Island, with water depth never exceeding 5 m. A thin veneer of sediments rests on a karstic surface of Pleistocene limestone produced during the last low stand of sea level. A line of coral reefs fringes part of the bank margin, and over the bank the deposits consist of carbonate sand and mud depending on the strength of water movements (Fig. 3.6). The carbonate sands consist either of ooids, concentrated in marginal areas where waters supersaturated in Ca^+ and HCO_3^- ions sweep over the bank from the surrounding ocean, or *grapestone*—composite particles of distinctive shape also produced by the aggregation of minute crystals of argonite. Locally wind-generated sand waves and dunes occur.

Mud composed of aragonite needles occurs in the more protected area west of Andros Island. There has been much dispute about whether this is an inorganic precipitate or a sediment produced by organisms. The argument now seems to be resolved in favour of the latter; the needles are evidently the disintegration products of green algae. Widely developed gelatinous algal mats serve to stabilize the sediment. Organic productivity is high only on the edge of the bank and the most conspicuous features of the more level surfaces are mounds produced by the activity of burrowing shrimps (Bathurst 1975).

The Persian Gulf differs in being a marginal sea in an arid region, with average depth about 35 m, reaching a maximum of 100 m near the Strait of Hormuz. Another difference is that tidal currents are important, although subordinate to wind-driven currents and waves as agents in controlling sediment transport (Purser 1973).

Carbonate sediments are deposited only on the more gently sloping southern side of the gulf, away from the influence of the Tigris–Euphrates river system. Coarse-grained deposits characterize the shallower water, more agitated areas down to about 20 m, below which the sediments are exclusively muddy. Tidal oolite deltas are found between barrier islands and coastal lagoons. Unlike on the Great Bahama Bank, calcareous green algae are not abundant and most of the carbonate mud appears to be a breakdown product of shell detritus, much of which is the result of organic boring activity. A feature of especial relevance to the interpretation of limestone hardgrounds is the occurrence of recently lithified surfaces produced by interstitial high magnesium-calcite precipitation at times of very slow sedimentation (Shinn 1969).

There is a moderately rich benthic fauna that diminishes in diversity with increased salinity, the biggest reduction occurring above 48%, with the loss of all corals, echinoids and most calcareous perforate and arenaceous foraminifera (Hughes Clarke and Keij 1973).

Coral reefs, or more accurately coral–algal reefs, occur widely in the tropics and have been well described in several places such as the Caribbean, Bahamas, Bermuda, the Seychelles and the Great Barrier Reef off Queensland. Because of their restriction to the photic zone the living reefs rarely extend to depths greater than 20 m. They have all been produced during the

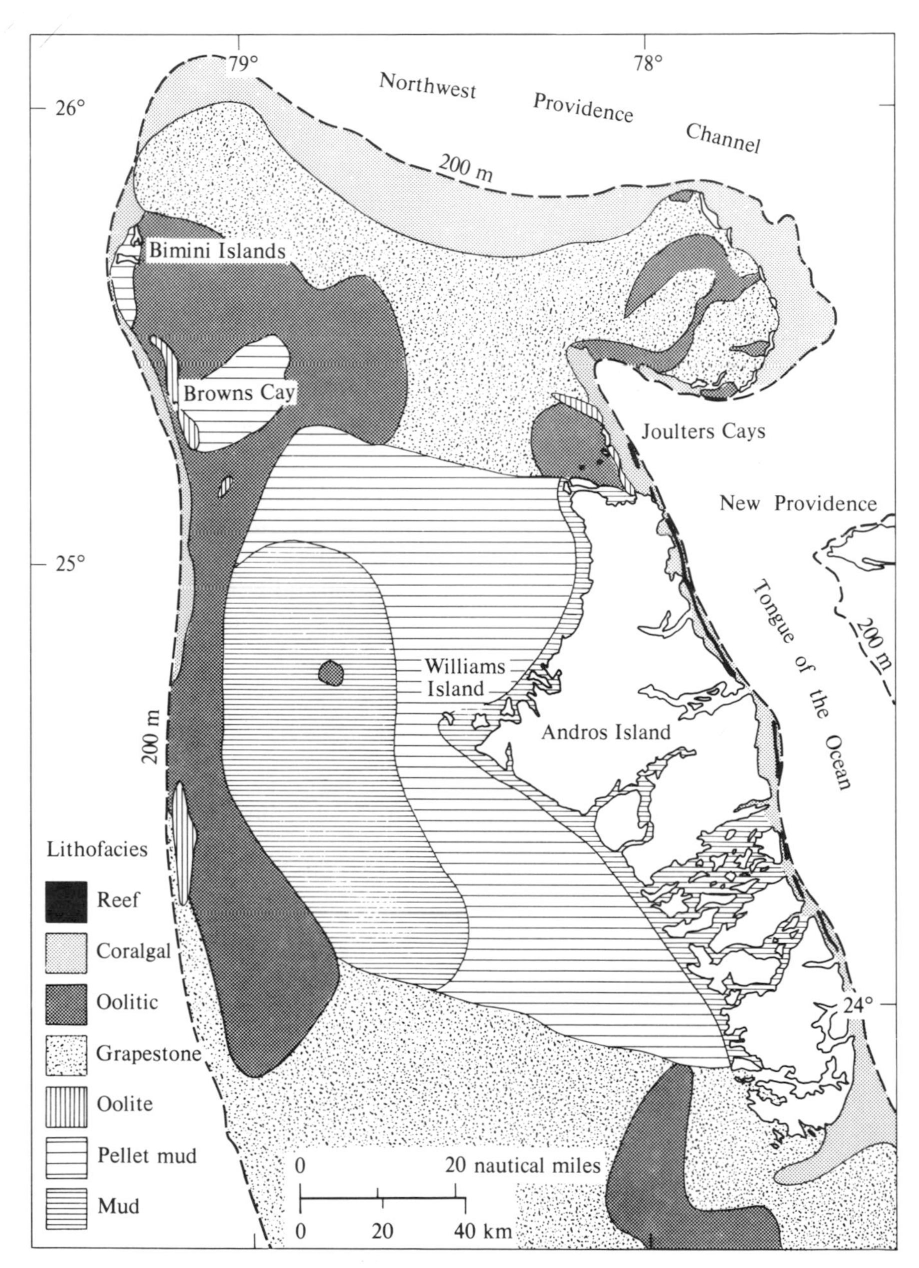

3.6 Major sediment distributions on the Great Bahama Bank. After Bathurst (1975).

Holocene transgression and rest on a karstic surface of limestone including coral rock, as in the case of the Bahamian deposits.

Such reefs are conventionally classified into fringing, atoll and barrier reefs, although the type example of the last category, the Great Barrier Reef, is a feature on the grand scale, with a whole complex of environments (Maxwell 1968). The most active growth tends to occur on the shelf edge because of the access of nutrients in the form of zooplankton.

From the point of view of making actualistic comparisons, however, *patch reefs* are equally important. Detailed descriptions from a geological point of view have been made of patch reefs in the shelf lagoons of the Bahamas and Bermuda respectively by Zankl and Schroeder (1972), Garrett *et al*. (1971) and Scoffin (1972). The Bahamian reefs range in length up to 300 m and contain large cave and tunnel systems up to 5 m high and several tens of metres long. The larger Bermudan reefs occurring at depths of 15–20 m, range from 75 to 400 m in diameter; the smaller ones, in waters less than 10 m deep, are usually less than 5 m wide but may range up to 2 m in height. The component coral species, which exhibit some degree of depth zonation, may build striking pillar structures coalescing laterally into cellular reefs. The associated fauna includes encrusters, borers and other vagile or sessile suspension feeders or grazers of various types.

The coral masses become progressively enveloped in sediment and cemented together as rock in early diagenesis. The interstitial sediment is composed largely of coral and algal fragments produced through a combination of organic boring activity and wave action and the original coralline frame builders may eventually constitute only a subordinate part of the rock so produced.

It has been argued that the high growth rates of living scleractinian corals leads to the creation of strong, upstanding, wave-resistant frameworks which provide a variety of crevices, and is probably a consequence of the symbiotic association with zooxanthellae (Coates and Oliver 1973). Similar growth forms should not therefore be expected with Palaeozoic rugose and tabulate corals.

Ancient carbonate platforms were enormously more extensive than modern ones, especially in the Palaeozoic (Wilson 1975). Particularly good and well-described examples of carbonate platform deposits occur in the Cambrian to Devonian sequence of the Central Appalachians (Laporte 1971) and the Upper Triassic and Lower Jurassic of the Mediterranean region (Bernoulli and Jenkyns, 1974). Shallow subtidal environments are inferred from oolites and shelly micrites, lagoonal environments from a variety of micrites, pellet and oncolitic limestones with a low diversity fauna and intertidal–supratidal environments from stromatolitic and birdseye limestones with horizons containing vadose pisolites and karstic erosion features (Fig. 3.7).

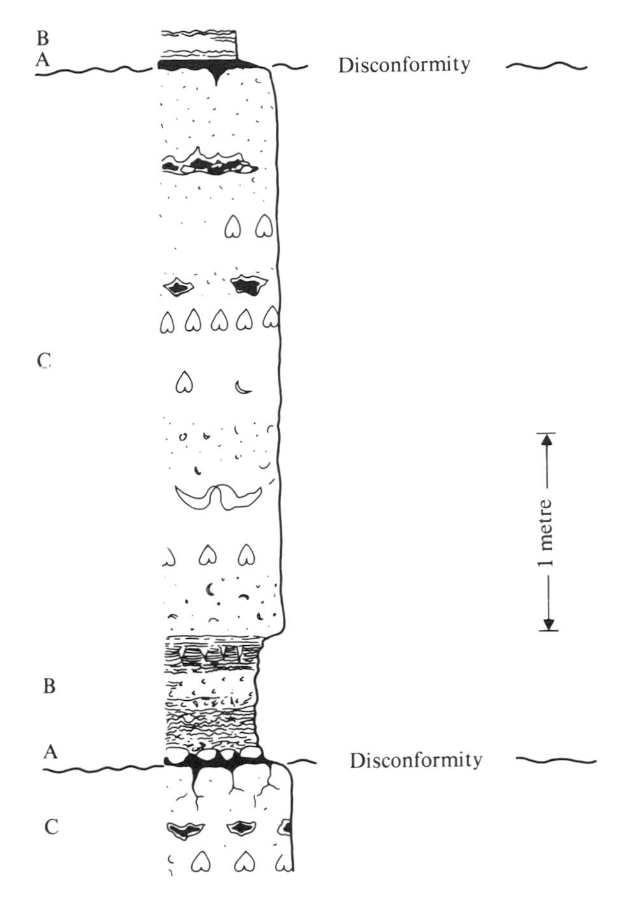

3.7 Diagrammatic representation of Fischer's (1964a) Lofer cyclothem in the Dachstein Limestone (Upper Triassic) of the Northern Calcareous Alps. A, Basal argillaceous unit representing reworked residue of weathered material (red or green), commonly confined to cavities in the underlying limestone; B, intertidal unit with dolomitic algal mats, desiccation pores (birdseye limestone) and mud cracks; C, subtidal megalodont limestone unit, with cavities produced by desiccation and solution during succeeding regression.

Many large lensoid bodies of massive, unbedded or poorly bedded limestone in the stratigraphic record have been termed reefs (Laporte 1974). However, the term is open to objection because it comes from the old Norse word *rif* meaning a rib of rock and has a special significance for navigators. Detailed studies suggest that many ancient 'reefs' might not have been firmly consolidated or reached wave base, hardly the sort of structures in fact that would constitute a hazard to shipping. Furthermore, there may be a conspicuous absence of organic skeletons. For these reasons the non-committal term *buildup* is to be preferred for these structures (Heckel

1974). Reef can then be defined as a biologically produced buildup that displays evidence of potential wave resistance or growth in turbulent water and of control over the surrounding sediment.

Whereas some buildups have a shape and facies association suggesting growth on the edge of a bank, as with most of the large modern reefs, others have a symmetrical mound- or dome-like shape suggesting isolated growth on a flat surface. Fossils vary from abundant to rare and vary widely in type. The lithology ranges from micrite to bioclastic limestone and the nearest to a common characteristic is the sparry calcite structure known as *Stromatactis*. This was formerly taken to be the recrystallized remains of an unidentified organism but is now generally accepted as a post-depositional cavity fill that could be produced by burrowing activity or decay of a soft-bodied organism in highly coherent sediment, followed by precipitation of calcite from circulating pore waters.

Some buildups may possibly have resulted from the entrapment of fine sedimentary particles by the baffling activity of benthic plants, as can be observed today in the shallow waters of Florida Bay and elsewhere. It is dangerous to assume, however, that all buildups necessarily formed in shallow water. Neumann *et al*. (1977) have, for instance, described large, steep-sided carbonate mounds up to 50 m high on the sea bed of the Florida Straits at 600–700 m depth. They have crusts of lithified rock which provide attachment surfaces for a dense and diverse community of corals, sponges and crinoids. Something resembling *Stromatactis* voids can also be produced. Therefore, as with so many other rock types, the facies context must be fully taken into account in environmental reconstruction.

Because of evolution it is only to be expected that faunal composition should change through time (Heckel 1974). Thus large and extensive stromatolite buildups are confined to the Precambrian and archaeocyathid buildups occur only in the Cambrian. The widespread Silurian buildups are dominated by tabulate corals and stromatoporoids, to which may be added rugose corals for the even more extensive and spectacular Upper Devonian buildups. Whereas the so-called Waulsortian buildups of the Lower Carboniferous are composed mainly of micrite with *Stromatactis* and subordinate bryozoans, the celebrated Triassic structures of the Alps are dominated by scleractinian corals, calcisponges and hydrozoans. Buildups with siliceous sponges in the Upper Jurassic of southern Germany were without much doubt formed in deeper water than the equally impressive coral buildups of similar age in the Paris Basin and Jura Mountains. Rudistid bivalves are an important component of some Cretaceous buildups. Many buildups have been intensively dolomitized and the fossils destroyed. What is loss for the palaeoecologist may be gain, however, for the petroleum geologist, for these frequently cavernous structures are among the best reservoir rocks known.

Fig. 3.8 illustrates three well-studied examples of large buildups that

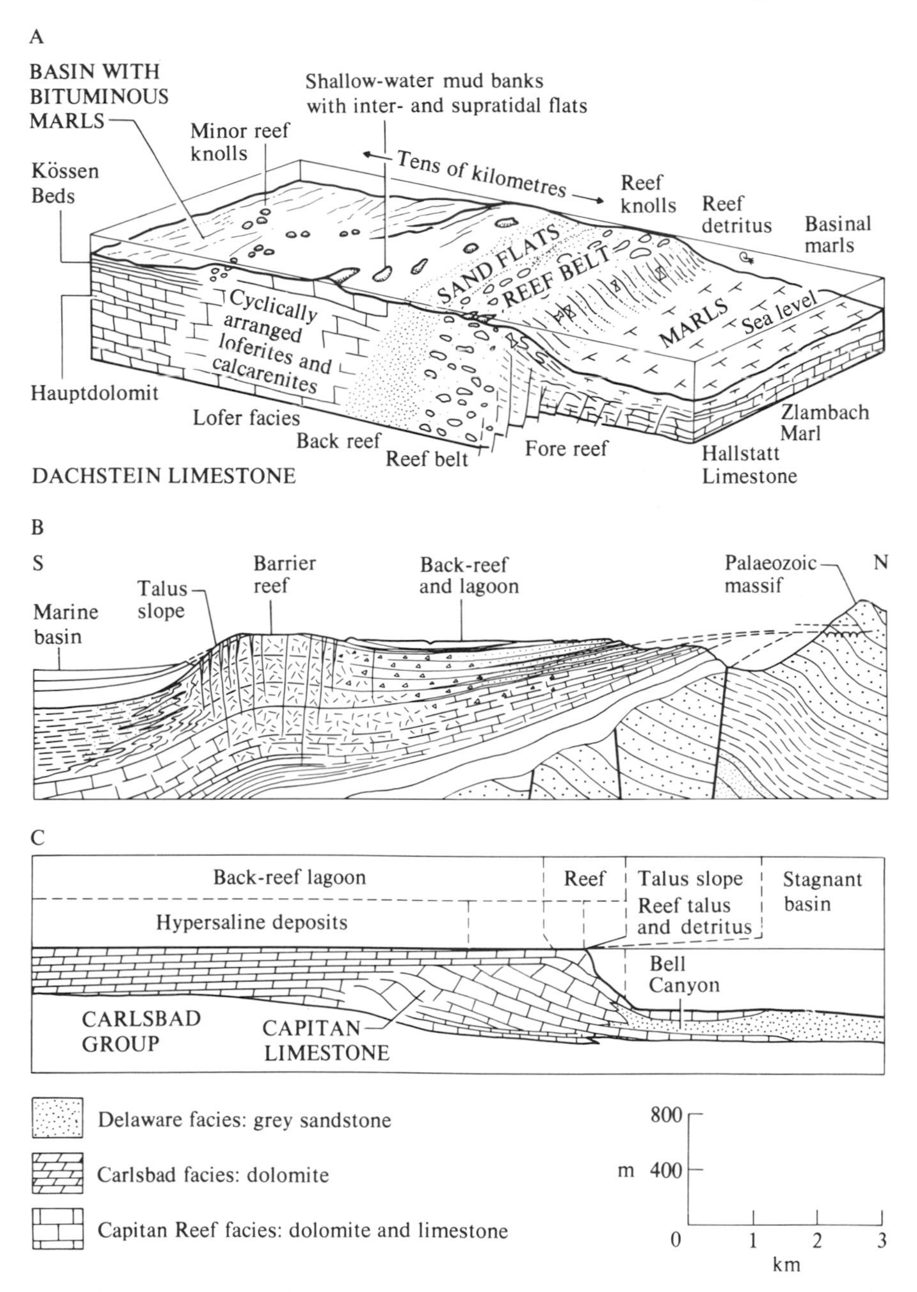

3.8 Schematic representations of three carbonate buildups interpreted as barrier reefs, separating a back-reef area from deeper sea. A, Upper Triassic of the Northern Calcareous Alps, after Zankl (1971); B, Lower Jurassic Bou Dahar reef complex of the Moroccan High Atlas, adapted from du Dresnay (1977); C, Permian reef complex of Texas and New Mexico, after Newell *et al.* (1953).

appear to qualify as genuine barrier reefs, with a characteristic sequence of environments: deep basin, fore-reef talus slope, reef front and back-reef lagoon. Of these perhaps the most celebrated is the Capitan reef complex. Perhaps it is not surprising that, because it has attracted the attention of so many geologists, several different environmental models have been proposed (Dunham 1972; Kendall 1969; Newell *et al.* 1953).

STAGNANT BASINS

Oceanic bottom waters are fully oxygenated at the present day except in a few regions, notably where land-locked basins are separated from the ocean by a narrow and shallow sill. In such '*barred basins*' the water tends to stratify as a result of restrictions to circulation combined with the development of temperature or salinity gradients with depth. Should stratification be maintained for extended periods of time, the water beneath the density boundary will become anoxic as dissolved oxygen is replaced by hydrogen sulphide, and this will be reflected in the bottom sediment. The best-known examples are the Black Sea, Cariaco Trench and certain Norwegian fjords. Normal shelf depths may be considerably exceeded, but it is convenient to consider them here because they occur close to land. Anoxic bottom conditions can also occur locally in more open oceans with highly fertile and productive surface waters usually associated with upwelling, where the oxidation of organic matter descending through the water column results in the formation of an oxygen-minimum later. In places where this layer impinges on the sediment–water interface, as in the north-west Indian Ocean and off Peru and Namibia, zero oxygen values are recorded (Thiede and Van Andel 1977).

In the ocean system as a whole only a minute fraction of the organic matter produced is incorporated into the bottom sediment (Hallam 1967b), but in stagnant basins a considerable percentage is preserved in the form of millimetre-thin laminae. The principal source of this organic matter is phytoplankton, but there is usually in addition a variable proportion of terrestrially derived pollen, spores and macroscopic plant debris. Direct evidence that the rhythmic couplets of organic and inorganic mineral matter are annual varves has come from investigations in the Clyde Sea of Scotland (Moore 1931), a bay of the Adriatic (Seibold 1958) and in the centre of the Gulf of California (Calvert 1966). The organic layers may correspond to heavy fall of phytoplankton following seasonal blooming in the surface waters but in the Gulf of California a more or less constant fall of diatoms is seasonally interrupted by siliciclastic influx from the Colorado River. Sedimentary carbonates may occur in the intervening inorganic sediment, as in the Black Sea, but generally carbonates tend to be dissolved in the deeper anoxic water layer (Degens and Stoffers 1976).

The Black Sea has frequently been taken as the type example of a barred or *euxinic* basin (the very term comes from the ancient Greek name for the Black Sea). Recent investigations of the sedimentary sequence have established, however, that anoxic conditions coincide with climatic optima and a fully oxidized water column was evidently in existence through much of the Quaternary (Degens and Stoffers 1980).

Examples of shales and micritic limestones with bituminous laminae abound in the stratigraphic record. The extent to which the barred-basin model is apposite will be discussed in chapters 5 and 8.

Deep marine regimes

PELAGIC DEPOSITS

The term *pelagic* relates to the open sea and does not necessarily signify deep water. The great bulk of what are usually known as pelagic deposits nevertheless occur at bathyal and abyssal depths. The predominant type of sediment is calcareous ooze, composed of coccoliths and planktonic foraminifera, and occupies nearly half the total oceanic area. Next in importance (38%) is the non-calcareous *red clay* of the abyssal plains, most widespread in the Pacific. Siliceous oozes consist mainly of diatoms in high and radiolaria in low latitudes (Fig. 3.9).

The distribution of these various types of deposit is controlled primarily by variations in surface water productivity and the *calcite compensation depth* (CCD). This is the depth below which calcite does not accumulate on the ocean floor, and corresponds to the level at which the rate of calcite solution is balanced by the rate of supply. It occurs on average at about 4500 m but the level fluctuates in relation to productivity, so that it is depressed in the high-productivity equatorial zone (the compensation depths for aragonite and opaline silica are shallower and deeper respectively). In contrast to shallow marine regimes, benthic organisms are usually extremely sparse, though most phyla are represented and the diversity may be quite high.

Jenkyns (1978) reviews pelagic sediments in relation to major physiographic features. On spreading ridges, typically rising to about 2700 m depth, the sediments directly above the basalt tend to be of distinctive type. They are characterized by smectite derived from the chemical breakdown of basalt and are characteristically enriched in metals, notably iron and manganese, but with subordinate copper, lead, zinc, nickel and cobalt. There is some dispute about the origin of these metals, whether they are a product of secondary leaching of basalt (Klinkhammer *et al.* 1977) or come directly from a primary magmatic source. There is extensive submarine lithification by high magnesian calcite and local ponding and redeposition that gives rise to pelagic turbidites. The pattern of sedimentation varies according to whether the ridge is of the fast- or slow-spreading type (Fig. 3.10).

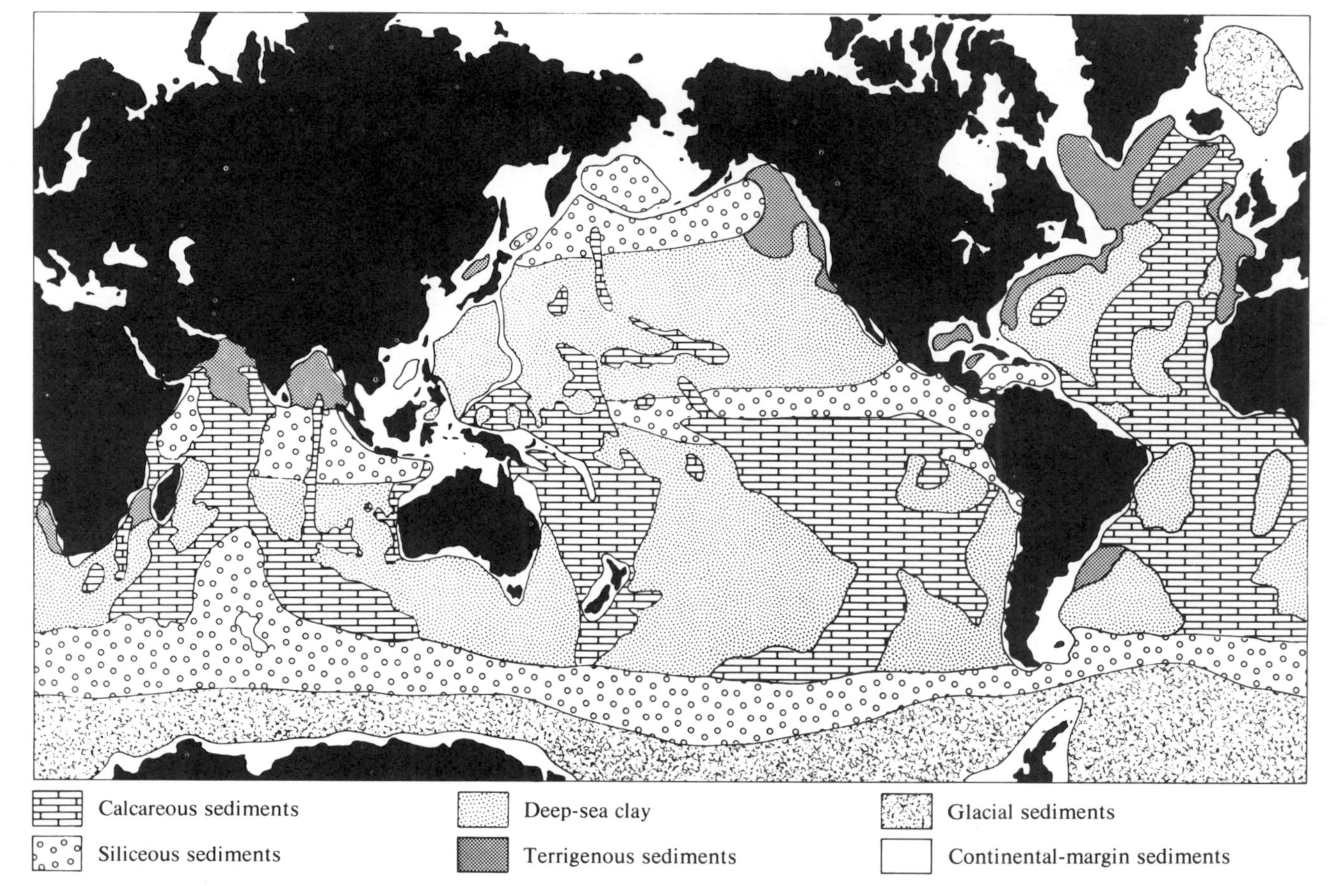

3.9 Global distribution of principal types of pelagic sediment on the ocean floors. After Jenkyns (1978).

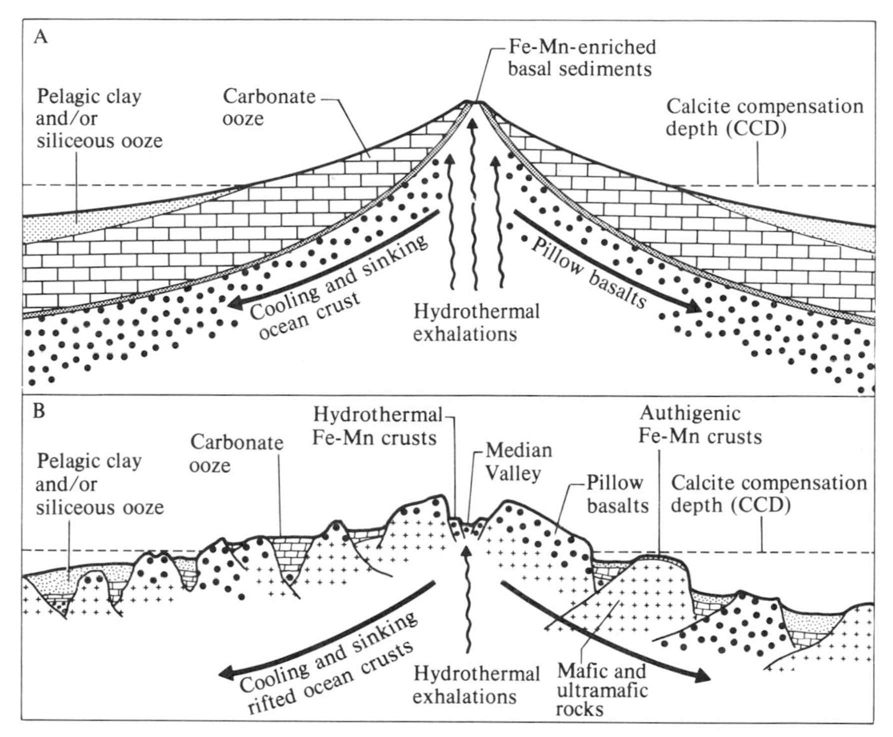

3.10 Sediment distribution on (A) a fast-spreading rise of East Pacific Rise type, (B) a slow-spreading rifted ridge of Atlantic type. After Jenkyns (1978).

Recent research utilizing submersibles has yielded some spectacular results. Thus massive ore-grade sulphides of zinc, copper and lead have been found associated with the basalt of the East Pacific Rise (Francheteau *et al.* 1979). Both on the Rise and in the Galapagos Rift Zone dense colonies of suspension-feeding benthos occur at sites of hydrothermal discharge. The fauna is dominated by bivalves of unusually large size for a deep-sea fauna (Lonsdale 1977). The existence of this unusual fauna poses the intriguing problem of food source. It seems likely that the organisms feed on bacteria that proliferate at these warm submarine springs.

Aseismic ridges such as the Rio Grande–Walvis Ridge in the South Atlantic differ from spreading ridges in lacking metal-rich deposits; also there may be evidence of shoal-water deposition. Volcanic seamounts have basal sediments chemically distinct from spreading ridge sediments and normal Pacific pelagic clays, having a generally lower content of trace elements. Many guyots in the West Pacific are capped by early Cretaceous limestones with a typical reef fauna of corals, rudists, algae etc. indicating considerable subsidence in the last 100 million years. Iron–manganese crusts are widespread on seamounts and there may often be a sessile epifauna of foraminifera, corals, bryozoans and serpulids. Those that rise into the photic zone may also have calcareous algae.

The oceanic plateaus rising to within 2–3 km of the surface, such as the Shatsky and Magellan rises of the West Pacific, are characterized generally by thick sedimentary sequences related to high plankton productivity. The deep ocean basins have calcareous sediments above the CCD and red clay below. Around topographic highs there may be redeposited volcanogenic and pelagic aprons. Siliciclastic terrigenous turbiditic sediments occur near the continents. The red clay, much of which is probably aeolian in origin, is composed principally of illite and smectite with subordinate authigenic zeolite minerals. The presence of manganese nodules, especially in the North Pacific, cosmic spherules and sharks' teeth, testifies to an extremely low rate of sedimentation. The dominant colour, brown rather than red, is due to the presence of iron oxides.

The siliceous sediments of the ocean basins are composed almost entirely of the remains of planktonic diatoms, silicoflagellates and radiolaria, which secrete skeletons of the relatively soluble opaline silica. The highest concentrations occur in the peri-Antarctic zone, where the sediments are overwhelmingly diatomaceous (some 80% of the world total) whereas radiolarian skeletons are predominant in the equatorial Pacific (Calvert 1974). In certain hemipelagic regimes where there is upwelling of nutrient-rich water, as off southern California, there may be concentrations of phosphatic sediments (Bromley 1967).

A large number of examples of ancient pelagic deposits have been recorded from rocks now exposed on the continents (Hsü and Jenkyns 1974; Jenkyns 1978). One of the most interesting is found in the Troodos Massif of Cyprus, which is the site of possibly the best documented ophiolite sequence and sedimentary cover. The basal sediments are brown mudstones of Campanian age rich in iron, manganese and other metals and known as *umbers*. These appear to have been deposited in ponds on the spreading ridge, and pass up into marls, chert and chalk (Fig. 3.11). The ophiolites notably contain pockets of metallic sulphide ores.

Other notable deposits include the Cretaceous red clays of Timor, which contain iron–manganese nodules and sharks' teeth, and the Palaeogene clays of Barbados rich in foraminifera, radiolaria and nannoplankton and also containing many fish teeth and cosmic spherules. The subject of ancient pelagic deposits will be dealt with more extensively in chapter 8.

DEEP-SEA TURBIDITE AND ASSOCIATED DEPOSITS

Because of the difficulty of making direct observations in modern seas the interpretation of turbidites and associated deposits leans heavily on sedimentological studies of rock sequences backed up by laboratory experi-

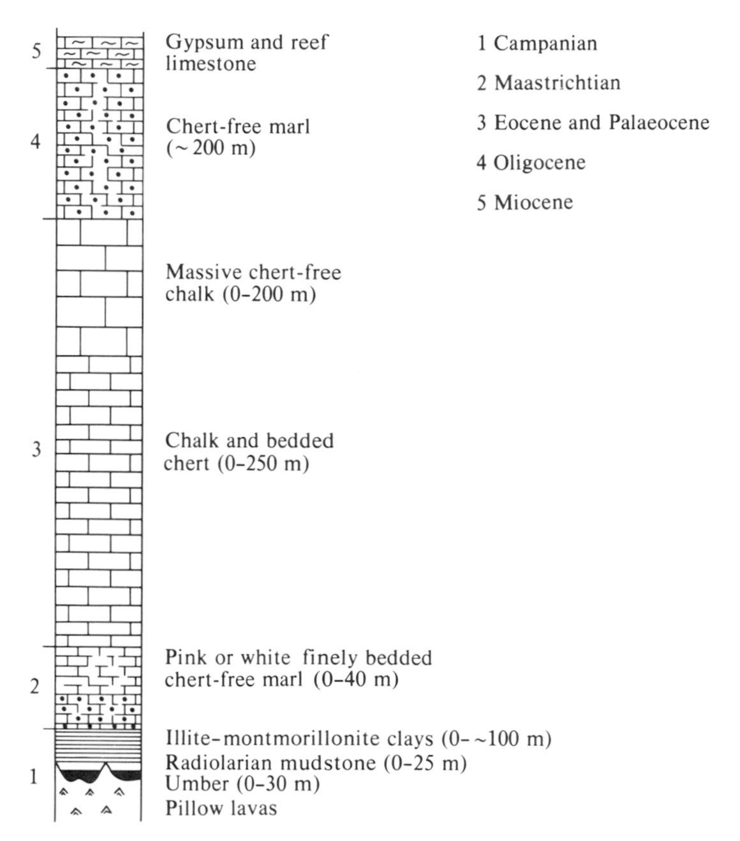

3.11 Composite section of Upper Cretaceous and Tertiary oceanic sediments of Cyprus, overlying Troodos ophiolite complex. After Robertson and Hudson (1974).

ments. Before application of turbidity current theory to the Macigno of the Apennines by Kuenan and Migliorini (1950), the numerous small-scale alternations of sandstone and shale in thick 'flysch' sequences of orogenic belts were generally attributed to frequent vertical tectonic oscillations. Since that classic paper was published there has been a major change in thought and *turbidite* has become deeply embedded in geological literature, even though it suffers from the disadvantage of being interpretative rather than descriptive.

There is no depth connotation and turbidites are deposited also in lakes, but by far the most important regime nowadays is in the deep sea beyond the mouths of submarine canyons. By no means all the sediments of deep-sea fans and their surroundings are to be attributed to deposition from turbidity currents. Rupke (1978) distinguishes three major categories of sediment transport that can give rise to characteristic types of deposit:

1. *Mass gravity transport.* This appears to be the predominant process transporting siliciclastic sediment to the deep sea at present. The sediment moves downslope only when the shear stress exerted by gravity exceeds the shear strength of the sediment. This may be induced by thickening of the sedimentary pile by deposition, by increase in pore fluid pressures leading to fluidization, or by thixotropic changes, converting a gel into a sol. The latter two categories tend to be triggered by earthquakes, tsunamis or storm waves.

Slumping involves the movement of a mass of semi-consolidated sediment along a basal plane of failure while retaining some internal coherence of bedding, although there may be some rotational deformation along slide planes. The term *debris flow* is used if the sediment contains matrix-supported clasts. Thick-bedded coarse-grained sediments may be the result of *grain flow* (Fig. 3.12). Slumping is largely restricted to areas of rapid deposition such as delta fronts and canyon heads and can occur on slopes as gentle as 1°. Transport can involve volumes of sediment of the order of hundreds of cubic kilometres.

TURBIDITY CURRENT

Rippled or flat top
Ripple drift micro-x-lamination
Laminated
Good grading ('distribution grading')
Flutes, tool marks on base

FLUIDIZED/LIQUEFIED FLOW

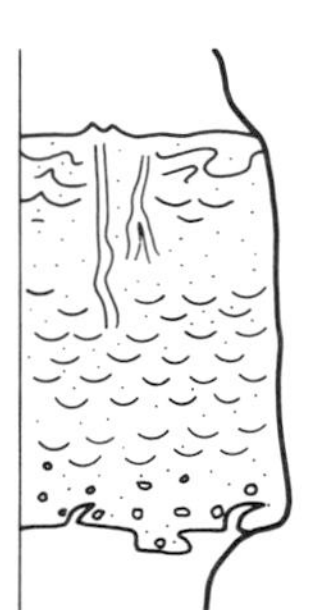

GRAIN FLOW

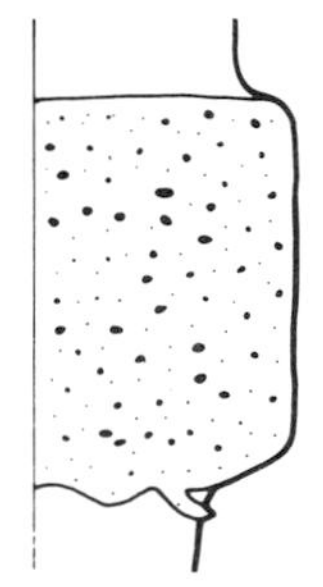

DEBRIS FLOW

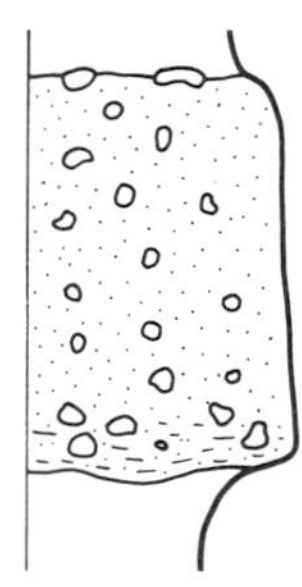

3.12 Structures and textures of deposits of mass-gravity flows. After Middleton and Hampton (1976).

2. *High-density turbidity currents.* These are suspensions of sand and mud with a specific gravity between about 1.5 and 2. When the flow decelerates deposition of graded beds may result (Fig. 3.12). Many deep-sea cable breaks are attributed to the action of fast-moving turbidity currents along channels. River-generated turbidity currents seem to be associated with times of high discharge. The characteristic *Bouma sequence* of massive graded beds passing up via laminated into rippled and then more laminated beds has been interpreted in terms of a waning flow regime (Fig. 3.13).

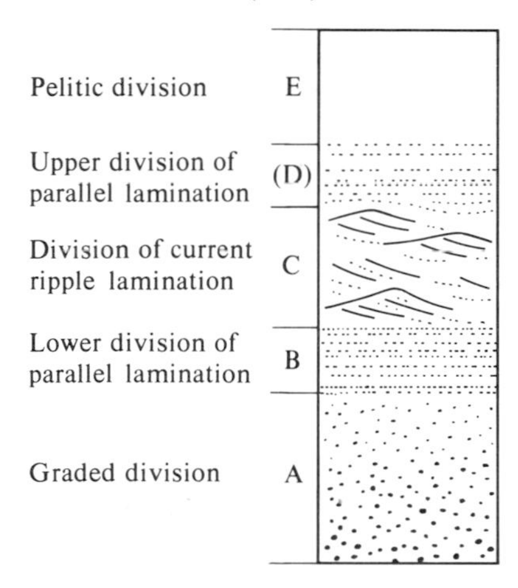

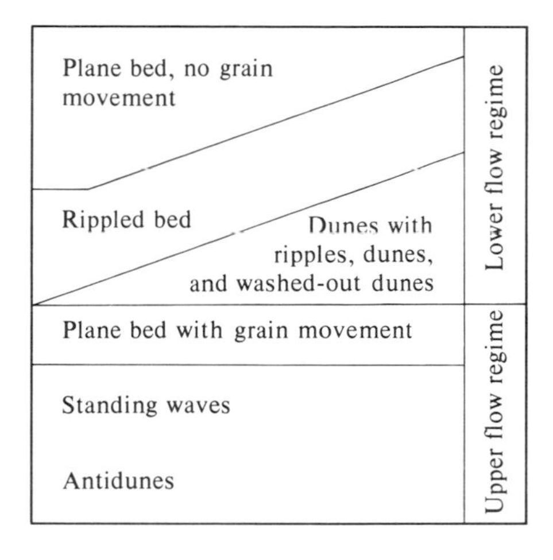

3.13 Interpretation of the Bouma turbidite sequence in terms of a waning flow regime. After R. G. Walker (1967).

3. *Low-density turbidity currents.* It has been recognized comparatively recently that many of the clay–silt layers between turbiditic sandstones may be the result of deposition from low-density turbidity currents rather than 'background' pelagic sedimentation. Close examination of these deposits may indicate such diagnostic features as fine lamination and grading.

Traction currents are also known to operate in the deep sea, as evidenced by ripple and small-dune forms (Heezen and Hollister 1971). Such currents may follow topographic contours but are relatively slow moving. They may give rise to rather fine-grained, well-sorted, thin sand beds with sharp tops and bottoms.

Within the deep Atlantic drilling has established the existence of laterally very extensive sandstone beds back from the present through the Tertiary. The Pacific is characterized in contrast by deep-sea fans related to canyon heads and by channelized flows. Turbidites are ponded by fault-bounded basins and trenches near the continents. The trenches have on average 10° slope towards the land and 5° towards the ocean. The proximity of volcanic land of high relief is signified by texturally immature volcaniclastic sediment. The direction of turbidity currents in the trenches is predominantly longitudinal, with slumping directions at right angles.

Modern turbidite fans are partial cones analogous to alluvial fans and may occur off major river deltas. It is possible to distinguish an inner fan, with thick immature turbidites and poorly developed Bouma sequences, from middle and outer fans with distally thinner and mature turbidites showing good Bouma sequences. The fans are traversed by channels margined by levées and interchannel areas where silts are the most typical sediments. Most such fans seem to be the result of resedimentation of siliciclastics during Pleistocene low stands of sea level, and at present a more slowly deposited mud blanket is forming.

The kinds of sedimentary sequence to be expected in deep-water slope, fan and basin plain environments are protrayed in Fig. 3.14. R. G. Walker (1967) makes an important distinction between *proximal* and *distal* turbidites, respectively nearer to and more distant from the sediment source. Proximal turbidites are characterized by thick, coarse-grained, poorly graded or ungraded beds often with scoured bases, while distal turbidites are thinner and finer grained, with more even surfaces. Scours and channels are rare, tool marks more frequent and grading good.

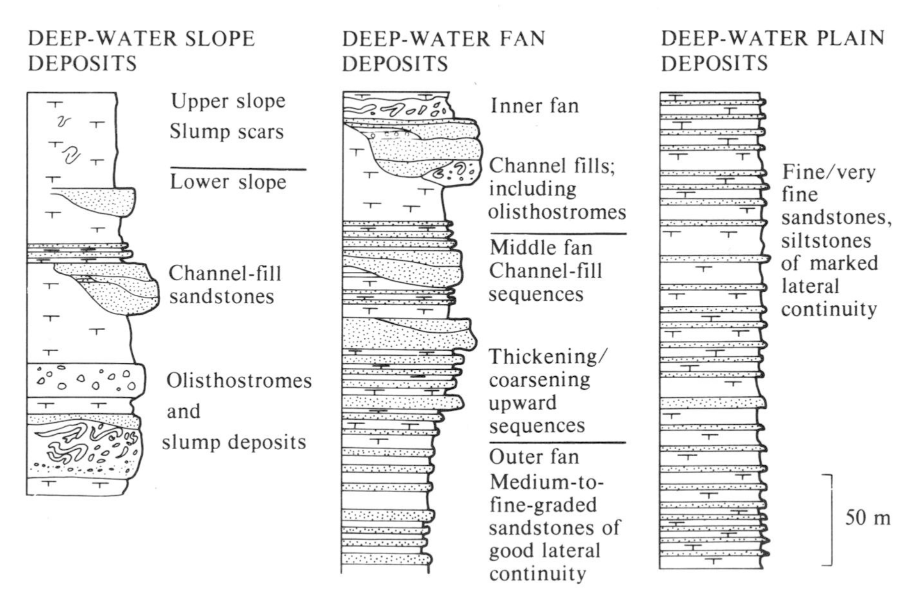

3.14 Sequences characteristic of slope, fan and basin plain facies. After Rupke (1978).

Numerous ancient turbidite sequences have been described, but are more appropriately dealt with in the next chapter. Two further important facts need, however, to be mentioned here. Whereas benthic organisms are sparse in such sequences, there may be a rich bedding-parallel ichnofauna of complicated crawling and feeding tracks, the so-called *Nereites* facies of Seilacher (1967). Many Palaeozoic sequences in orogenic belts are typified by the poorly sorted mixed sand–clay rock known as greywacke, which frequently exhibits typical turbidite features like graded bedding and sole marks such as flute and groove casts. Cummins (1962) has shown convincingly that the 'clay' component may be derived from the chemical breakdown during diagenesis or mild metamorphism of unstable volcaniclastic grains of sand grade. Thus many such greywackes might originally have been as well sorted as many modern deep Atlantic sands or sandstones in the Alpine flysch.

4. Sedimentation and tectonics

Many stratigraphic units exhibit striking lateral changes in thickness and facies over short distances. Substantial water depths and exceptionally high sedimentation rates may be inferred for some deposits, which may also contain clasts of a wide size range indicating the proximity of land of high topographic relief. Slump and slide deposits may indicate the former existence of appreciable depositional slopes, while in extreme cases enormous masses of strata ranging in size up to a kilometre or more, might have been detached and incorporated into younger deposits. There may be interlayered lava and ash beds, or volcaniclastic detritus in the sediments, to indicate contemporary volcanicity. Frequently the strata might have been intensely deformed by fold movements.

It is by such characteristics as these that we infer significant control of sedimentation by tectonics of the sort that create orogens or faulted horst and graben structures. By far the most influential concept over the last century relating sedimentation and tectonics has been that of the *geosyncline* (Dott 1974; Mitchell and Reading 1978). Although the concept has today been largely superseded by plate tectonics it is instructive to trace its history in outline and thereby learn something about the evolution of thought.

The recognition that elongate zones of comparatively thick sedimentary sequences, implying high subsidence rate in a trough, may broadly coincide with orogenic belts was first made by James Hall (1859). The rocks he was concerned with were shallow-water Palaeozoic deposits in the Appalachians; he believed that sedimentation was the sole cause of subsidence and that such zones of subsidence were more liable to be folded subsequently than more 'stable' zones of low sedimentation and subsidence rate, such as occur further towards the North American continental interior. The term geosyncline was proposed not by Hall, however, but by Dana, whose views differed insofar as he could not accept that sedimentation alone caused subsidence. Instead he postulated that tectonic downbending of *geosynclinals* (Dana's original term) was complemented by upwarped *geanticlinals* (that could incidentally be a source of geosynclinal sediment) during the progressive contraction of the Earth.

Towards the end of the nineteenth century the geosynclinal concept was taken up by leading European geologists such as Haug, Bertrand and Suess to account for significant features of the Alpine orogen. A fundamentally different viewpoint emerged on the European Continent. To Haug, a geosyncline was an elongate trough of considerable water depth receiving pelagic sediment. Steinmann and others saw the geosynclinal sequence beginning with thick ophiolitic igneous rocks followed by deep-water radiolarian cherts and shales. American geologists failed to recognize such rocks in their geosynclinal sequences and expressed scepticism about the so-called deep-sea deposits.

Several attempts have been made this century to subdivide geosynclines, the most important of which was by the German tectonicist Hans Stille. His *orthogeosynclines* are those linear feature between cratons associated with orogens and were divided into two categories, eugeosynclines and miogeosynclines, respectively with and without igneous rocks in the sequence. His *parageosynclines* embraced other zones of significant subsidence and sedimentation, often associated with block faulting, commonly ovate rather than linear, and comparatively short-lived.

Stille's terms *eugeosyncline* and *miogeosyncline* became widely adopted whereas parageosyncline, or its various subdivisions proposed by Kay (1951), did not. This is probably because there is frequently only a tenuous association with orogeny, if any, and because the simple term *basin* (with some adjectival qualification) is usually perfectly adequate as a descriptive term. It is certainly more euphonious and less of a mouthful than Kay's terms exogeosyncline, autogeosyncline and zeugeosyncline.

Two important developments of Stille's ideas were by Kay (1951) and Aubouin (1965), respectively concerned with North America and the circum-Mediterranean region. Kay introduced the idea of paired geosynclinal belts bordering a central craton, the eugeosyncline being more distal than the miogeosyncline. Rather than having to invoke tectonic borderlands to provide sources of sediment, uplift of island arcs *within* the eugeosyncline could provide sufficient sediment. Aubouin recognized, primarily on the basis of his own studies in Greece, that both types of orthogeosyncline could be divided into longitudinal ridges and furrows.

A different way to classify geosynclinal deposits is in terms of their temporal relationship to orogeny. Facies research in the Alps led to the concept of a pre-orogenic phase characterized by sediment starvation in deep basins, the *leptogeosyncline* of Trümpy (1960). Similarly in the Alps, the synorogenic phase is characterized by *flysch* facies and the post-orogenic phase by *molasse* facies. Flysch and molasse have proved such useful terms that they have become international currency, despite the attempt by some Swiss geologists to exercise proprietorial rights (an attempt that has been as futile as that of French purists who would wish to exclude 'weekend' and 'camping' from their language).

Flysch was introduced by Studer in 1827 as a lithological term for a Lower Tertiary formation of alternating sandstones and shales in the Simmental and by late last century was recognized as a widespread facies in mountain belts. It is now generally understood to signify a thick succession of sandstones, bioclastic limestones or conglomerates alternating with shales and mudstones, interpreted as having been deposited by turbidity currents or mass flow in the deep-water environment of a geosynclinal belt.

Studer was also the first geologist to use the term molasse, applying it to the whole Cenozoic sequence of the Swiss plateau. Bertrand and Haug used it in a more general sense and it is now usually applied to thick siliciclastic sequences dominated by sandstones and conglomerates as well as to alluvial deposits deposited in a subsiding trough or basin after the principal orogenic episode, the sediment source being the newly risen mountains (Van Houten 1974). The fact that terms like flysch and molasse can be used in more than one sense precludes any rigorous definition, which would be too limiting. No great harm is done if the rocks in question are well described elsewhere.

The weakness of geosynclinal theory was less that the term geosyncline was used in a different sense by different people than that it was not tied to an adequate theory of global tectonics; hence much of the confusion surrounding Stille's parageosynclines. With the advent of plate tectonics in the late 1960s this situation has changed drastically for the better (e.g. Dickinson 1974; Dott and Shaver 1974), but there are still problems. A major difficulty in relating sedimentary facies to plate tectonics is that sedimentation is only indirectly related to the geophysical processes on which the theory is based. One simple classification that suggests itself is into *extensional* and *compressional* regimes but both subduction zones and transform/strike-slip fault systems may contain regions of extension or compression, or regions which are not well characterized by either term. A more useful distinction is to attempt to relate sedimentary facies to general tectonic settings specified by plate theory, such as spreading-related, subduction-related, strike-slip fault-related and continental collision-related settings. This is the classification proposed by Mitchell and Reading (1978) and is adopted here.

Spreading-related settings

These systems are characterized by sedimentation in graben or half-graben created by extensional tectonics or *taphrogeny*. They may be intracontinental, as in the case of the huge East African and Baikal rift systems containing lacustrine and alluvial sediments. Alternatively they may be oceanic ridges, in which case the sediments will be pelagic (see chapter 3) or marginal to an incipient ocean like the Gulf of California or Red Sea.

The western margin of the Red Sea shows a set of highly characteristic structures, namely fault blocks tilted away from the axis of spreading, forming for example the Danakil Alps and Depression (Hutchinson and Engels 1970). Up to 5 km of Oligocene and Miocene evaporites and siliciclastic deposits occur in association with volcanics related to tensional faulting along the margin of the coastal plain (Fig. 4.1). Such tilted fault blocks are found along many continental margins, although the relationship to spreading axes is not always clear. It has even been suggested that there was a worldwide phase of continental margin taphrogeny during the

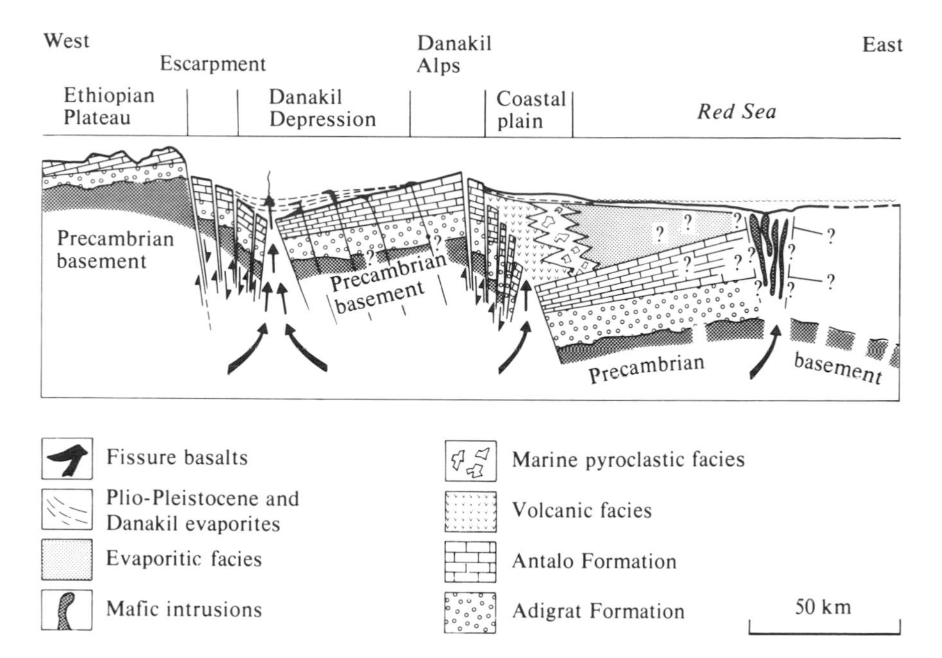

4.1 Section across the Red Sea and Danakil Depression. After Hutchinson and Engels (1970).

Permian–early Cretaceous interval, after which there was a tendency to downsagging without faulting (Kent, P. E. 1977).

The kind of control that the faulting of tilted blocks can exert on sedimentation is well illustrated by Surlyk's (1978) detailed account of end-Jurassic submarine fan deposits in East Greenland. These deposits change rapidly in facies from breccias laid down at the base of the fault scarp, via inner-fan conglomerates to mid-fan turbidites and outer-fan shales (Fig. 4.2). This is a facies association reminiscent of deep-sea fans, but the depth of sea in this case was probably quite modest. One important difference is that the sedimentary sequence wedges out rapidly eastwards as the crest of the next fault block is approached. A further feature that is characteristic of fault-block sedimentation is the great thickness of rudaceous deposits close to the fault scarp, which may often be greater in vertical than in lateral dimensions.

Surlyk recognizes several fining-upward megacycles about 100 m thick, which he interprets as corresponding to major phases of faulting, with a gradually diminishing supply of sediment following rapid erosion and retreat of borderlands. Smaller-scale fining-upward cycles probably result from the progressive filling and abandonment of inner- and mid-fan channels. Remarkably similar deposits of approximately the same age occur in north-

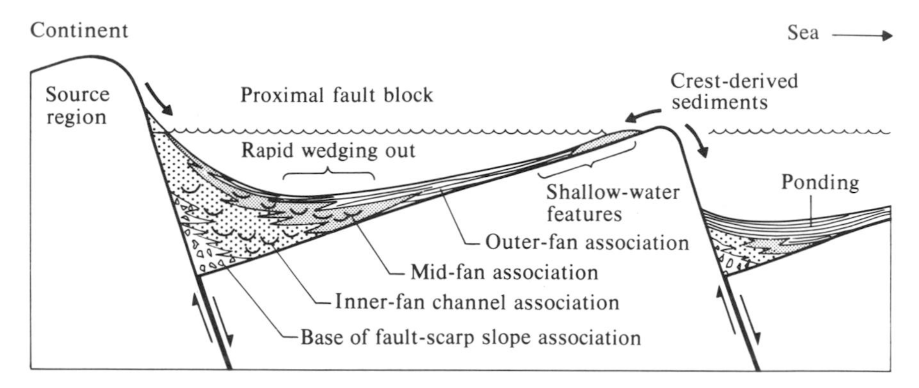

4.2 Model for submarine fan sedimentation along scarps in a tilted fault-block situation, based on Jurassic–Cretaceous boundary beds in East Greenland. After Surlyk (1978).

east Scotland and the northern North Sea (Selley 1976b), where westward-tilted fault blocks are widespread and indeed line up, on a pre-drift reconstruction, with East Greenland to the north.

Naturally fault-block sedimentation is not confined to continental margins, and a number of intracontinental examples have been described, notably from the Permo-Triassic of the two sides of the North Atlantic. The Stornoway Formation of the Outer Hebrides of Scotland provides a good case history (Steel and Wilson 1975). This consists of about 4000 m of mainly conglomeratic deposits thought to belong to the Permo-Triassic, and interpreted as alluvial fan, mudflow, streamflood and braided stream deposits with subordinate flood plain, channel and overbank deposits. Fining-upward fan sequences are interpreted as signifying basin-margin faulting of gradually decreasing intensity while coarsening-upward sequences suggest increasing faulting intensity through time.

Fault-block sedimentation in a different environmental setting from either of the two previous examples took place in the circum-Mediterranean region in Jurassic times (Bernoulli and Jenkyns 1974). In the late Triassic and early Liassic the region was occupied by an extensive, shallow-marine carbonate platform, which began to break up increasingly during the Pliensbachian and Toarcian, with the rapid collapse of certain sectors into at least moderately deep water. Thick, marly deposits accumulated in the newly created graben, with many interlayed turbidites.

Some of the intervening horsts or 'seamounts', such as the Trento Swell, were the sites of long-continued extremely slow sedimentation, and are characterized by the pinkish nodular limestone known as Ammonitico Rosso. From the great rarity of normal benthic fauna and concentration of iron–manganese nodules and crusts, the Ammonitico Rosso and its facies equivalents, such as the Adnet Limestone of the Salzburg Province of

Austria, have long been regarded as deep-water pelagic deposits. However, the occurrence of stromatolitic horizons with well-developed, laterally-linked hemispheres throws some doubt on this, as do some other criteria (Hallam 1975). At any rate the succeeding Middle to Upper Jurassic radiolarian cherts and the overlying coccolith limestones known as Maiolica and Biancone are almost certainly pelagic deposits of quite deep water (Fig. 4.3).

The presence of neptunian dykes and sills, together with evidence of minor volcanicity, confirm the existence of a taphrogenic regime in the Jurassic, which is probably bound up with the early opening phase of the Atlantic Ocean.

The name *aulacogen* has been given to large linear troughs that extend at a high angle from orogens far into the interior of cratons. Such rift structures were first described from the Russian Platform but have since been recognized in all the major continents, extending back in age to the early Proterozoic (Burke and Dewey 1973; Hoffman *et al.* 1974).

The Southern Oklahoma Aulacogen is a good American example (Fig. 4.4). This is a NW-trending faulted and folded Palaeozoic trough extending from the Ouachita Geosyncline across a foreland platform. It began as a graben underlain by Precambrian granites and was filled in the early and mid-Cambrian by siliciclastic sediments and volcanics, followed in the late Cambrian to late Ordovician by carbonates. Then follows a much thinner Siluro-Devonian sequence comparable to that deposited elsewhere on the platform. The late Palaeozoic was a time of folding and faulting but locally thick sequences of clastics were deposited. The transition through time from a fault-bounded graben to a broad downwarp that later became compressed is characteristic of many aulacogens.

Extensional tectonics are often associated with domal uplift, possibly the result of the rise of mantle plumes, in which case triple junctions may be generated. One arm of such a junction usually ceases to develop before the spreading stage while the other two arms form a divergent plate boundary leading to the creation of new oceanic crust. Many such *failed arms* have been recognized (Burke and Dewey 1973) and embrace the category of aulacogen. One of the best examples is the Benue Trough of West Africa, which extends NE from the Gulf of Guinea re-entrant. It began as an early Cretaceous graben developed contemporaneously with the opening of the South Atlantic, forms the site of the Niger Delta and contains over 10 km of Cretaceous and Tertiary submarine fan, deltaic and alluvial sediments.

Subduction-related settings

This is the setting most relevant to the interpretation of what most geologists have had in mind when they used the term eugeosyncline. The sedimentary sequences are characteristically thick, with abundant turbidite and mass

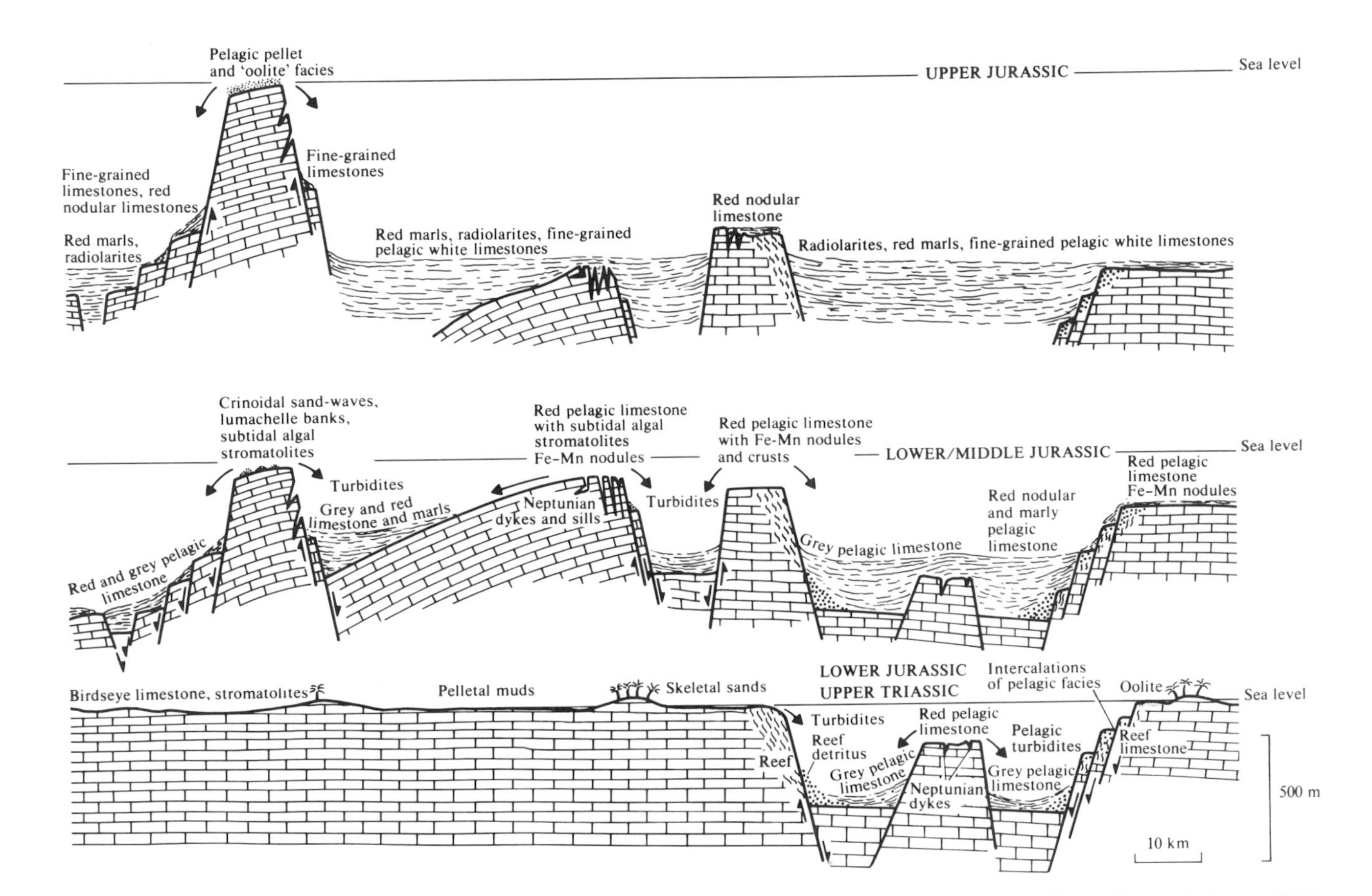

4.3 Schematic representation of the Jurassic collapse of the Mediterranean Tethys carbonate platform. After Bernoulli and Jenkyns (1974).

Late Proterozoic–Middle Cambrian

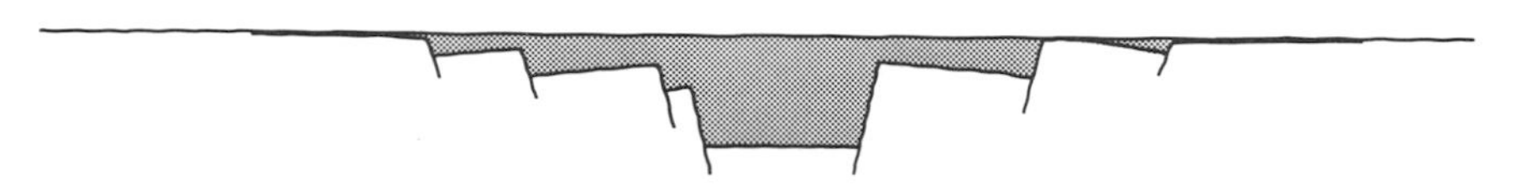

Late Cambrian–Early Devonian

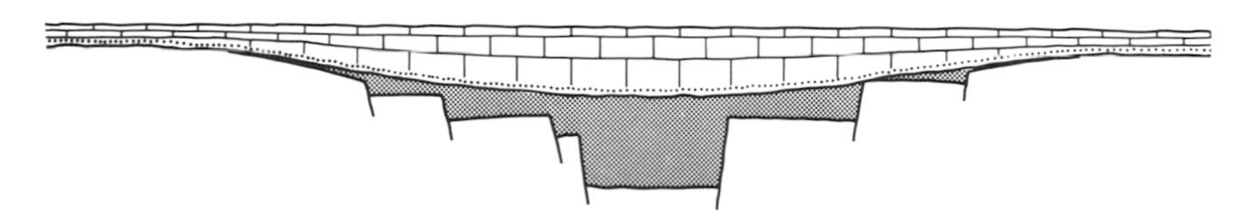

Late Devonian–Early Carboniferous

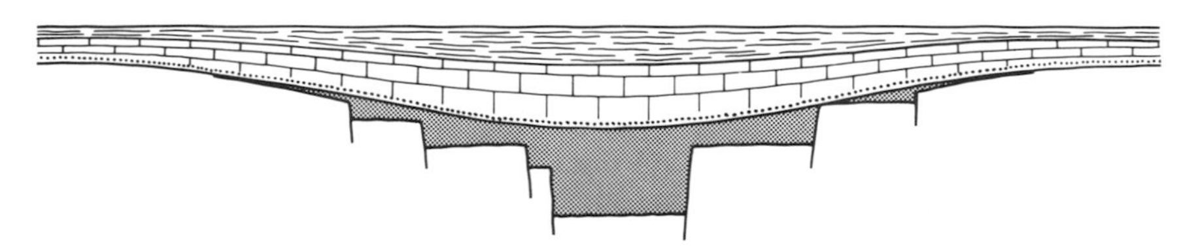

Late Carboniferous–Permian

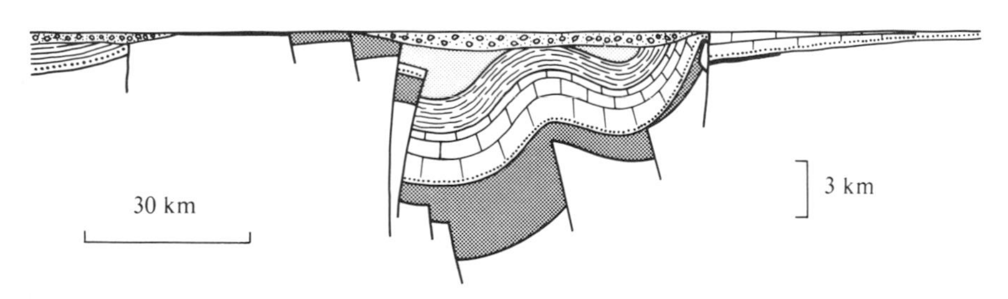

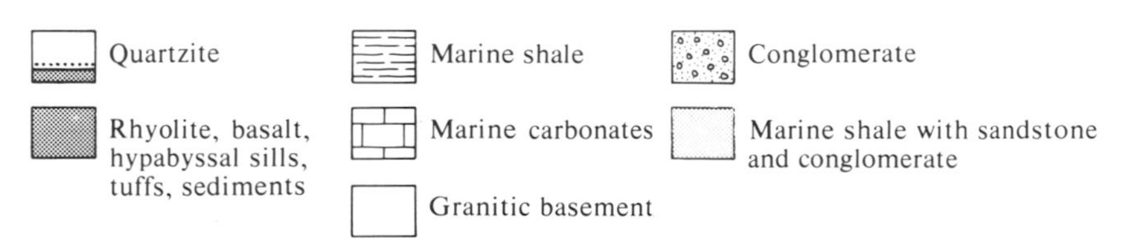

4.4 Schematic cross-sections through the Southern Oklahoma Aulacogen. After Hoffman *et al.* (1974).

flow deposits. There are associated igneous rocks and often much volcaniclastic debris and the strata have subsequently been deformed during the creation of orogens. Long before the emergence of plate tectonics it had become customary to seek modern analogues in the intra-oceanic island arc-trench systems most extensively developed on the western side of the Pacific Ocean, nowadays of course related to subduction zones.

It is usual to distinguish several physiographic units in an intra-oceanic island arc system (Fig. 4.5). From the ocean towards the continent occur in succession the trench, arc-trench gap, volcanic arc and back-arc basin. Subduction, and hence compression, only concerns the trench-volcanic arc system and the back-arc basin is typically a zone of spreading.

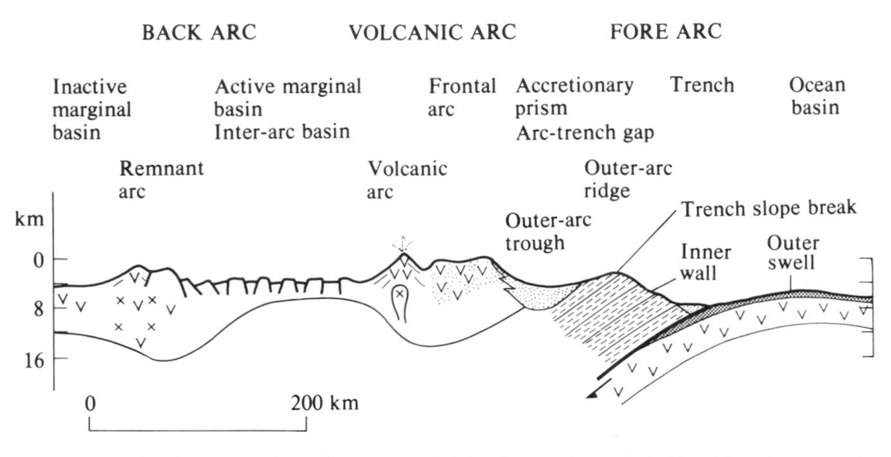

4.5 Generalized cross-section of intra-oceanic island arc. After Mitchell and Reading (1978).

There has been a widespread belief that the deep Pacific trenches correspond to eugeosynclines forming today. Modern work hardly bears this out (Scholl and Marlow 1974). Not only do they appear to be too narrow to fulfil this function, being only 50–100 km wide, but they could not accumulate more than 2–3 km of sediment without overspill on to the adjacent deep-sea floor. Through most of the Cenozoic the trenches accumulated only comparatively thin sequences of pelagic and hemipelagic deposits and thick turbidites did not enter them until the unusual conditions of the Pleistocene low stands of sea level.

Scholl and Marlow consider that the 'missing' deposits were lost through subduction but, more recently, seismic reflection profiling combined with other data suggests an alternative—that they have been plastered against the fore arc as an *accretionary prism* (Figs 4.5, 4.6). This consists of a series of steeply inclined successive wedges of sediment younging towards the ocean, but there is an overall younging of the sequence within each wedge towards the continent. Confirmation of the accretionary prism model has recently been obtained by deep-sea drilling off southern Mexico (Moore *et al.* 1979).

Intra-oceanic island arcs such as Tonga, the New Hebrides and the Lesser Antilles consist of subaerial and submarine volcanics, predominantly low potash basalts or andesites with subordinate calc-alkaline siliceous lavas and tuffs. The associated sediments are mainly derived from erosion of the volcanic rocks but reef carbonates may also occur. Mass flow gives rise to aprons of volcaniclastic and carbonate boulder breccia, conglomerate and sandstone. Continental margin arcs as in the Andes and Sumatra tend to have more siliceous and potash-rich dacites, andesites and rhyolitic ignimbrites; basalts are rare. Volcanogenic and maybe granitic debris are often found in associated fault troughs.

The back-arc region consists of both ridges and basins. The sediments include volcaniclastic debris from the arc, biogenic ooze and varied clays, often rich in smectite. The sedimentation patterns can be highly varied and complex depending primarily on the amount and nature of the terrigenous input. Both thick turbidites or pelagics may directly overlie ocean crust. This seems a more likely environmental setting for ancient 'eugeosynclines' than the much narrower and comparatively sediment-starved trench systems.

Ancient subduction zones are most readily recognized by the presence of paired metamorphic belts and slivers of obducted ophiolite, and there may be considerable difficulty in assigning given sedimentary units to precise physiographic or tectonic regimes. In the older work on flysch terrains, reviewed by Dzulynski and Walton (1965), it was found that turbidity currents characteristically moved along the length of troughs parallel to the tectonic strike. More recently it has been argued that many turbidite sequences represent deep-sea fan deposits (Mutti 1974; Nelson and Nilson 1974). Turbidite fills in ancient trenches should be signified by long, linear channel facies parallel to the trench axis, with small fans built out at right angles to this direction.

It is possible that regions like the Burma Orogen and the Sunda Arc provide better models for interpreting some ancient orogens with thick turbidite sequences than regions such as the Andes, Japan and intra-oceanic island arcs, where much less of the tectonic history of the subducting plate is preserved. A feature of special interest is that thick delta-fan turbidites on the deep-sea floor extend for some 3000 km south of the Bengal Delta. Thus this huge body of sediment has a longitudinal relationship relative to the strike of the Indo-Burman ranges, and thereby invites analogy with the older flysch examples cited by Dzulynski and Walton (1965). A plausible comparison can be made between the Indo-Burman fold ranges passing into the Eastern Highlands, and the Scottish Caledonides (Mitchell and McKerrow 1975). For both regions active subduction can be inferred beneath a continental margin bounded by a strike slip fault. More recently Leggett *et al.* (1979) have made a case for an accretionary prism of turbidites in the southern part of the Scottish Caledonides, adjacent to the presumed subduction zone.

The most spectacular rock assemblages associated with subduction zones are known as *olisthostromes* and *melanges* (Hsü 1974). Both consist of a chaotic jumble of clasts ranging from pebbles up to huge blocks several kilometres in length, in a usually fine-grained matrix, often consisting of a distinctive scaly clay. Olisthostromes are generally regarded today as deposits emplaced by debris flow or some other mass gravity process, containing large exotic clasts termed *olistholiths* which are older than the enclosing sedimentary matrix. Classic examples of olisthostromes include the Palaeogene *Wildflysch* of the Swiss Alps and the *Argile Scagliose* of the Apennines. They may be associated with incipient phases of thrusting and hence form a sole zone to major thrusts, as in the Taconic ranges of New England and at the base of the *Coloured Melange* zone of the Zagros Mountains, Iran.

The term melange may be used comprehensively to embrace olisthostromes when distinction is difficult. In the stricter sense melange refers to bodies of deformed rocks with tectonically mixed clasts that may be younger than the pervasively sheared matrix. The Franciscan of Northern California is one of the best-studied examples. This consists of blocks of greywacke, greenstone, chert, serpentinite and rare blue schist and eclogite in a matrix of sheared mudstone containing rare fossils of Tithonian to Valanginian age. It is thought to be the result of long-continued Cretaceous and Palaeogene eastward-subduction, and probably represents the random offscrapings of a variety of oceanic terrains extending over a large area (Hsü 1971).

A further instance of a melange zone, this time formed by Neogene subduction, has been recognized in the Indonesian region (Hamilton 1977). Seismic reflection profiling and other data indicate that the outer-arc ridge between Java and Sumatra and the active Java Trench is the top of a wedge of melange and imbricated rocks whose steep to moderate dips are sharply disharmonic to the gently dipping, subducting oceanic plate beneath (Fig. 4.6). The wedge has grown by scraping off of oceanic sediments and basement against and beneath its toe, and also by internal imbrication which is a gravitationally driven counter to subductive dragging at the base.

Further east, the Banda Arc records the collision of an island arc with the Australia–New Guinea continent, and the shoving up on to the continent of an imbricated pile of material derived partly from the continent itself, and partly from deep-water melange in front of the advancing arc. Whole islands such as Roti, Timor and Seram are composed of melange together with imbricated rocks. Scaly clay forms the matrix of a chaotically distributed assemblage of olistholiths of all sizes up to tens of kilometres long, of shelf

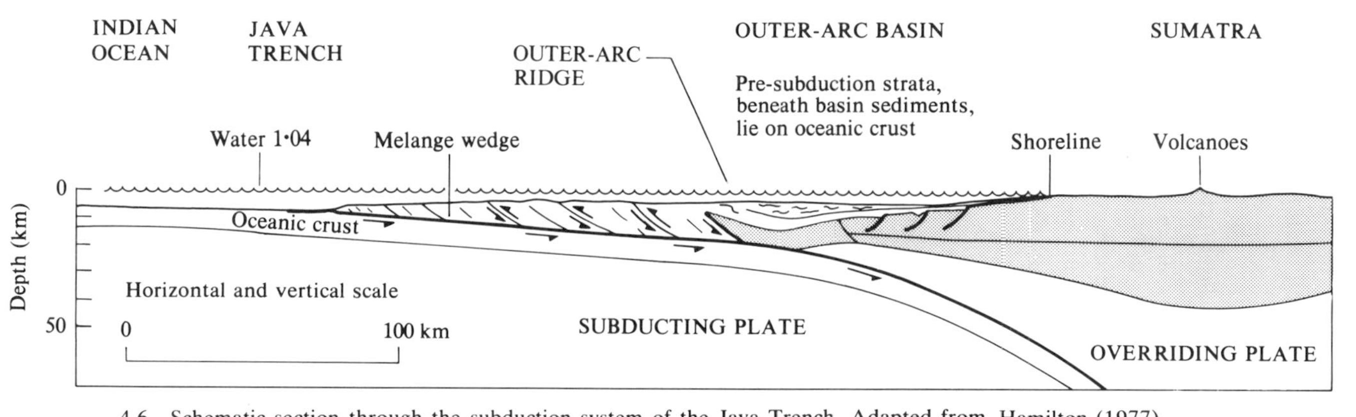

4.6 Schematic section through the subduction system of the Java Trench. Adapted from Hamilton (1977).

(Permian and younger), slope and deep-water (Cretaceous and younger) sediments and of igneous and metamorphic rocks, mainly of mafic and ultramafic composition.

Transform/strike-slip fault-related settings

Transform plate boundaries can be studied today in California, Alaska, Venezuela, the Middle East and New Zealand. Their most characteristic sedimentary feature is the development of small, tectonically active basins in which accumulate locally thick successions exhibiting rapid lateral and vertical facies changes, with abundant indications of contemporaneous gravity movement. Another typical feature is alternating uplift and subsidence in different localities, which may indicate local changes in the tectonic forces, from extension to compression. Many basins may lie along lines of secondary faulting and ancient transforms may be hard to recognize.

The basins occur where strike-slip faults diverge and are known as *pull-apart basins.* An outstanding example is the Ridge Basin associated with the San Andreas Fault system (Crowell 1974). This has dimensions of 50 by 20 km and has a thick Neogene siliciclastic sequence that was deposited as the basin margins were progressively pulled apart. The most striking rock formation is the Violin Breccia, some 10 000 m thick but much more restricted laterally, and interpreted as a fault-margin alluvial fan conglomerate. It passes laterally into varied alluvial, lacustrine and marine sediments.

A further case of a Cenozoic basin associated with a major transform boundary is described from New Zealand by Norris *et al.* (1978). A much older example of basins associated with strike-slip faulting occurs in the Cantabrian Mountains of northern Spain, which belong to a Hercynian orogen (Reading 1975). There are rapid vertical and lateral facies variations, with sediments including mass flow deposits, and pronounced vertical movements have induced gravity sliding (Fig. 4.7).

Continental collision-related settings

When two continents collide it is unlikely that they will do so along a straight line. The initial contact will instead tend to be irregular, with local gaps acting as sediment traps, eventually to be destroyed as the continents continue to converge. The Bay of Bengal is one such *remnant ocean basin* that is gradually closing as a result of eastward-subduction beneath the Indo-Burman ranges and Andaman–Nicobar and Sunda outer arc. The Bengal fan has been deposited above pelagic sediments of the ocean floor. With continued subduction the fan turbidites are scraped off to form an outer arc. At a later stage such deposits are overthrust on to the continental margin.

South of the rising Himalaya a series of late orogenic basins developed in the Neogene, in which were deposited the thick alluvial fan Siwalik sediments. Far into the Asian interior a number of intramontane troughs have

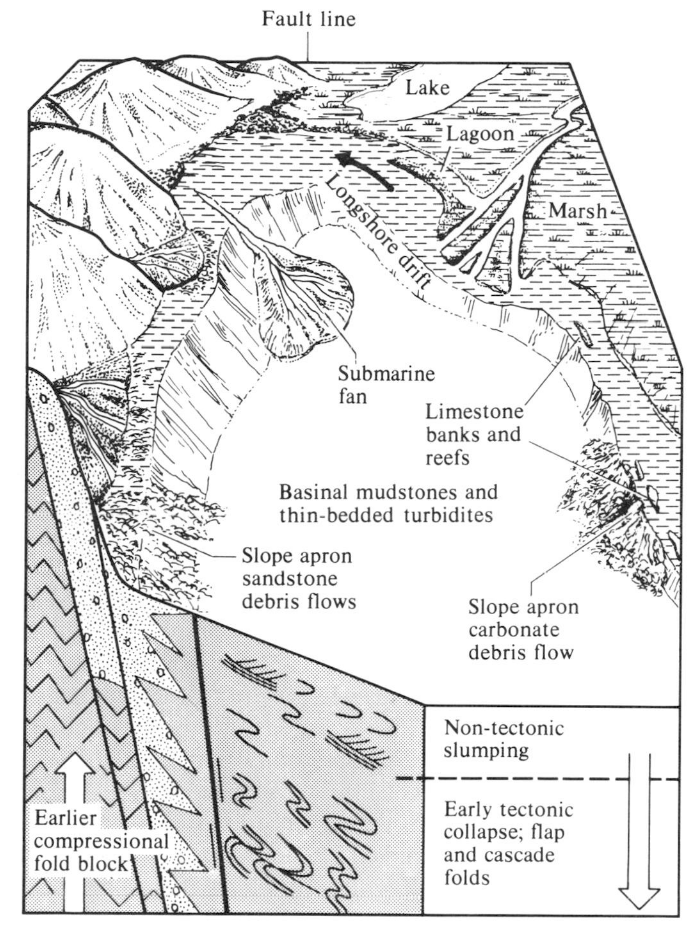

4.7 Sedimentation and tectonic model for a strike-slip orogenic belt, based on the Cantabrian Mountains, Spain. After Mitchell and Reading (1978).

been created by strike-slip faulting related to the continued northward push of India (Molnar and Tapponnier 1975).

With regard to the Alps, Milnes' (1978) detailed analysis suggests that the final stages of continental collision, leading to locking together of the separate landmasses, had been completed by the early Oligocene. Thereafter significant uplift commenced concomitant with subsidence of a foredeep to the north. This is the classic late orogenic (sometimes called post-orogenic) trough of the Swiss Plateau in which were deposited 3000–6000 m of molasse (Van Houten 1974). The most distinctive rock type is *Nagelfluh*, a thick fanglomerate. Several large alluvial fans have been recognized, up to

1000 m thick and 40 km wide. Northwards the sediments thin and become finer grained, being mainly alluvial plain deposits but with shallow marine intervals of mid-Oligocene and mid-Miocene age. Several coarsening-upward sedimentary cycles reflect uplift of the orogen to the south, northward translation of nappes and faulting of the basement.

Van Houten also describes typical molasse deposits from the Ebro and Aquitaine Basins, whose fanglomerates have been derived from the uplifted Pyrenees. Similar thick fanglomerates in the Lower Devonian of Central Scotland appear to represent analogous deposits of a late orogenic trough developed shortly after the closure of the Iapetus Ocean.

The Wilson Cycle

A major weakness of classical geosynclinal theory was that no account was taken of significant lateral movements, which could bring together rocks formed over an extensive area and perhaps far distant from the eventual orogen. It is indeed ironic that Stille based his concept of the eugeosyncline primarily on Alpine ophiolite complexes, at a time when they were considered an integral part of the geosyncline, in effect the result of *in situ* intrusions and extrusions. This concept persisted until the advent of plate tectonics (e.g. Aubouin 1965) but nowadays ophiolite complexes in the hearts of orogens are generally considered to have been obducted, and were created by seafloor spreading elsewhere. In fact Steinmann's famous association of ophiolites and pelagic sediments is now seen to have nothing directly to do with the compressional tectonics that create mountain belts.

In a similar way, little remains of Trümpy's leptogeosynclinal phase in the Alps, because the deposits of his deep starved basins of the Southern and Eastern Alps are now considered to be seamount deposits, perhaps laid down on the southern side of the Tethys, which have since been transported a considerable distance northward. More generally, at least some melange deposits may be a chaotic jumble of material scraped from a large area of sea floor.

Clearly a new type of concept is required which takes into account plate tectonics. Let us first consider briefly three orogenic case histories.

The northern Appalachians have been a classic region for the study of geosynclines since the pioneer work of Hall and more especially since Kay's distinction of a westerly miogeosyncline and easterly eugeosyncline in the Palaeozoic of New England, separated by the volcanic uplift of the Green Mountains. Bird and Dewey (1970) were the first to apply a plate tectonic model to the Appalachian orogen. Attention will be confined to the northern area extending from New England to Newfoundland, where the absence of a late Palaeozoic Alleghanian orogenic overprint makes it possible to study more easily the effects of the older Taconic and Acadian orogenies.

In western New England an area traditionally known as 'Logan's Zone' corresponds to Kay's miogeosyncline, with a series of Cambrian and Lower Ordovician carbonate platform deposits overlying a Lower Cambrian orthoquartzite, the Potsdam Sandstone. The sequence thickens eastwards and a similar facies extends along the length of the Appalachians. In western Newfoundland the famous Cow Head Breccia is interpreted as a slide breccia on the south-eastern margin of Logan's Zone, where the platform was abruptly replaced by a zone of deeper water.

The facies changes abruptly in the Middle Ordovician, the shallow water carbonates passing up into a thick argillaceous formation, the Normanskill Shales. Westwards this passes into the Trenton Limestone, but in the other direction there are intercalations of greywacke, clearly derived from the east. Then follows the allochthonous unit of the Taconic Mountains, with several westerly directed thrust sheets. A distinctive wildflysch facies forming the sole is interpreted by Bird and Dewey as a melange generated by the dumping of erosion products from the westward-advancing klippen. This undoubted compressional event, abruptly disturbing a long history of slow continental margin subsidence and comparative tectonic stability, is dated as late Ordovician and marks the Taconic Orogeny (Fig. 4.8).

To the south-east is the Piedmont Zone, with late Precambrian to Ordovician clastics and volcanics intensely deformed and metamorphosed during the Taconic Orogeny. This corresponds to Kay's eugeosyncline and is now considered to have been situated on the continental rise at the margin of a spreading ocean. Late Precambrian basalts suggest an early tensional phase with the formation of graben, while the evidence of other volcanics suggests the establishment in early Ordovician times of an island arc system associated with a newly created westward-dipping Benioff Zone (Fig. 4.9). The subsequent history is one of a progressively narrowing ocean (Iapetus) as crust was subducted, with eventual continental collision in the Devonian being signified by the Acadian Orogeny and westward spread of molasse facies (the so-called Catskill Delta).

On the western side of North America the Palaeozoic sequence exhibits a mirror image of that east of the central cratonic region. Broadly speaking three parallel facies belts can be recognized, from east to west as follows:

1. Shallow-water carbonates with orthoquartzites, traditionally interpreted as a miogeosyncline.
2. Shales with chert horizons (graptolitic in the Lower Palaeozoic), generally considered to be a deep-water facies.
3. Greywacke–volcaniclastic–volcanic association.

Zones 2 and 3 have been taken to signify a eugeosyncline containing volcanic island arcs, and a long history of intermittent eastward subduction has been inferred. A marginal basin zone is believed by Churkin (1974) to have existed through most of the Palaeozoic, with several alternating phases

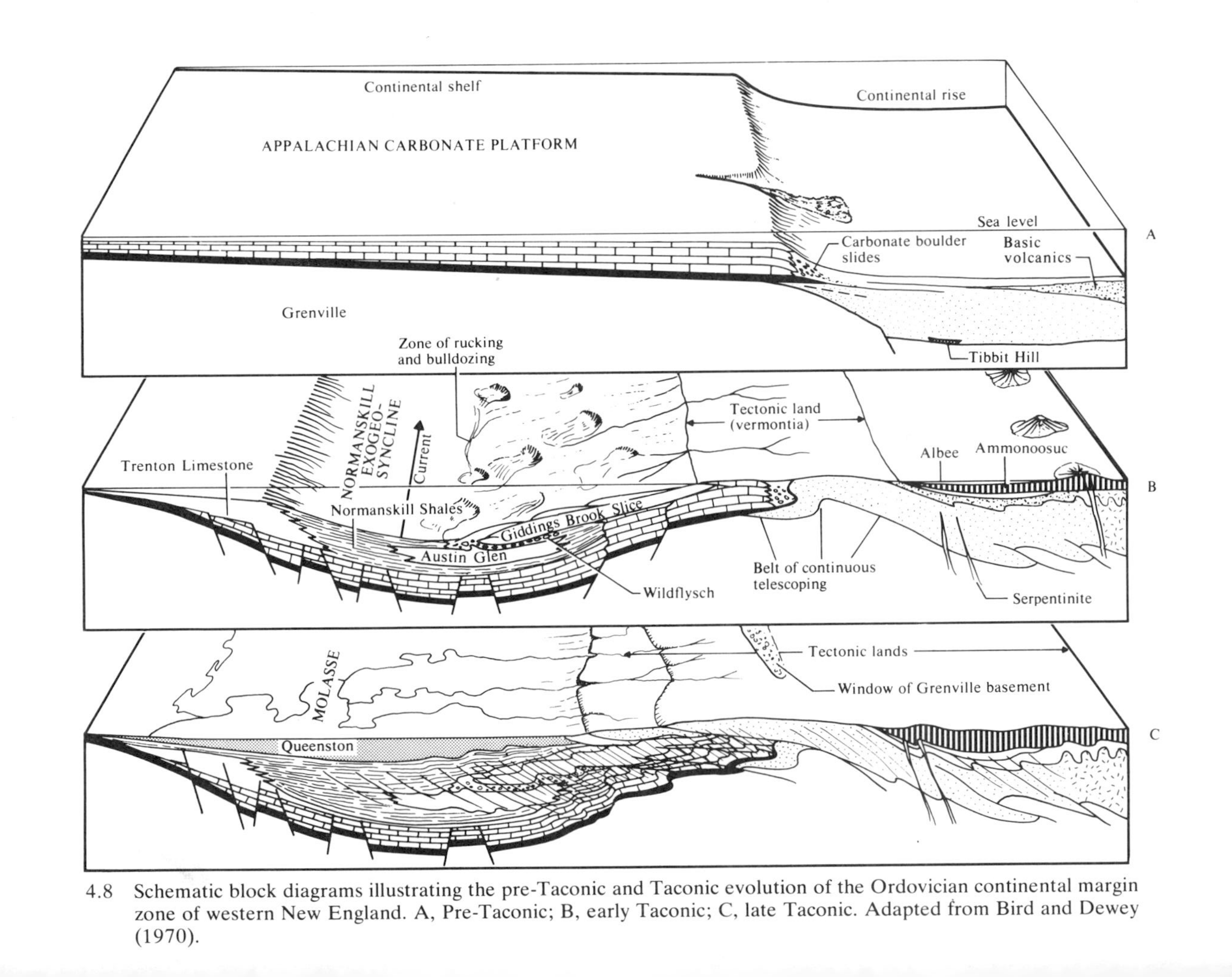

4.8 Schematic block diagrams illustrating the pre-Taconic and Taconic evolution of the Ordovician continental margin zone of western New England. A, Pre-Taconic; B, early Taconic; C, late Taconic. Adapted from Bird and Dewey (1970).

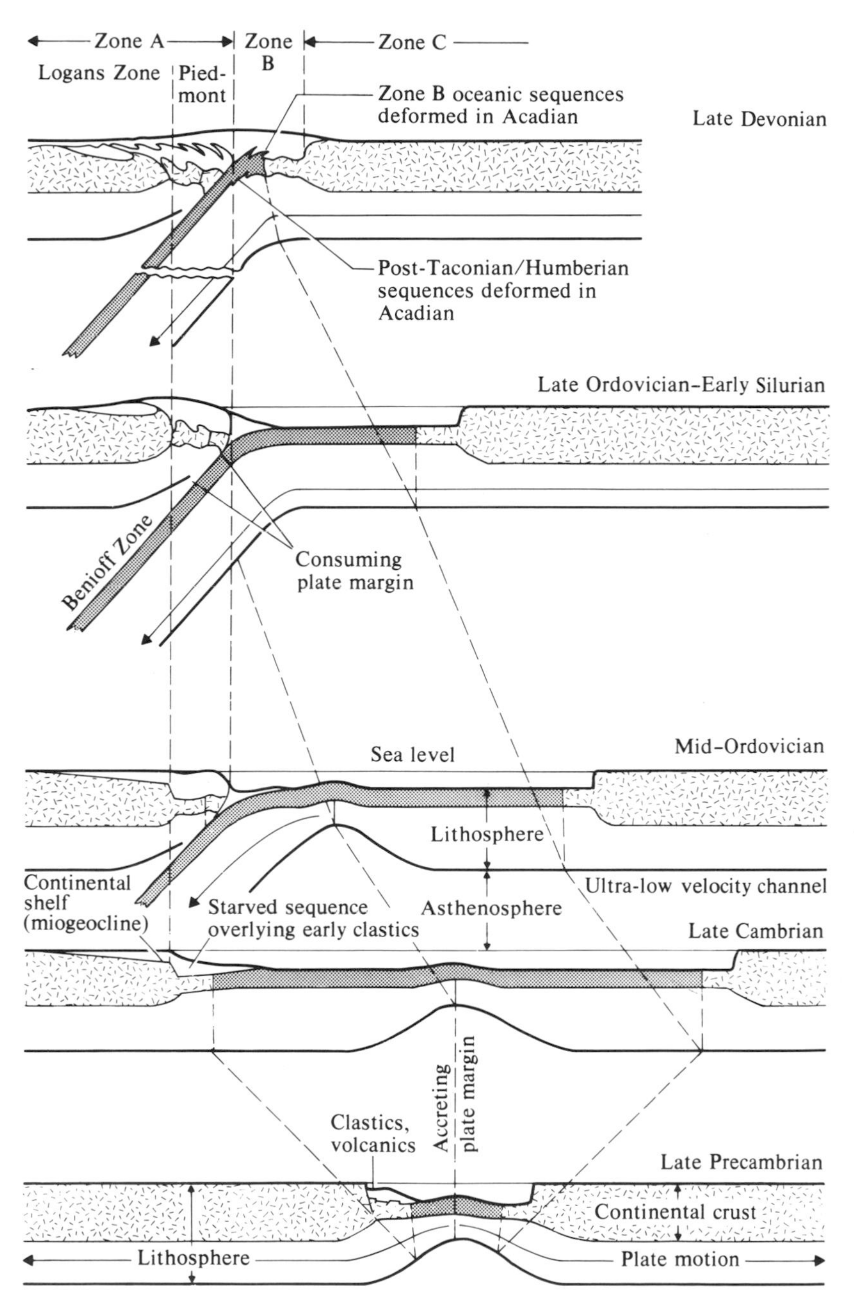

4.9 Interpretation of New England geology in terms of the Palaeozoic closure of the Iapetus Ocean following an earlier taphrogenic episode. After Bird and Dewey (1970).

of opening and closure. Thus the late Devonian to early Carboniferous Antler Orogeny is attributed to collision of a marginal basin and volcanic arc, with concomitant uplift and overthrusting of pelagic sediments; a huge clastic wedge of conglomerates and sandstones derived from the west unconformably overlies deposits of zone 2. Renewed rifting caused the creation of a new volcanic arc-marginal basin system towards the end of the Palaeozoic (Fig. 4.10).

Churkin drew the obvious analogy with the behaviour of the Western Cordilleran system of North America in the Mesozoic and Cenozoic, but it should be noted that the inferred relatively simple pattern of alternating east–west extensional and compressional phases has been complicated by the recognition that large sectors of the coastal belt might have been translated for considerable distances northward (Jones *et al.* 1977).

Turning attention now to the Old World, a line of ophiolite complexes and associated melange deposits extending from the Oman through southern Iran to southern Turkey and Cyprus marks a major continental collisional event that took place in the late Cretaceous. The geology changes little along the strike for great distances and the resemblances between the Oman and southern Iran are especially close (Glennie *et al.* 1973; Hallam 1976). A thick series of shallow-water carbonate platform deposits ranging in age up to the Cenomanian are overlain by a wildflysch unit with a chaotic jumble of limestone blocks in a shaly matrix. Above this in thrust contact is a thick series of radiolarian cherts (radiolarites) with turbiditic limestone intercalations containing fossils of Triassic to Cretaceous age. Several thrust units can be recognized and near the top is a unit containing enormous olistholiths of Permian and Triassic limestone. A major ophiolite unit has in turn been thrust over the radiolarites and the whole complex is unconformably overlain by Maastrichtian and younger shallow-water limestones, giving an upper-age limit for the collisional event (Fig. 4.11).

The interpretations proposed for the two areas are broadly similar. At some undetermined time in the Mesozoic new ocean crust was generated to the north of the Arabian carbonate shelf, which was downfaulted at the margin as a result of extensional tectonics. Calcareous sediment was transported seaward into a deeper water zone of siliceous pelagic muds deposited directly on oceanic basalt, to give rise to turbidite interbeds and occasional horizons of exotic blocks. During the subsequent late Cretaceous compressional episode these deep-water deposits were thrust southwards and south-eastwards over the carbonate platform, the initiation of the thrusting being marked by subsidence of the outer margin of the platform and deposition of the wildflysch olisthostrome at the sole of the advancing nappe. Obduction of the ophiolite nappe marks the final closure of the old seaway (Fig. 4.12).

There are some striking resemblances with the early Palaeozoic deposits

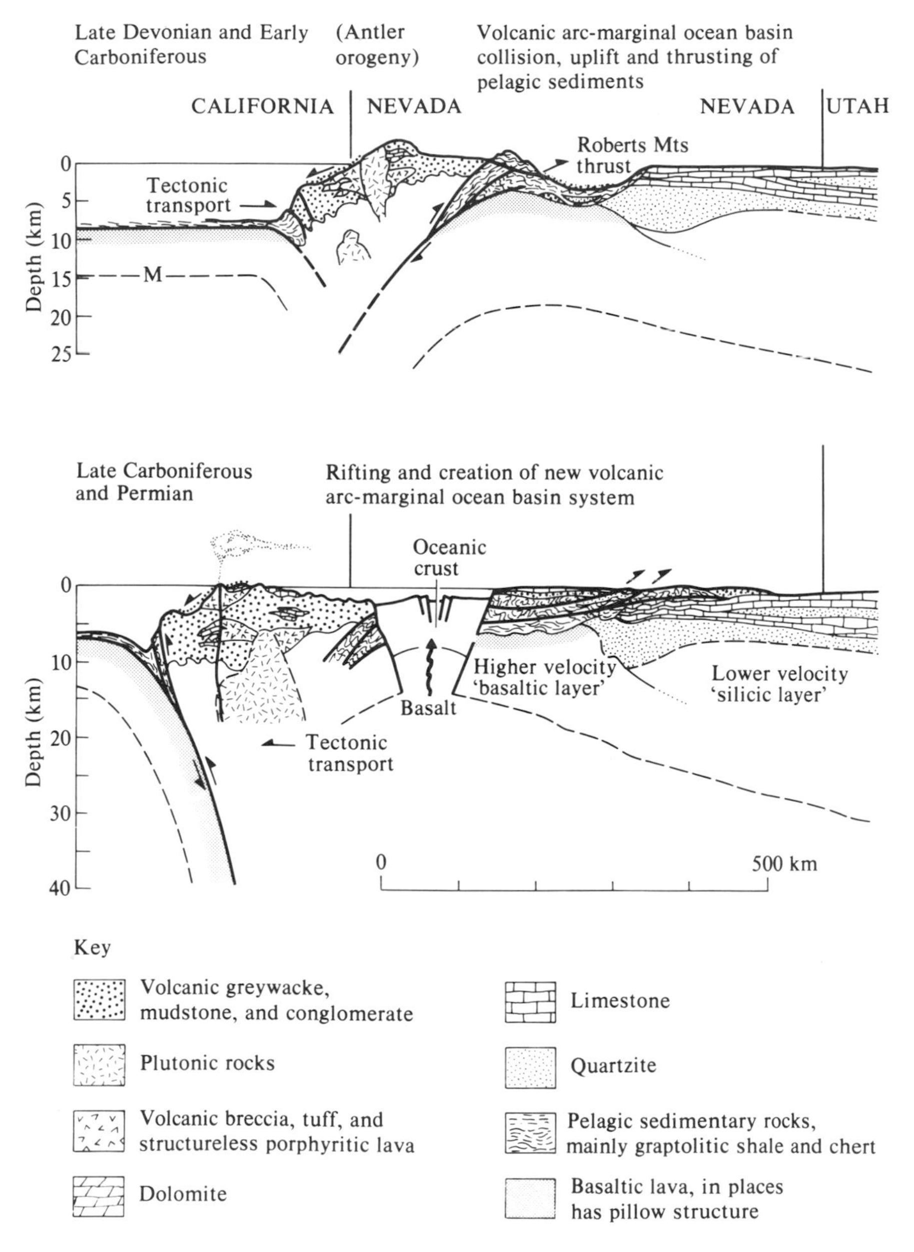

4.10 Model of late Palaeozoic extensional and compressional tectonics in the Western Cordillera of the United States. Adapted from Churkin (1974).

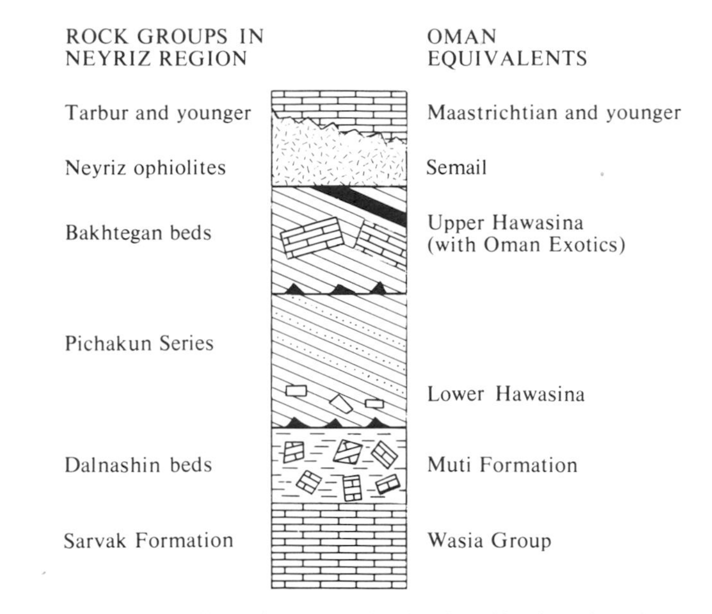

4.11 Comparison of stratigraphic-structural units of the Neyriz region of southern Iran and the Oman Mountains. After Hallam (1976).

of the northern Appalachians. The Cow Head Breccia of Newfoundland recalls the small olistholiths of the Lower Hawasina and Pichakun Series, derived from the carbonate platform. The Normanskill Shale is similar in tectonic situation to the Dalnashin and Muti shales and in both cases there is a wildflysch sole to the thrust complex.

A very similar history of structural telescoping has been worked out for rocks of near-identical facies in the Othris Mountains of Greece (Smith *et al.* 1979). Notable differences are that the carbonate platform lay to the north-east and the thrusting took place somewhat earlier, in late Jurassic to early Cretaceous time.

These and other case histories provide a general model for tectonic evolution of successive ocean opening and closure, with an early open ocean stage, a later remnant basin stage and a final stage of continental collision. Since the model was first proposed by J. Tuzo Wilson it is known as the Wilson Cycle (Fig. 4.13).

Basin subsidence within cratons

Despite the numerous complications and uncertainties that persist, many of the geological events, including sedimentation, that take place at plate boundaries are thought to be comparatively well understood, at least in outline. The formation of sedimentary basins within the interior of cratonic

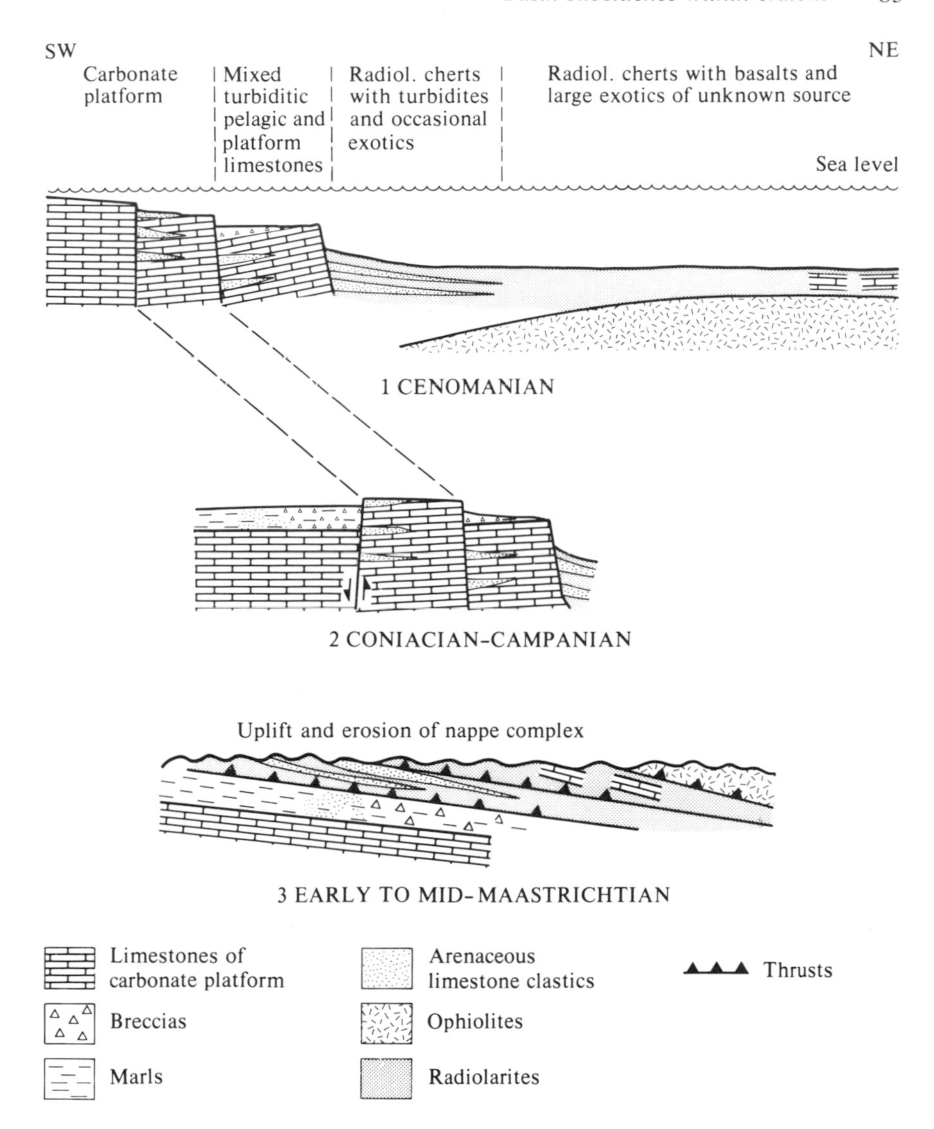

4.12 Model of the structural evolution of the Neyriz region of southern Iran during the late Cretaceous. After Hallam (1976).

regions is much more enigmatic (Bott 1976). By no means all such basins can be regarded as aulacogens or failed arms, and hence tensional by-products of plate-margin events elsewhere.

Consider for instance the case of the Michigan Basin of the United States interior. The sedimentary sequence of this basin records a history of continued subsidence throughout most of the Palaeozoic, but it has a nearly

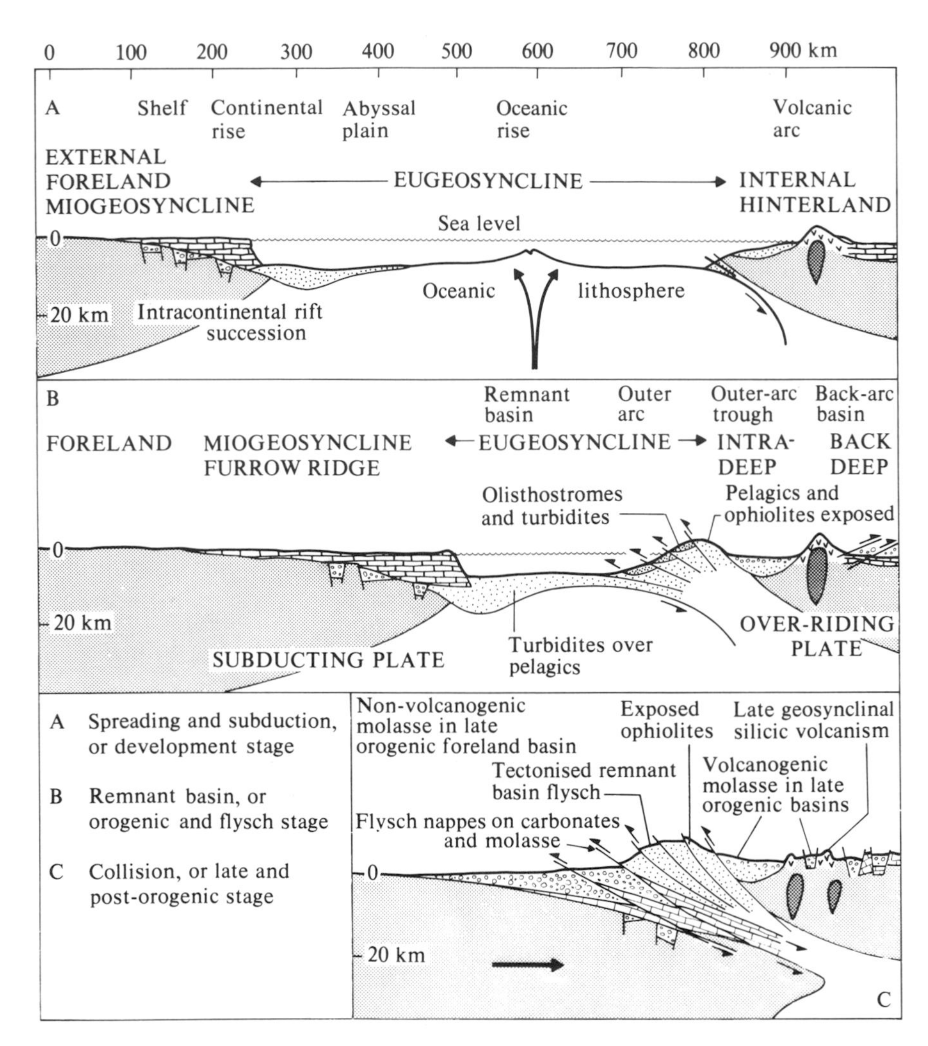

4.13 The Wilson Cycle of oceanic opening, subduction and continental collision. After Mitchell and Reading (1978).

circular outline and has remained undeformed. According to Haxby *et al.* (1976) the size and shape provide strong evidence of flexure of the lithosphere under a load of small horizontal dimensions compared with the radius of flexure. They propose a thermal model of evolution whereby a mantle diapir initially penetrated the lithosphere. Gabbro was transformed to eclogite and, as cooling took place by conduction, the basin subsided under the load of eclogite.

This model is rejected by McKenzie (1978), who considers that the required phase change is implausible. If thermal contraction occurred it should have been preceded by an initial period of uplift on a large scale, with significant erosion in consequence, and there is no evidence for this in either the Michigan or other basins, such as the North Sea Basin or Pannonian Basin of Hungary.

McKenzie prefers instead a model involving an initial phase of stretching associated with block faulting and subsidence. The lithosphere then thickens as a result of heat conduction to the surface and further slow subsidence without major faulting occurs. The principal objection to this model, as McKenzie admits, is the large amount of extension required, because extension by a factor of two is needed for a mere 4.5 km of sediment accumulation. On the other hand it appears to accord well with the history of many aulacogens insofar as they show an early phase of graben development followed by downsagging.

It is beyond the scope of this book to evaluate fully these or other geophysical models, but it cannot be too strongly emphasized that much of the necessary evidence for such evaluation must come from thorough stratigraphic and facies analysis of basin formation and evolution. Only this kind of research can provide an adequate answer to a whole series of important questions, such as: What was the shape and size of the basin? How long did it persist? At what rate did it subside? Did the rate vary systematically through time? What was the association, if any, with faulting, either of the basement or basin margins?

Sedimentation and tectonics in relation to petroleum occurrence

Although an extended treatment of economic geology is beyond the scope of this book it is pertinent to cite the recent discoveries in the northern North Sea as an outstanding example of how the interaction of sedimentation and tectonics has controlled the occurrence of petroleum in what has become established as a major oil province.

The North Sea is now known to be underlain by a large Mesozoic taphrogenic zone with a centrally located graben system extending along its length characterized by tilted fault blocks. The oilfields are intimately associated with this graben system (Selley 1976b; Ziegler 1977) so that some kind of causal relationship is indicated. The northern North Sea is a major petroleum province because of a combination of factors. There is a deep sedimentary basin in which organic shales are interbedded with, or brought into faulted juxtaposition with, porous sands, and the requisite temperatures for conversion of organic matter into petroleum have been achieved by a combination of deep burial and a high geothermal gradient along the rift axis, where the continental crust is thinned. Numerous oil traps developed because of structural movement synchronous with sedimentation.

Petroleum is currently being extracted from three types of reservoir:

Palaeocene sands (in the Forties Province); Maastrichtian and Danian chalks (in the Ekofisk Province); and Jurassic sands (in the Brent and Piper Provinces). Each type of occurrence has features of particular interest. The Palaeocene 'play' is in massive sands interpreted as deep-water grain-flow channel deposits, interbedded with a less porous turbiditic sand–shale complex, produced as a result of uplift of the Scottish mainland and Shetland Platform Caledonian basement. Unlike in other major oilfields, such as those of the Mississippi and Niger deltaic complexes, there have been no major oil discoveries yet from the delta front zone where sands pass seaward into shales. The chalk occurrences are unusual in that, though chalk is often very porous (up to 35%) it is normally highly impermeable, so that pore fluids cannot be extracted. Indeed, in some Middle East oilfields chalk acts as a cap rock. Probably the most important factor creating high permeability as well as porosity is tensional fracturing produced as a consequence of domal uplift of chalk deposits above diapirs of Permian salt, though other post-depositional factors may also have contributed.

By far the most abundant occurrences, however, are in Jurassic sands of the Viking Graben between the Shetland Platform and Norway, namely the topmost Triassic-basal Jurassic Statfjord Sand, laid down variously in an

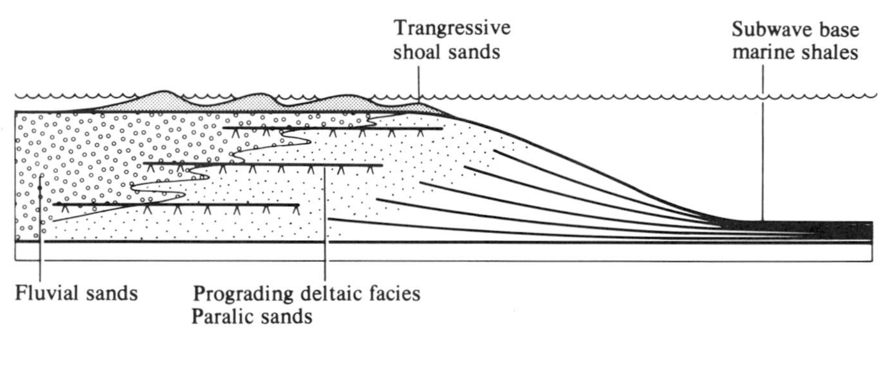

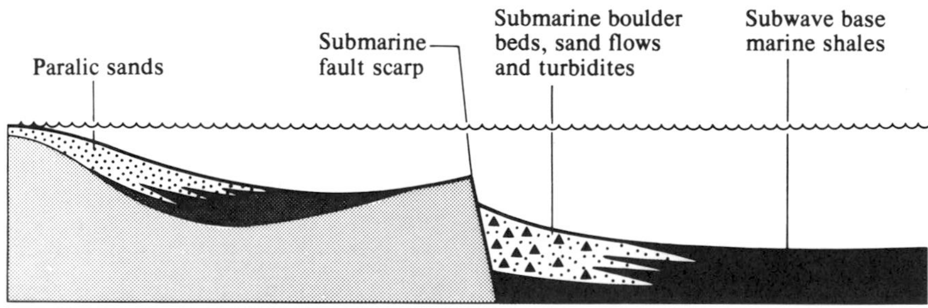

4.14 The various types of sand facies in the Jurassic of the northern North Sea. After Selley (1976b).

alluvial-deltaic and shallow-marine environment, and the most important alluvio-deltaic Bajocian–Bathonian Brent Sand. A third, Oxfordian–Kimmeridgian sand, the Piper Sand, produces oil in the E–W elongated Moray Basin, which intersects the main graben system. This is a shallow-marine sand produced by reworking of the underlying Brent Sand at the start of a late Jurassic transgression.

The various types of Jurassic sand occurrence are illustrated in Fig. 4.14 and the three types of trapping mechanism in Fig. 4.15. The tilted fault block occurrence is the most important. In the Brent oilfield end-mid-Jurassic and end-Jurassic faulting has brought the Brent and Statfjord Sand reservoirs into unconformable contact with Upper Jurassic and Lower Cretaceous shales, which have acted as cap rocks. Without much doubt the source rock is the strongly bituminous Kimmeridgian shale.

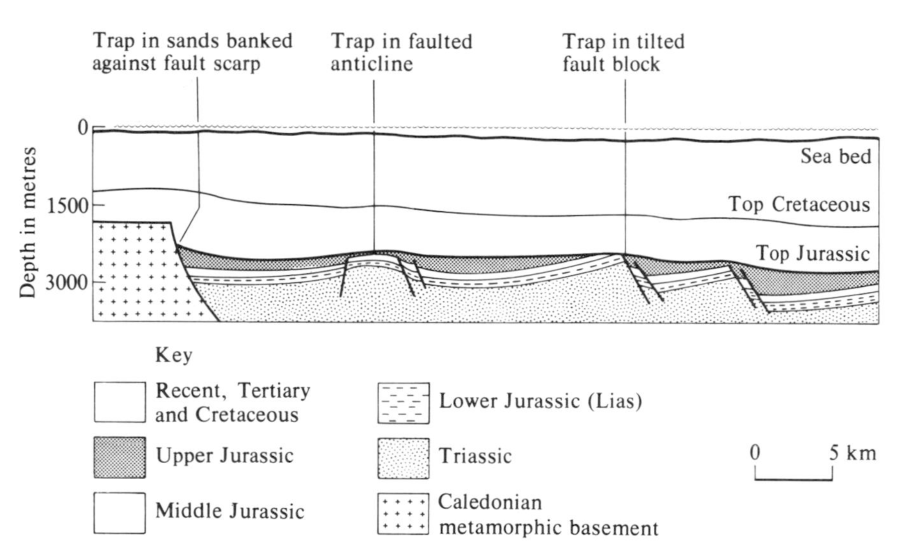

4.15 The three main types of oil-trapping structure found in the Jurassic rocks of the North Sea. After Selley (1976b).

The oilfields were originally located by seismic reflection studies and so far it has only proved economically profitable to exploit these structural traps, but as the price of oil contines to rise increasing attention will be paid to possible stratigraphic traps, which at present are only of secondary importance. The full resources of the facies analyst will then be brought into play, for instance in the prediction of the occurrence of delta or submarine channel systems, and the zones of interfingering of sand and shale.

5. Ancient epicontinental seas

Nowadays shelf seas occupy only a limited area at the margins of continents and are appropriately termed *pericontinental*. In contrast, during long periods in the past shallow seas extended far into continental interiors. The term shelf seems inapposite for such seas, which are usually described as *epicontinental* or *epeiric*. The facies associations that record the existence of epicontinental seas are strikingly different from those of 'eugeosynclines'. They are characterized predominantly by orthoquartzites, carbonate platform deposits and shales and mudstones containing generally abundant benthic fossils that testify to shallow-water depths.

Another important difference from deposits associated with intense tectonism is that individual beds or bed groups often extend laterally for great distances with little or no change of facies. I can illustrate this with a personal anecdote. For my PhD thesis I made a detailed facies study of the Blue Lias Formation of southern England, a thin-bedded sequence of limestones, marls and shales of Lower Jurassic (Hettangian and Sinemurian) age. The classic section on the Dorset coast was already well-studied stratigraphically, and distinctive beds of limestone were known by such colourful quarrymen's names as 'Mongrel' and 'Pig's Dirt'.

Some years later I was being shown a series of beds of the same age in the eastern Paris Basin, 430 km to the east. My French guide was astonished that I could pinpoint the Hettangian–Sinemurian boundary to within half a metre, purely on the basis of the lithological resemblance with Dorset. My subsequent experience has persuaded me what I did was unremarkable for anyone who has made detailed examination of this kind of facies, and I am prepared to risk the assertion that lateral persistence of distinctive horizons over hundreds of square kilometres is almost commonplace, and by no means unusual even over thousands of square kilometres.

In dealing with the deposits of extensive epicontinental seas we are faced with a prime example of a situation in which there are no modern analogues, so that we are more than usually dependent upon circumstantial evidence and reasoning based on a full understanding of stratigraphic and facies relationships. It also helps to model the situation, as was attempted by Shaw (1964) and Irwin (1965) with special reference to the Palaeozoic deposits of the North American craton.

There were times, notably in the Ordovician, when it would have been possible to cross the continent for thousands of kilometres without seeing exposed land. Assuming that the maximum depth of sea attained that of the modern outer shelf (200 m), oceanward gradients are estimated not to have exceeded about 1 in 50 000, compared with modern shelf gradients of between 1 in 500 and 1 in 2500. Moreover, water depths over extensive regions were without much doubt appreciably less than 200 m.

Now such a physiographic situation as this would have had important hydrographic consequences. Major ocean current systems transferring large bodies of water could hardly have existed. Keulegan and Krumbein (1949) have shown mathematically that, given a sufficiently low oceanward gradient of the sea bed, a condition will be attained whereby waves generated by wind some distance offshore will dissipate their energy before reaching the shore. Not only would the formation of coastal cliffs be prevented but extensive stretches of sediment on the sea bed would be subjected to very little wave disturbance. In a similar way, tides would be reduced by friction with the bottom and tidal range reduced almost to zero at the shorelines remote from the open ocean. Effectively the only water energy within extensive regions of epicontinental sea would be that of waves generated by local winds.

Because of the extreme shallowness over huge areas minor topographic irregularities of the sea bed would have a disproportionately large effect in restricting circulation. In areas of high temperature and low precipitation those parts of the sea beyond the limits of tidal exchange and ocean currents would not be able to replace water lost by evaporation and there would be a tendency towards hypersalinity. Conversely, freshwater runoff from the land in humid regions would tend to render the sea slightly brackish for considerable distances offshore.

One might also expect organic productivity to be high in such seas, because nowadays phytoplankton productivity is appreciably increased in continental shelf waters (Fig. 5.1). Thus the waters of the shelf off New York have an annual production of about 120 g carbon m^{-2}, about four times as high as in the tropical ocean; the productivity of the North Sea is almost as high. Ryther (1963) relates this to two factors: firstly the increased supply of nutrients from the land; secondly the fact that in shallow water, the depth of the wind-mixed layer does not usually exceed that of the euphotic zone sufficient to reduce phytoplankton growth through lack of light. In addition, benthic algae play an important, if not a dominant, part in contributing towards total plant productivity. Land-derived organic detritus may be the most important component in some shelf regions. Thus one study in the northern Gulf of Mexico showed that there were only 0.032–0.096 mg l^{-1} phytoplankton compared with 0.2–0.5 mg l^{-1} of detritus (Jørgensen 1955).

Since most modern carbonate sediments are evidently biogenic in origin the presumed high productivity of ancient epicontinental seas may help to account for the large volumes of carbonates that occur, that have evidently been deposited at a far higher rate than carbonate oozes in the deep ocean. An extreme example is provided by the late Triassic carbonate platform deposits of the southern and eastern Alps, for which a sedimentation (and

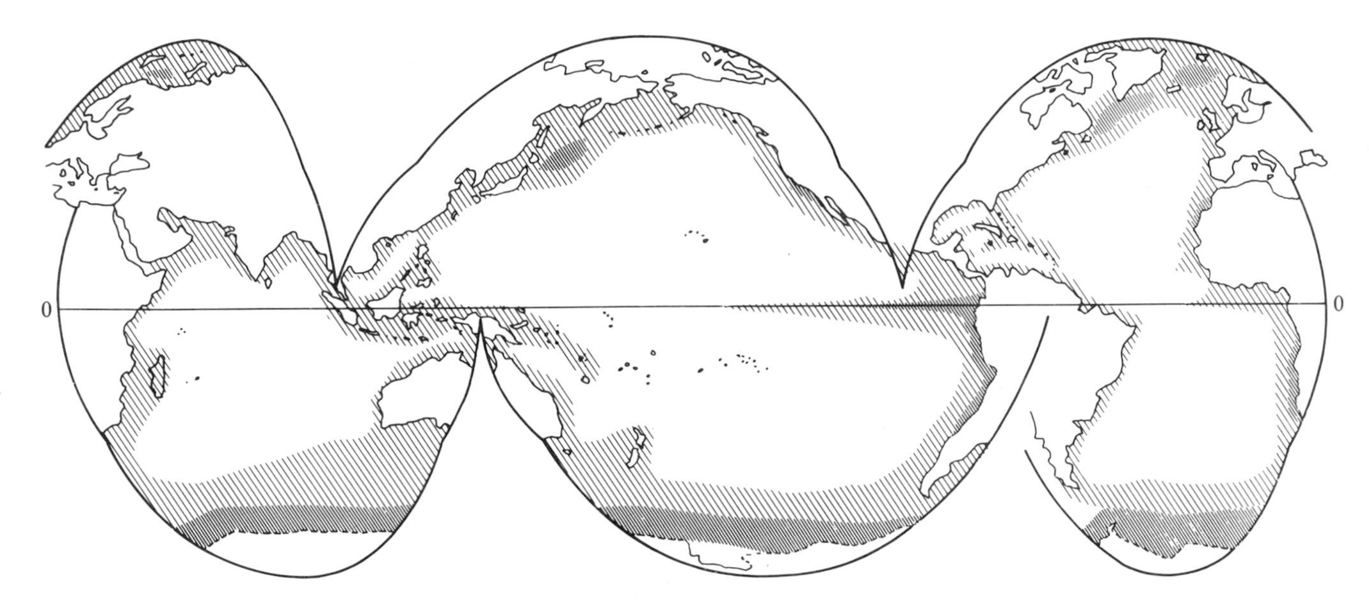

5.1 Generalized picture of phytoplankton productivity in the oceans. Density of ornamentation proportional to productivity. After Hallam (1967b).

subsidence) rate of about 100 m per 10^6 years has been estimated (Garrison and Fischer 1969).

Irwin (1965) proposes an epicontinental sea model with three zones of varying hydraulic energy (Fig. 5.2):

1. A wide, low-energy zone occurring in the open sea below wave base (Zone X).
2. A narrow, intermediate, high-energy belt beginning where waves first impinge on the sea floor and thus expend their kinetic energy on the bottom, extending landward to the limit of tidal action (Zone Y).
3. An extremely shallow, low-energy zone, which may be very wide, occurring landward of Zone Y, in which there is only limited water circulation, where tides are negligible and in which the only wave action is that produced by storms (Zone Z).

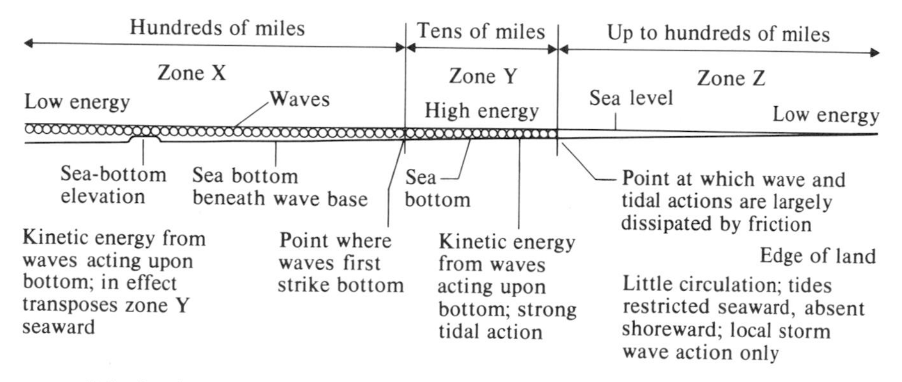

5.2 Section showing energy zones in epicontinental seas. After Irwin (1965).

Let us explore some of the possible consequences of this model. In a carbonate regime organic reefs might be expected to occur on the seaward side of Zone Y, and perhaps a barrier sand in a siliciclastic regime. Depending on the relationship of evaporation to precipitation, the salinity in Zone Z might achieve values appreciably above or below that of the ocean. Fine-grained sediments indicative of little water disturbance would be characteristic and might extend with little change over large areas. The combination of high productivity and restricted circulation in Zone Z would increase the tendency towards stagnation compared with the open sea.

The general model outlined here, which is an extension of those of Shaw and Irwin, can be tested in a number of ways. Clearly the most crucial part is the geographically extensive Zone Z because this lacks close modern analogues. It may be considered in terms of the lack of evidence of significant tidal activity and the tendency towards stagnation and abnormal salinities.

Negligible tidal range

In modern environments a variety of diagnostic features have been recognized in tidal regimes where siliciclastic sediments are being deposited. As noted in chapter 2, deltas may be classified as fluvial-, wave- or tide-dominated, and that whereas there are many modern representatives of the latter two categories the overwhelming majority of examples described from the stratigraphic record bear the characteristics of fluvial dominance. Among the most distinctive features of modern tidal flats are tidal channels that erode laterally, but there are very few well-documented ancient examples of their effects in the deposits that provide the best candidates for such environments. In fact diagnostic scour-and-lag concentrate features appear to be rare.

With regard to shallow-marine sand bodies, purportedly diagnostic features such as herringbone cross bedding, flaser bedding and clay drapes do not appear to be unequivocal criteria of tidal activity, and there is doubt about how reliably former sand waves can be identified, as indicated in chapter 3. In fact in the complicated and imperfectly understood hydrographic regime of shallow seas the respective importance of tidal currents and storm-generated waves is often quite difficult to sort out.

Now it may be objected that some shallow seas such as the North Sea and the Yellow Sea have an appreciable tidal range, and indeed tidal currents must travel faster in shallow water to transport a given volume of water. Such seas can hardly be compared, however, with the enormously more widespread epicontinental seas of the past. Because of the numerous indications of very modest water depth over vast areas, tidal transfer would have to take place at millrace speed, but there are no indications of the extensive scour that would occur in consequence.

A further argument is that free tidal exchange with the ocean would guarantee the maintenance up to the shoreline of normal marine salinities, but the evidence to be cited below does not always support this.

It would be unreasonable to maintain that tidal range was invariably negligible. Certain topographic configurations and the phenomenon of resonance would guarantee at least some tidal activity, and Zone Z might have been narrow or non-existent in certain cases. It is all the more remarkable therefore that so little convincing or unequivocal evidence for notable tides has been brought forward. Bearing in mind the argument based on overall geometric configuration and the evidence of widespread evaporite and lagoonal facies to be discussed below, the burden of proof appears to lie firmly with those who would maintain that tidal currents were of much importance, at least in those parts of epicontinental seas far distant from the ocean.

The relief of stromatolite heads has been regarded as an indicator of tidal range, based on study of the structures in Shark Bay and elsewhere, but the reliability of this criterion has been questioned by Scrutton (1978), who also points out that there is no firm evidence that the extensive Proterozoic stromatolites are intertidal. He indicates, however, that careful study of growth increments of bivalve shells might yield useful information on tidal patterns.

Tendency towards stagnation

One of the most characteristic types of epicontinental marine deposit, though not volumetrically the most abundant, is finely laminated bituminous shale (often called black shale) which often weathers to produce distinctive paper-thin laminae. Such rocks have a much higher organic-carbon content than normal shales, though it rarely exceeds more than a few per cent. The bulk of the organic matter consists of structureless kerogen but there is also a variable content of recognizable fossils in the form of dinoflagellates, pollen and spores, together with subordinate quantities of macroscopic terrestrial plants. Typically it occurs with slightly thicker clay interlayers, together forming rhythmic couplets averaging about 20–30 μ in thickness. Finely dispersed diagenetic pyrite crystals are usually common, and the terrigenous sand/silt content ranges from low to moderate.

The most striking faunal feature is the almost total absence of trace- or body-fossil burrowers. Indeed, substantial bioturbation would easily destroy the thinly interlayered bedding. Nektonic or nektobenthic organisms such as fish and ammonoids may occur frequently and in a very well-preserved state, but benthic organisms, if present at all, are almost always low in diversity and generally small in size. Faunal density can be high, however, with bedding planes crowded with a single species. Low diversity and high density are characteristic of an environment of high physiological stress. The species are usually epibionts that lived either at or just above the sediment/water interface.

There is general agreement that such bituminous shales were deposited below normal wave base in quiet, poorly aerated water, but dispute persists about the extent to which the bottom waters were completely anoxic, a matter to which we shall return later.

Selection of the most appropriate facies model has been even more controversial. Traditionally the most popular has been the deep, barred-basin model (Fig. 5.3A) based primarily on analogy with the Black Sea (see chapter 3). This, however, runs into some difficulties. Many bituminous shale horizons are extremely widespread, even extending across whole continents, and the associated facies is usually of shallow-water type. Most significantly, they seem to occur most frequently at or close to the base of sequences deposited during marine transgressions (Hallam and Bradshaw 1979).

(a) Barred basin

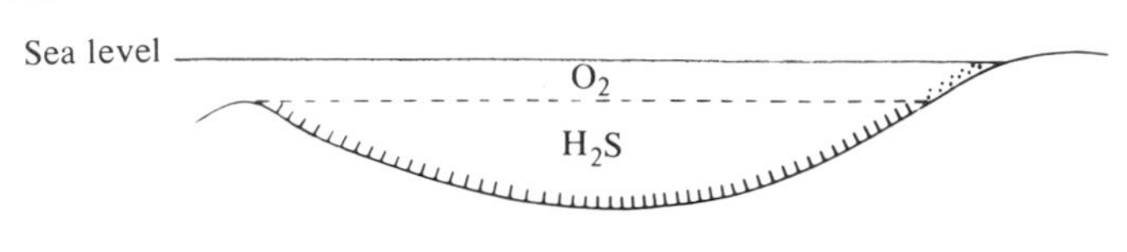

(b) Shallow shelf sea

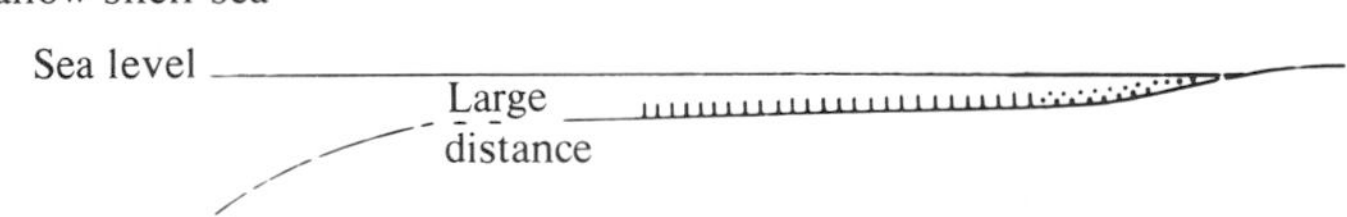

(c) Irregular bottom topography

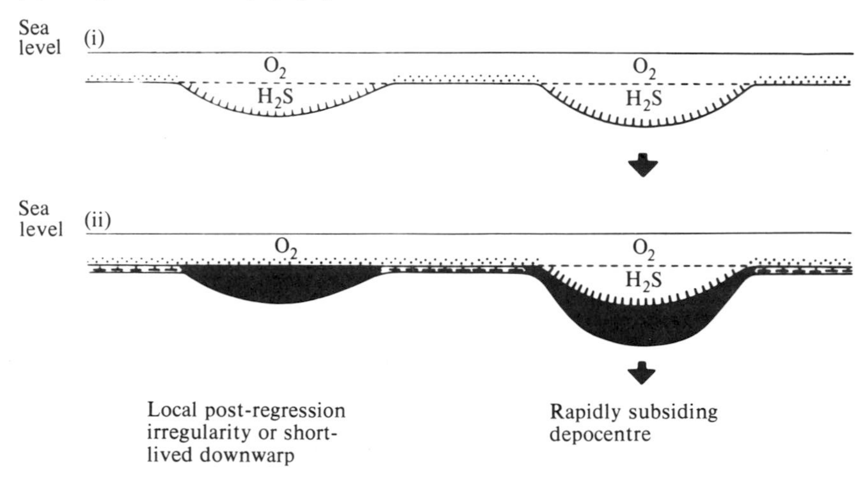

Bottom environment

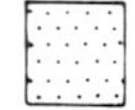 Aerobic

 Anaerobic

Deposits

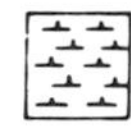 Non-bituminous, bioturbated shale

 Bituminous shale

5.3 Proposed models for the depositional environments of Jurassic bituminous shales. After Hallam and Bradshaw (1979).

Thus the well-known Chattanooga Shale and its lateral equivalents elsewhere in the Appalachians and Mid-West occur near the base of a Devonian–Carboniferous boundary transgressive sequence (Conant and Swanson 1961). Marine horizons within the Upper Carboniferous Coal Measures of Europe and North America often consist of bituminous shale. The *Kupferschiefer* of the German Upper Permian Zechstein directly overlies the basal conglomerate of the marine sequence, and its lateral equivalent in north-east England, the Marl Slate, rests upon the Yellow Sands, a large-scale, cross-bedded, aeolian or transgressive shallow-marine deposit (see chapter 2).

Another bituminous shale, the Rhaetian Westbury Beds of England, overlie a condensed bone bed which in turn rests on non-marine or marginal marine Keuper. The overlying Lower Hettangian, a time of renewed transgression after end-Rhaetian regression, contains a series of widely developed, thin bituminous beds. The thicker Lower Toarcian *Posidonienschiefer* of Germany and its French and Yorkshire equivalents (*schistes cartons* and Jet Rock respectively) occur only a short stratigraphic distance above widely developed sandstones, ironstones and marls displaying evidence of very shallow-water conditions, and correspond to a time of major transgression elsewhere in the world; bituminous shales at this horizon are also recorded from Alberta and Japan. The Middle Callovian Lower Oxford Clay of England is similarly close to the base of a major transgressive sequence traceable widely across Europe. Further Jurassic examples are cited in Hallam and Bradshaw (1979).

With regard to younger rocks, the widespread Albian and Cenomanian transgressions are often associated with bituminous shales (Schlanger and Jenkyns 1976), and in the Eocene of the Gulf Coast region Fisher (1964) described sedimentary cycles with laminated clays containing thin-shelled bivalves near the base of transgressive sequences, overlying condensed units rich in glauconite and phosphorite.

By no means all black shales are confined, of course, to epicontinental seas, and one recalls the graptolitic black shales of Lower Palaeozoic geosynclinal regions. It is intriguing to learn therefore that the deposition of graptolitic bituminous mud was initiated on the floor of the Iapetus Ocean in the British Isles region at about the start of the Caradoc, coinciding with extensive transgression in adjacent shelf areas. There are also widespread epicontinental marine black shales in the British Isles both at this time, the early Upper Cambrian and the Lower Llandovery; the Llandovery examples are also associated with significant transgression (Leggett 1980).

Because of this association with marine transgressions and shallow-water deposits, the wide extent and the general absence of evidence of a palaeosill for the depositional basin, I rejected the barred-basin model in favour of one involving deposition in a shallow sea where stagnation occurred because of a combination of high organic productivity, equable climate and extremely low oceanward gradient that cause restriction of circulation and the damping of tide- and wind-generated wave activity (Hallam 1967b and Fig. 5.3B). Freer circulation ensued as the sea deepened during the course of the transgression. This gave rise to more aerated bottom waters and the formation of non-bituminous, bioturbated shales containing a moderately diverse infauna.

Unfortunately this model also has shortcomings, notably in that it fails to explain why bituminous shales are not found near the tops of regressive sequences, nor why such shales are more fully developed in thicker basinal sequences (Hallam 1975, 1978a). A third model has therefore been proposed based on analysis of Toarcian bituminous shales in Western Europe, which incorporates elements of models A and B (Hallam and Bradshaw 1979).

This model (C) is based on the fact that anoxic, or near-anoxic, conditions were maintained longer in regions with greater subsidence rates. It invokes topographic lows on the sea floor that would locally have inhibited bottom circulation, allowing isolated pockets of stagnant water to persist while aerobic sediments were being deposited around them (Fig. 5.3C). As some depressions filled up, circulation improved and with it bottom aeration, but the more rapidly subsiding depocentres kept pace with sedimentation and maintained a bottom relief. One significant implication of this model is that in an epicontinental sea, bituminous shales may well be deeper-water deposits than the contemporary non-bituminous deposits surrounding them, but they are not necessarily different in depositional depth from the non-bituminous deposits stratigraphically above or below; much depends on local topographic configuration.

Further insight into the conditions of formation of bituminous shales can be obtained by considering the distribution of the European Jurassic deposits as a whole (Hallam 1978a). The facies is not uniformly developed throughout the system, being commoner and more widespread in the Lower than in the Middle and Upper Jurassic. Thus a number of thin bituminous horizons are widely distributed in the Hettangian and Lower Sinemurian in England, France and Germany; there is a more restricted development in part of the Pliensbachian of northern Spain and locally elsewhere on the continent, and a very widespread Lower Toarcian horizon. For the post-Lower Jurassic, bituminous shales are effectively confined to the Middle Callovian and Kimmeridgian of parts of England and Scotland.

This general pattern of distribution with respect to time may be tentatively accounted for as follows. As the Jurassic sea progressively spread across Europe and subsiding areas accommodated the deposited sediment, initial topographic irregularities, of the sort that tend to restrict water circulation in a shallow sea, were gradually reduced or eliminated. Thus the likelihood of partial or complete stagnation in particular areas diminished through time as water circulation became freer, to recur only as a consequence of appreciable deepening of the sea, either through eustatic rise or a marked basinal subsidence (Fig. 5.4). Support for this hypothesis is provided by Hallam and Sellwood (1976), who demonstrate for southern England that the amount of lateral thickness variation for all deposits, which might be expected to correlate with topographic irregularities and differential subsidence rates, diminished from the early and early–mid-Jurassic into the later Jurassic; also that the Kimmeridgian marked a time of increased tectonic disturbance and basinal subsidence in the British area.

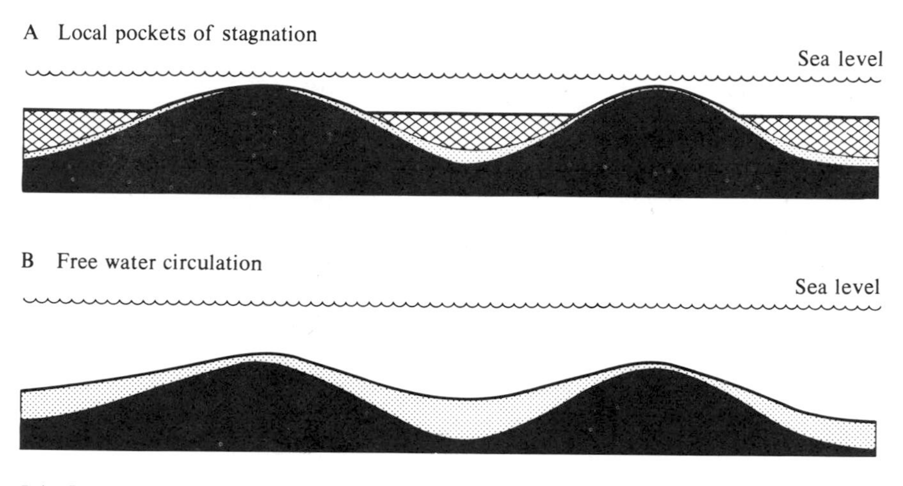

5.4 Interpretation of conditions controlling formation of bituminous or normal shales in European Jurassic. A, Early and B, later stages of deposition in a shallow-marine area of irregular topography. Stippling indicates sediments, black represents basement and cross-hatching symbolizes stagnant water. After Hallam (1978a).

It should also be borne in mind from the general model outlined earlier that the shallower the sea, the more marked would be the influence of topographic irregularities in restricting circulation and hence the greater the likelihood of stagnation. This may help to account for the occurrence of many bituminous shales at or close to the base of transgressive sequences.

Detailed examination of the Jurassic bituminous shale sequences shows that they are rarely uniform for more than short vertical distances, and that the change of facies can be interpreted in terms of varying degrees of stagnation (Hallam 1975). This has been borne out by more recent work.

Thus in the type Kimmeridgian of Dorset a series of small-scale cycles has been recognized by Tyson *et al.* (1979), which are interpreted in terms of varying degrees of oxygenation of bottom waters related to cyclic ascent and descent of the O_2/H_2S interface (Fig. 5.5). Thermoclines are thought to have probably been the principal agent in controlling stratification and stagnation of the water column. Similarly, Morris (1980), in his comparison of the principal British Jurassic bituminous shale formations, considers that the Kimmeridge Clay probably accumulated in an environment that periodically fluctuated between mildly oxygenated and totally anoxic. In contrast, the Callovian Lower Oxford Clay signifies deposition in mildly oxygenated bottom waters, while the Toarcian Jet Rock was probably laid down in very poorly oxygenated bottom waters, with reducing conditions extending up to the sediment surface.

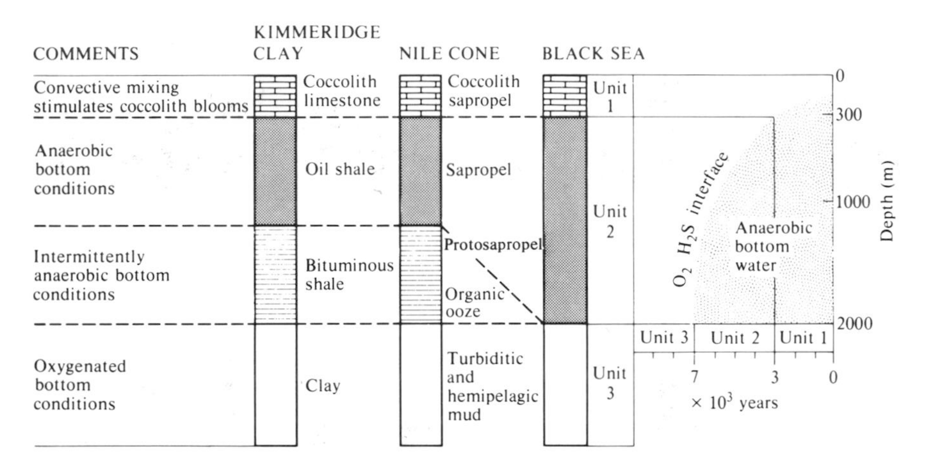

5.5 Lithologies and environments of the Kimmeridge Clay compared with the Nile Delta cone and the Black Sea. After Tyson *et al.* (1979).

One of the most interesting interpretations comes from a detailed palaeoecological analysis by Kauffman (1978) of the stratigraphic equivalent of the Jet Rock in southern Germany, the *Posidonienschiefer*, long celebrated for its superbly preserved vertebrate fauna.

The conventional opinion concerning this much discussed formation has been that it was deposited in totally anoxic conditions, based on the fine organic laminae, the presence of pyrite, the exquisite fossil preservation including skin, squids with tentacle hooks in place, entire arthropod skeletons and articulated fish. Bivalves and crinoids associated with driftwood have been considered to have been pseudoplanktonic in habit.

Kauffman, however, brings forward arguments against this interpretation:

1. There is a rare but undoubted benthos including foraminifera, bivalves and echinoids.

2. The eponymous bivalve '*Posidonia*' (= *Bositra*?) has shell characteristics that argue for a benthic mode of life.

3. Byssate and cemented bivalves are abundant but driftwood rare. Floating driftwood is a poor substrate for modern bivalves.

4. Vertebrates are only well preserved on their *undersides* and the upper sides have evidently been scavenged.

The middle Posidonienschiefer, with coccolith limestone horizons, is considered to represent the deepest, most stagnant water. '*Posidonia*' only occurs near the top and the bottom of the formation and is thought to have required more oxygen than *Inoceramus*, which occurs throughout. The largest ammonites have encrusting epibionts on their upper surface, suggesting that they rose above the anoxic layer. The crinoids were *not* attached to floating driftwood.

Kauffman's interpretation (Fig. 5.6) invokes frequent periodic small-scale fluctuations in degree of anoxicity, implying a condition of delicate balance, and speculates on the existence of an algal–fungal mat that trapped the $O_2/-O_2$ boundary below the mat while allowing clay sediment to pass through it. The kind of environment inferred by Kauffman is decidedly not the sort of environment to be expected in a stagnant basin with an extensive anoxic zone in the lower water column.

Abnormal salinities

Let us firstly consider the evidence of hypersaline regimes provided by evaporite basins within continental interiors. One of the best studied of such basins is the Devonian Elk Point Basin of Western Canada, extending from Alberta into Saskatchewan. Laminated sediments composed of calcite, dolomite, anhydrite and organic matter form the basal unit of the Muskeg and Prairie Formations and pass up into thick anhydrite and halite beds. These deposits are restricted to intermount or interreef areas between carbonate buildups of the Winnipegosis and Keg River Formations (Fig. 5.7).

Shearman and Fuller (1969) invoke a sabkha model to account for the laminites. The organic laminae are thought to have been produced by the growth of algal mats in a shallow subaqueous or intermittently exposed intertidal environment, with displacive nodular anhydrite being introduced during subsequent exposure in a supratidal setting.

This interpretation is disputed by Davies and Ludlam (1973), who point out that the laminae, ranging in thickness from a fraction of a millimetre to about 1 cm, are composed of mineral layers of calcite, dolomite or anhydrite alternating with films of structureless organic matter containing dinoflagellates. Interbedded with these laminites are clastic carbonate graded beds, which are thicker and commoner near the carbonate buildups.

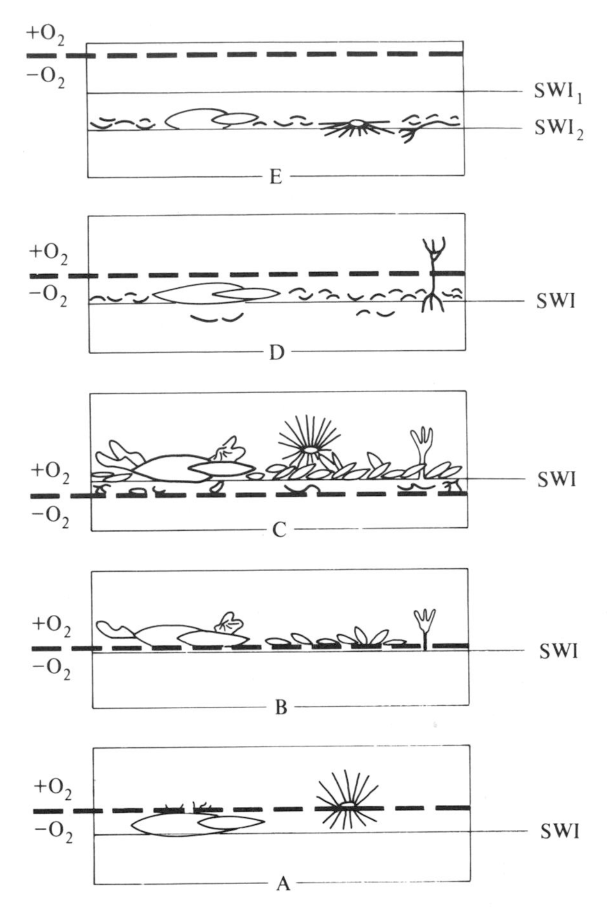

5.6 Model showing delicate balance and small-scale fluctuations of aerobic ($+O_2$)–anaerobic to oxygen-depleted ($-O_2$) boundary (heavy dashed line) on the Posidonienschiefer sea floor, and response of benthic biota. After Kauffman (1978).

Several arguments are brought forward against the algal mat–tidal flat hypothesis of Shearman and Fuller:

1. There is no close similarity between the Devonian laminites and modern algal laminated sediments of Shark Bay and the Trucial Coast. Thus there are no indications of subaerial exposure and the laminites are laterally extensive, individual laminae being traceable over distances of 25 km.

2. The sabkha model is inadequate to account for the overlying thick anhydrite and halite deposits.

3. The graded beds indicate subaqueous deposition and transport by turbidity currents from the topographically higher buildups.

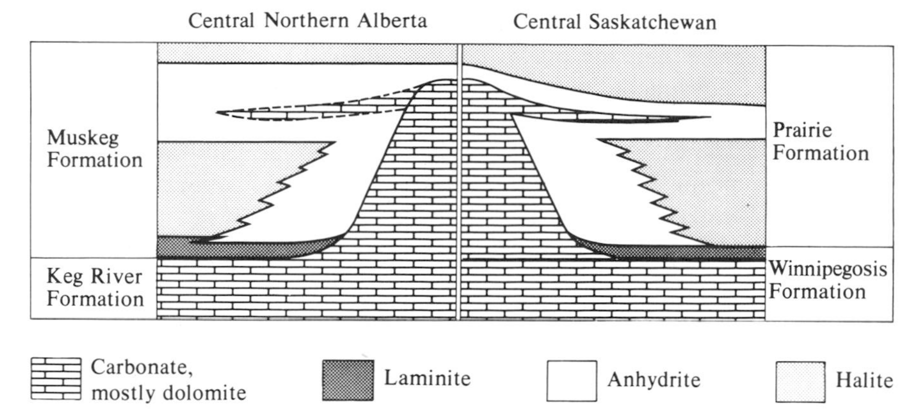

5.7 Middle Devonian stratigraphic relationships for the deposits of the Elk Point Basin. Adapted from Davies and Ludlam (1973).

Comparisons are made with laminites forming today in stratified water bodies and it is noted that in the Black Sea individual laminae are correlatable over distances as great as 1000 km. The lack of benthos and bioturbation point to stagnant bottom conditions.

The deposits are thought to have been laid down below wave base, because of the absence of any indication of scour, in a shallow, partly enclosed epicontinental sea open to the north-west. The carbonate buildups helped to restrict circulation and were the source of the graded beds. Evaporation exceeded precipitation and in parts of the basin more distant from the ocean, the upper-water stratum became hypersaline. A return underflow of brine occurred, as recorded by facies belts passing from anhydrite via halite to potash salts. Both the nodular anhydrite and enterolithic folds are thought to be diagenetic features unrelated to periodic emergence.

Davies and Ludlam estimate the depth of water to have been 50 m at maximum, based on the height of the buildups, some of which have vadose features at their top. If the mounds grew during deposition of the laminites the depth of water would have been less than 50 m. A deep-water (300 m) environment based on a total 300 m thickness of evaporites is excluded because the formation of the laminites was controlled by the distribution of the buildups, which would have been drowned.

The interpretation of the Elk Point Basin evaporites by Davies and Ludlam probably has relevance also to other major evaporite bodies, especially as bituminous laminae have frequently been recorded. Thus the celebrated Upper Permian evaporites of Texas and New Mexico show a broad similarity in that an older sequence of laminated bituminous calcite and anhydrite deposits of the Castile Formation is overlain by a younger formation, the Salado, with more soluble salts such as halite, sylvite and polyhalite.

These evaporites have been generally interpreted as having been deposited in a barred marine basin in a back-reef situation with respect to the underlying carbonates including the Capitan Reef, discussed in chapter 3 (Stewart 1954). Anderson and Kirkland (1966) have demonstrated an excellent correlation of the Castile laminae over distances exceeding 14 km, and also record the existence of comparable laterally extensive laminae in the late Jurassic Todilto Formation of New Mexico.

With regard to the Zechstein evaporites of north-west Europe, which are broadly correlative with the Ochoa Series, including the Castile and Salado Formations (Borchert and Muir 1964), Richter-Bernburg (1960) has claimed to be able to correlate with confidence laminae over distances of nearly 300 km. It is worth recalling that a marine bituminous laminite forms the base of the Zechstein sequence. By analogy with modern examples mentioned in chapter 3, many of the organic laminites may well be varves. This has also been widely assumed for many other types of laminae composed of various evaporite and carbonate minerals (Stewart 1954).

We turn now to other groups of deposits suggesting the existence of laterally extensive shallow seas passing into lagoons with salinities appreciably different from the ocean, in this case lacking or almost completely lacking evaporite minerals. Examples will be selected from the Mesozoic of north-west Europe.

The so-called Rhaetic beds are broadly correlated with the topmost Triassic (Rhaetian) Kössen Beds of the Austrian Alps, passing southward into the upper part of the Dachstein Limestone, by the common occurrence of the distinctive bivalve *Rhaetavicula contorta*. They consist principally of a sequence of laminated bituminous shales (Westbury Beds in England, *contorta* beds in Germany) passing up into marls and limestones in England and sandstones in Germany.

The Kössen Beds and their Carpathian equivalents (Gazdzicki 1974) contain a rich and diverse fauna with many species of bivalves (including large, thick-shelled megalodontids), foraminifera and ostracods. Articulate brachiopods, reef-building corals and various types of echinoderm are common, and ammonites, calcareous sponges and hydrozoans occur rarely. These deposits are generally regarded as having been laid down on the margin of the Tethys Ocean. In contrast, the north-west European 'Rhaetic' has a low diversity fauna confined largely to a few species of bivalves (without megalodontids) and ostracods, suggesting abnormal salinity. In view of the absence of evaporites and the presence of kaolinite and abundant plant material suggesting a humid climate on the neighbouring land, the environment is plausibly regarded as one of brackish water (Will 1969). The Rhaetic 'sea' occupied an extent of hundreds of thousands of square kilometres (Fig. 5.8), implying in effect an enormous lagoon for which there is no modern analogue.

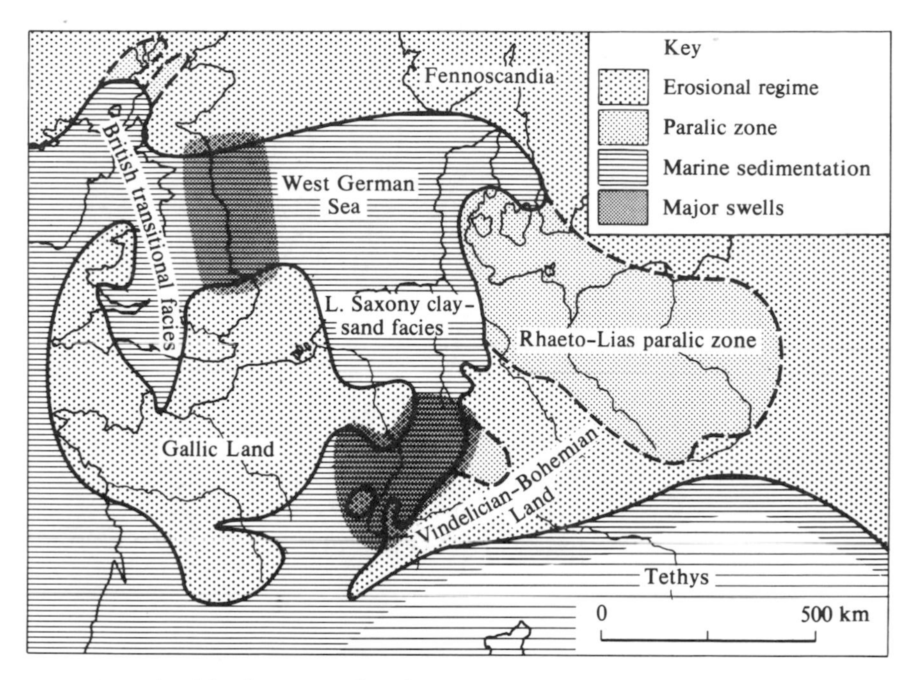

5.8 'Rhaetic' palaeogeography of western Europe. Adapted from Will (1969).

After a minor end-Triassic regression the sea returned at the beginning of the Jurassic, but fully marine conditions were not established immediately. Thus nowhere do ammonites occur exactly at the base of the Hettangian sequence, which is characterized by a high density fauna of presumably euryhaline oysters, unless it is very condensed, and the faunal diversity increases progressively through the stage (Hallam 1960). As with the 'Rhaetic' the facies indicates the persistence of quiet, shallow-water conditions across much of the continent.

The second example is taken from the Bathonian deposits of Great Britain. This stage marks a widespread regressive interval in the Jurassic marine sequence, with a zone of deltas centred on the North Sea passing southwards and westwards into a lagoonal facies (Fig. 5.9). Fully marine deposits are known only in south-west England and in the Viking Graben far to the north (Poulton and Callomon 1977).

Fine-grained marls and limestones in central England with a low diversity, bivalve-dominated fauna have been interpreted as lagoonal deposits comparable with those forming today in Florida Bay (Palmer 1979). Part of this succession can be traced northwards into a sequence of sands, silts and clays with horizons of truncated rootlets, interpreted by Bradshaw (1978) as signifying a series of prograding coastal marsh and swamp deposits inter-

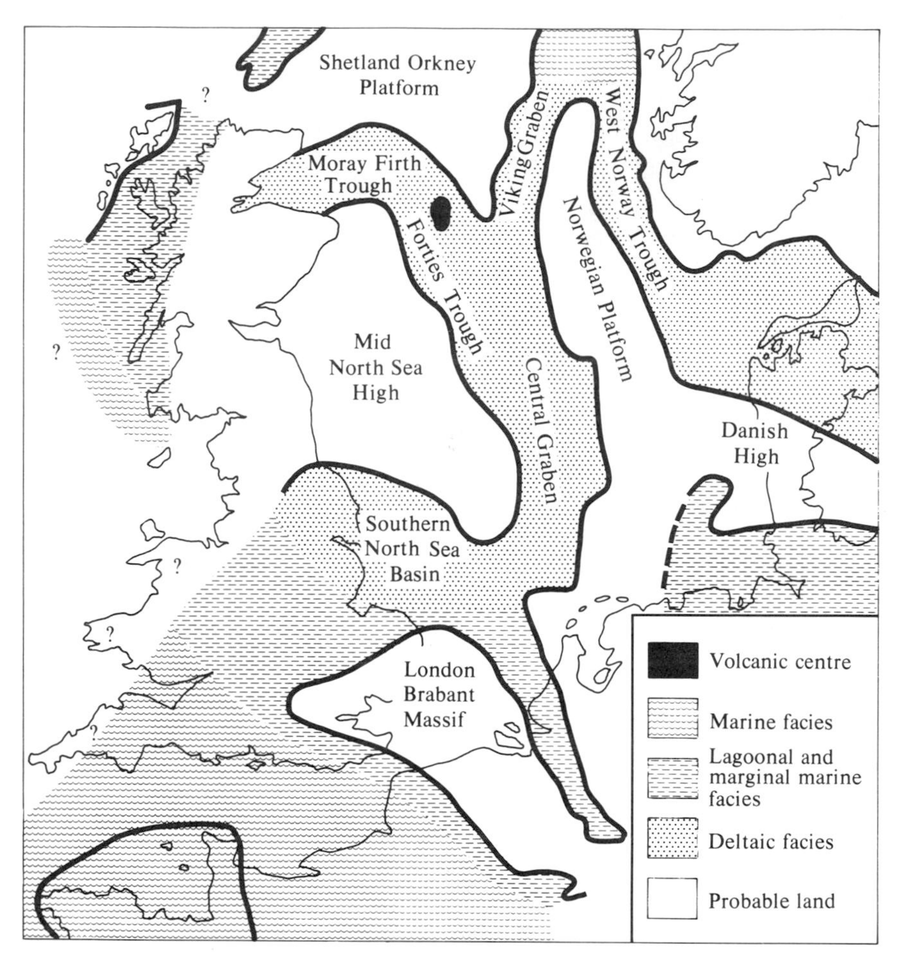

5.9 Bathonian palaeogeography of north-west Europe. Adapted from Sellwood and Hallam (1974).

rupted by periodic marine transgression. Faunal diversity in the marine beds decreases northwards and Bradshaw argues that fully marine conditions were never established.

In north-west Scotland the stage is represented by the so-called Great Estuarine Series. Hudson (1963, 1966, 1970) has recognized a series of faunas characterized by low diversity and high density, including estheriids, ostracods and the molluscs *Neomiodon, Liostrea* and *Viviparus*, and inferred considerable fluctuations of salinity, subsequently confirmed by carbon and oxygen isotope analysis (Tan and Hudson 1974). The closest modern analogues appear to be the deposits of the coastal lagoons of Texas and the bays of the Florida Everglades.

Although the zone of lagoonal facies is less extensive than that of the Rhaetic it is nevertheless much wider than any equivalent Recent environments. The coastal lagoons with which Hudson makes comparison are restricted by barriers, but in one place in the open Gulf of Mexico some 16 km offshore *Crassostrea* reefs occur in waters with a summer salinity of only 30% because of dilution by fresh water from the Atchafayala River (Coleman 1966). The winter salinity is presumably less. Bearing in mind that the Gulf of Mexico is a microtidal regime, we may treat this as a hint of what to expect with an extensive epicontinental sea in an area of humid climate.

Finally, the Purbeck Beds of southern England, which straddle the Jurassic–Cretaceous boundary, have traditionally been interpreted as the deposits of a brackish-water lagoon. Once again, the Purbeck facies is very widespread, extending across the Paris Basin as far as the Swiss Jura and southern Germany. Characteristic features include low diversity, high density faunas, algal and pellet limestones and horizons of sabkha-type evaporites and soils. Modern research reviewed by Hallam (1975) suggests a series of extremely shallow, quiet-water environments of strongly fluctuating salinity varying from brackish to hypersaline, with occasional emergence. As with the previous examples the tidal range is likely to have been negligible.

Other distinctive facies

It is appropriate to round off the chapter by considering briefly a few other distinctive types of epicontinental marine deposit that have no close modern analogue, although the extent to which they owe their origin to the special conditions of epicontinental seas may be debatable.

OOLITIC IRONSTONES

These so-called minette-type ores typically consist of ooids of the iron aluminium silicate chamosite in a matrix of chamosite mud or siderite microspar; the ooids may alternatively be goethitic or haematitic and the matrix calcite spar. They are commonest in Europe in the Ordovician and Jurassic and North American examples include the Ordovician Wabana Ironstone of Newfoundland and the Silurian Clinton Ironstone of the Appalachians.

Although chamosite occurs in modern marine sediments, for instance as altered faecal pellets off the Niger and Orinoco deltas, no modern analogue of these oolitic ironstones has been discovered, and their mode of formation is correspondingly intriguing. Most discussion has been concerned with European Jurassic examples (Hallam 1975). Although a lagoonal environment of deposition has sometimes been claimed, marine fossils are usually abundant and commonly include stenohaline elements such as brachiopods, ammonites and crinoids. The rich benthos and the presence of some horizons of cross bedding, scour surfaces and reworked pebble beds signify a well-oxygenated, shallow-water environment with at least episodes of high water energy.

Most argument has been concerned with the origin of the ooids and the source of the iron. Comparison of the chamosite ooids with aragonite ooids forming at the present day should not be pressed too far, because the chamositic examples tend to be ellipsoidal rather than spheroidal in shape and are often found in a matrix of chamosite mud, suggesting *in situ* formation in a little disturbed environment. Where the ooids are predominantly goethitic, as in the Sinemurian Frodingham Ironstone of central England and the Toarcian ores of Lorraine, the matrix is coarse calcite spar, rather than siderite microspar or chamosite mud, implying reworking in a more agitated environment, with winnowing away of fine sediment and concomitant oxidation of the ooid chamosite. Kimberley's (1974, 1979) attempt to resurrect Sorby's old idea that the chamosite ooids were formed by diagenetic replacement of originally calcareous ooids ignores the overwhelming petrographic evidence in favour of primary origin (Taylor 1949). Moreover, his interpretation requires that the ironstones be overlain by a thick deltaic sequence to provide a source of the required replacive ions by downward percolation. However, among the many well-known Jurassic ironstones or sandstones with chamosite ooids, only one, the Dogger Formation of Yorkshire, has a thick deltaic cover, and the stratigraphic evidence in the other cases does not favour the possibility that deltaics were subsequently removed by erosion.

The problems involved in chamosite ooid formation are really more subtle than this. Is the chamosite an alteration product of, for instance, land-derived kaolinite or was it formed *in situ* directly from ions in the sea water rather like glauconite? If the ooids formed by 'snowball' accretion in a muddy environment, why are not ooids of other clay minerals more frequent? Chamosite presumably required reducing conditions for formation (cf. Curtis and Spears 1968), in which case why does it occur in deposits with a rich benthos? Was it because it formed within fine-grained sediment but remained stable in mildly oxidizing conditions on the sea floor? What is the origin of the peculiar distorted ooids, often with hooked junctions, known as *spastoliths*? Rather than attempt to answer these and related questions I am content merely to pose them, because there is as yet insufficient evidence to solve them decisively.

As far as the more general problem of the source of the iron is concerned, the evidence from facies associations supports the traditional view that it came from nearby land rather than being derived from contemporary anoxic muds (Hallam 1975). A necessary prerequisite seems to be some form of pre-concentration in the form of laterite, produced by weathering under tropical conditions, and transport by sluggish rivers containing abundant organic matter in a topographically subdued land with a thick vegetation cover. The iron would be transported in the ferrous state possibly as part of organic-metallic complexes, and precipitated on entering the more alkaline environment of the sea. Alternatively, it might have travelled to the sea as ferric oxide particles coating the surface of clay micelles and afterwards reduced to the more soluble and hence more mobile ferrous form (Carroll 1958). Perhaps the iron concentration in the water of some epicontinental seas distant from the ocean was much higher than in present-day sea water, in which it is extremely low.

A further problem is how the iron, in whatever form, became separated in transport from terrigenous sand and clay. Without such a separation the iron minerals would be considerably diluted and no exploitable ore could form. The most popular solution has been to invoke some sort of *clastic trap* (Huber and Garrels 1953). whereby ironstones formed on slight topographic rises while terrigenous clastics were, as it were, siphoned off into surrounding depressions (Fig. 5.10A). Support for this comes from the fact that ironstone formations can be demonstrated by stratigraphic data to be condensed relative to laterally equivalent sequences of sand and clay.

The clastic trap hypothesis was challenged by Brookfield (1971), who argued that ferruginous ooids and siliciclastic sand grains have different hydrodynamic properties and can be separated on the sea bed by processes of suspension, saltation or rolling (Fig. 5.10B). This alternative idea seems implausible for the reasons given by Knox (1971), but a further problem remains: in some ironstone deposits the purest chamosite oolites may be thicker than their lateral sandy equivalents (Hallam and Bradshaw 1979; Fig. 5.10C).

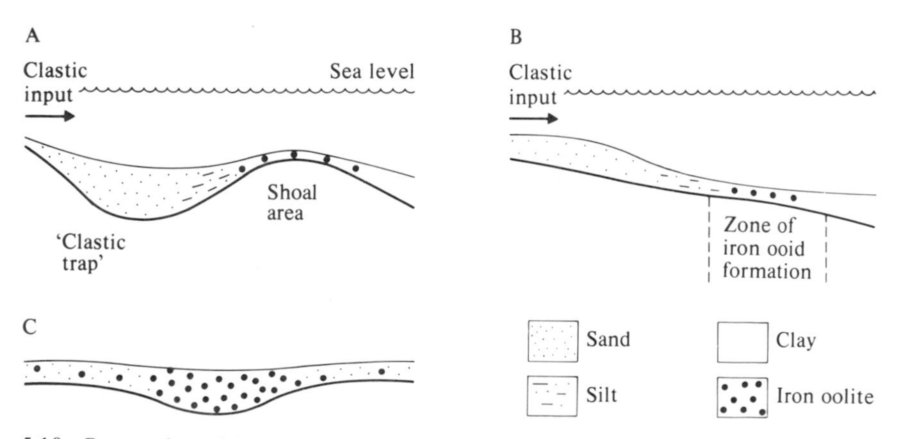

5.10 Proposed models for the depositional environment of oolitic ironstones. A, The classical 'clastic trap'; B, Brookfield's (1971) alternative; C, generalized facies relationships of the Marlstone Rock Bed (Pliensbachian) and Northampton Sand Formation (Aalenian) of the Midlands. After Hallam and Bradshaw (1979).

Consideration of the facies sequence may help to resolve this problem, because Jurassic ironstones characteristically occur towards the top of a major or minor regressive sequence, signified by marine shales passing up via sandy shales into sandstone, which is in turn capped by a thin ironstone (Hallam and Bradshaw 1979).

It seems difficult to eliminate entirely the need for topographic irregularities on the sea floor to account for the separation of chemical and siliciclastic components. Such irregularities could have been very slight, because there are many beds in which both sand grains and chamosite–goethite ooids occur in abundance together. Only in very restricted regions and for limited periods of time was separation of the two types of material total, leading to the formation of an exploitable ore.

Such conditions would have been favoured by regression, when any topographic high, of whatever origin, would have exerted a greater hydrographic influence. In deeper water finer silt and clay particles would have tended to settle as a more or less uniform carpet of sediment over the whole sea floor. As the sea shallowed coarser sediment would have replaced the finer, and the dominant mode of sediment transport would have changed from suspended- to bed-load. Bed-load transport is less efficient in dispersing sediment than suspended-load, and sand bodies would have tended to occur as restricted bars rather than extensive sheets. Sand grains might only have been dispersed more widely during occasional storms. In conditions of very shallow water not strongly affected by tidal currents even the most modest topographic rise might have resisted being covered by sand. In addition there could have existed small depressions within extensive shoals kept free from sand encroachment, within which a pure oolitic ironstone could have formed within comparatively quiet water.

Oolitic ironstones are not volumetrically abundant, because the ironstone formations rarely exceed a few metres in thickness and a few thousand square kilometres in lateral extent. They tend to pass laterally into thicker sequences of sandy shale or mudstone with nodules of concretionary siderite rather than the generally ubiquitous calcite, with subordinate horizons of sandstone, which implies that iron-rich waters were considerably more extensive than the ironstones themselves. Furthermore, chamosite or goethite ooids are dispersed through considerable volumes of siliciclastic or carbonate sediments (the so-called iron-shot limestones), indicating transport from the ooid-generating localities. Clearly ironstones only formed when a particular combination of chemical, sedimentological and physiographic conditions was realized.

PHOSPHORITE

Phosphorite is another chemical sediment that forms deposits of economic importance. The brown phosphatic mineral has traditionally been known as collophane but chemical and X-ray diffractometry analysis shows it to be a carbonate apatite. In the exploitable deposits it normally occurs in pelletal form, but it also occurs as phosphatized nodules associated with glauconite in some condensed horizons (Kennedy and Garrison 1975a).

Unlike in the case of oolitic ironstones Recent phosphorites are known, though it cannot uncritically be assumed that they are strictly analogous to the ancient pelletal phosphorites that are found in continental interiors. They occur at depths of a few hundred metres within 40° of the equator, predominantly off the west coasts of continents. This occurrence is usually related to zones of upwelling of phosphorus-rich oceanic waters, which are generally found in the trade-wind belt where surface waters are blown offshore by the trade winds and where the offshore current is augmented by the Coriolis force (Bromley 1967; Sheldon 1964a).

Geologically-young phosphorite is also found in rocks adjacent to areas of modern oceanic upwelling. Thus the phosphorite in the Miocene Monterey Formation of California is located next to the modern upwelling associated with the California Current and the Miocene phosphorite of the Peruvian coastal Sechura Desert occurs near the modern upwelling associated with the Peru Current. An origin bound up with the upwelling phenomenon appears in these cases to be highly plausible, but they hardly qualify as the deposits of epicontinental seas.

One of the best known of the older phosphorite deposits occurs, however, well within a continental interior, namely in the Permian Phosphoria Formation of western Wyoming and adjacent states. The question naturally arises as to what extent an oceanic upwelling model can be applied.

The pelletal phosphorite beds occur in a cyclic sequence of a variety of sediments. According to Sheldon's (1963) interpretive movel, the sequence represents successive transgressive and regressive phases of areally-zoned depositional environments, with dark bituminous phosphate-bearing and poorly fossiliferous mudstone representing the deepest-water phase. Successively shallower-water phases are represented by chert, carbonates including dolomite and bioclastic limestone, and light-coloured mudstone. These last pass into evaporite-bearing continental red beds. The regional facies distribution indicates land to the east.

Sheldon considers that the evaporitic red beds signify a low-latitude west-coast desert, implying offshore trade winds, and that the phosphorites were deposited as a consequence of upwelling on the edge of the ocean to the west (Fig. 5.11). The palaeogeographic setting is, however, on the borders of the North American Palaeozoic craton and western miogeosyncline and there is no convincing evidence of ocean or marginal ocean basin until the western coastal zone is reached (Churkin 1974). Therefore either Sheldon's model or the palaeogeographic reconstruction would appear to require some amendment.

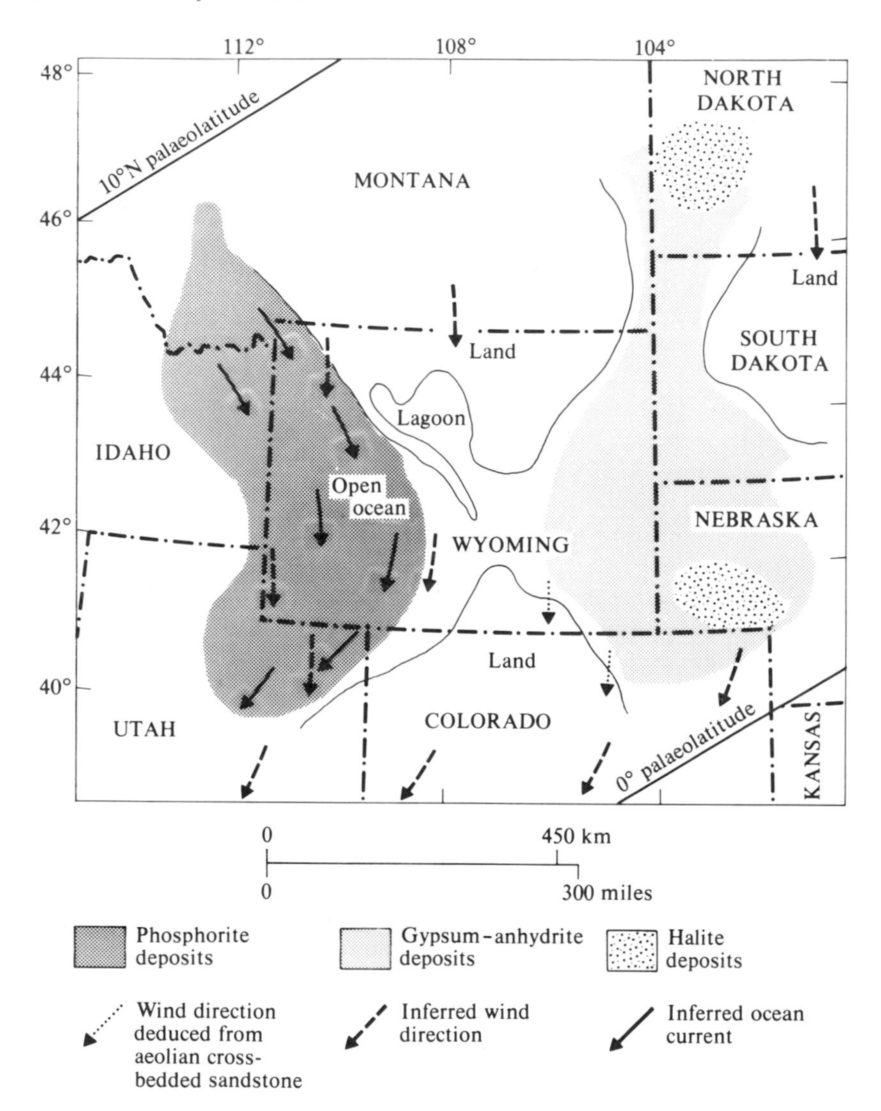

5.11 Palaeogeographic setting of the Permian phosphorites of the western United States. After Sheldon (1964a).

Other important phosphorite deposits, of late Cretaceous to Eocene age, occur in a zone extending from Morocco to southern Turkey and Iraq, and they are similarly associated with chert, black shale and carbonate (Salvan 1960; Sheldon 1964b). The modern Mediterranean does not exhibit upwelling, but in the very different palaeogeography of the time some at least of these areas would have been situated on the edge of a craton on the southern side of the Tethys. Since this was situated in the northern part of the trade-wind belt Sheldon (1964b) considers that the winds probably caused a strong east–west current with strong upwelling on the southern side of the sea caused by offshore winds and the Coriolis force.

Further insight into the conditions controlling phosphorite deposition in what was undoubtedly an epicontinental marine environment may be obtained by a study of the phosphatic black shales that occur within Pennsylvanian (= Upper Carboniferous) deposits of the United States mid-continent region, extending from Kansas to the Appalachians (Heckel 1977). The deposits in question are typically anoxic, being black and fissile, rich in organic matter and conodonts but characteristically devoid of benthic fauna; there is also an enrichment in heavy metals such as copper, nickel, vanadium and zinc. The abundant non-skeletal phosphorite occurs both as laminae and nodules. The black shales, which are of considerable lateral extent, are interbedded with limestones and normal shales with a benthic fauna and are interpreted by Heckel as the deepest water, most offshore deposits of a transgressive–regressive sequence.

Heckel presents a model (Fig. 5.12) in which a thermocline was developed to give a vertical density stratification. This was strong enough to prevent local wind-driven cells of vertical circulation from replenishing oxygen on the sea bottom. If the depth were great enough in an epicontinental sea anoxic or near-anoxic conditions could be established without requiring the presence of a restrictive barrier or sill, a point that has been argued earlier in this chapter.

The top of the oxygen-minimum zone (and the thermocline) is high in the eastern tropical Pacific as a consequence of upwelling produced by the trade winds blowing offshore. The upwelled deeper water is both enriched in phosphorus and low in oxygen, and provokes a rich growth of plankton which provides large quantities of organic matter. Deposition of phosphorite is favoured in this region both by the decrease of pressure and rise of temperature as the colder water wells up from the deep, as well as by rise of pH. The presence of phosphorite suggests deposition in water greater than 50 m deep because of the rapid assimilation of phosphorus by phytoplankton in the photic zone. The greatest concentration of phosphorus in modern marine muds is between 30 and 200 m water depth. Heckel infers a tropical latitude for the Pennsylvanian mid-continent sea, probably connecting to the open ocean westwards through the basins of West Texas. A depth of deposition of between 100 and 200 m is thought to be the most likely.

This intriguing association of phosphorite and black shale, which recurs throughout the stratigraphic column, may provide the crucial clue to phosphorite formation. It could well be that many other relatively deep-water black shales, virtually devoid of benthos, may prove to be enriched in phosphorite, though the occurrence of economically exploitable phospho-

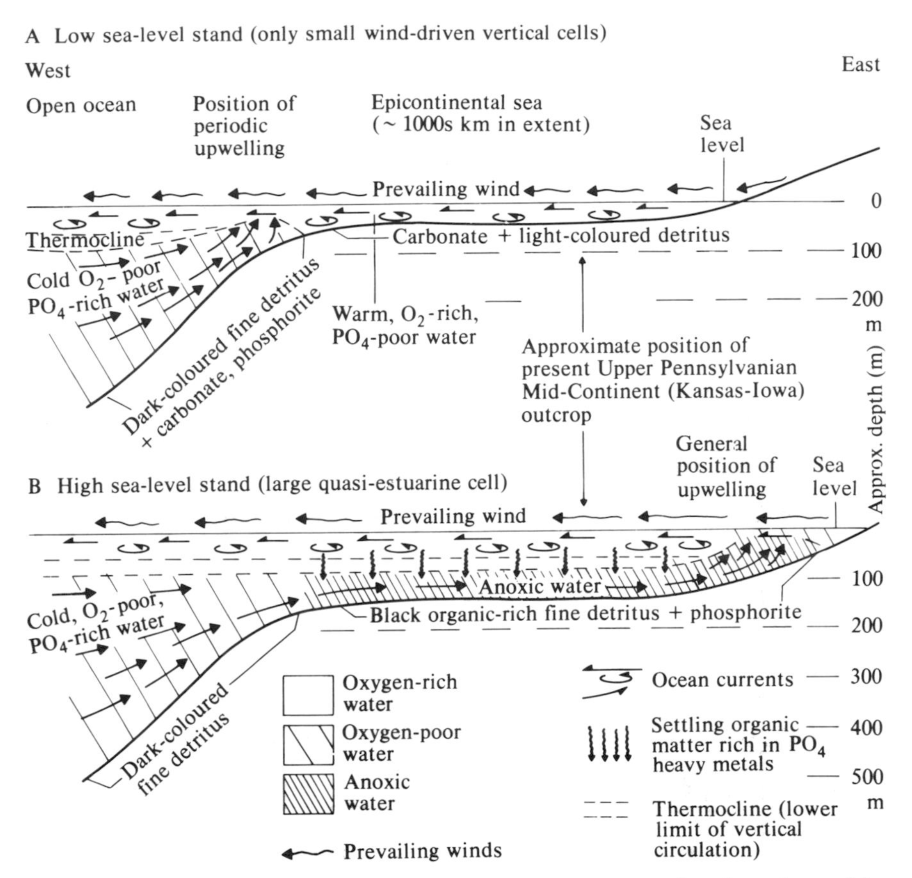

5.12 Model for the deposition of phosphatic black shales in an epicontinental sea. After Heckel (1977).

rite beds is apparently much more restricted. However, it is probably always necessary to invoke some form of upwelling, but it looks as though bodies of upwelled, phosphorus-enriched water have in the past extended far into continental interiors.

CHALK

One of the most distinctive deposits of ancient epicontinental seas is the soft, friable white micritic limestone known as *chalk*, which was laid down during the late Cretaceous over a vast area of northern Europe extending from Ireland to the Russian Platform and in a more areally limited belt in the eastern part of the North American mid-continent seaway. A major breakthrough in the understanding of chalk came with the advent of electron microscopy, whereby it was demonstrated that the great bulk of rock was composed of coccoliths and coccolith fragments, with a coarser fraction including *Inoceramus* prisms, foraminifera (mainly planktonic) and calcispheres (Hancock 1975). The paradox of why such a pure limestone had remained only partly consolidated and therefore soft was consequently resolved, because the fine components were all calcitic and there had accordingly been only a limited amount of post-depositional cementation.

Coccolith–foraminiferal muds are being deposited in the deep ocean at the present day, but it would be extravagant to maintain accordingly that large parts of continental interiors were submerged beneath thousands of metres of water during the late Cretaceous. The chalk is undoubtedly a pelagic deposit, however, and its widespread occurrence on the continents relates to an unusual event, namely the biggest marine transgression and episode of sea deepening since the mid-Palaeozoic, which will be discussed in the next chapter.

Soft, uniform white chalk with horizons of nodular diagenetic flint (= chert) is the predominant deposit in the northern European Upper Cretaceous. Much of it is not especially rich in macrofossils but a considerable fauna has nevertheless been collected, of which the burrowing irregular echinoids such as *Micraster* are perhaps the most familiar. The surface sediment was evidently soft, as indicated by the special adaptations of various bivalves to prevent sinking, such as spines on *Spondylus* and the development of a shell with a large surface area in some *Inoceramus.* At depths of no more than 0.5 m, however, the sediment was firm enough to preserve in clear outline such ubiquitous trace fossils as *Chondrites, Thalassinoides* and *Zoophycos.*

There has been much speculation about the depth of deposition. Within the white chalk, as opposed to some hardground horizons, there is no evidence of algal activity, which suggests deposition beneath the photic zone. The occurrence of a diverse fauna of hexactinellid sponges suggests, using an actualistic comparison, a depth of between 200 and 600 m and Hancock (1975), taking all evidence into account, proposes a depth range of 100–600 m. Hardground horizons on the other hand probably signify relatively shallow depths and, taking these into account, Kennedy and Garrison (1975b) suggest a range of 50–300 m.

The rate of sedimentation obviously varied between basins and swells but was commonly 10–50 mm per 1000 years and even as high as 150 mm per 1000 years (Funnell 1978). Funnell points out that this rate is high compared with the ocean floor beneath the most productive phytoplankton zone of the equatorial Pacific at present, where the rate of coccolith–foraminiferal ooze deposition is only 10–30 mm per 1000 years, and much less elsewhere. This implies a higher calcareous phytoplankton productivity within the chalk-depositing epicontinental sea compared with the deep ocean, which is consistent with the general model outlined at the beginning of this chapter.

One of the most interesting phenomena of the European Chalk is the frequent occurrence of *hardgrounds* and *nodular chalk*, which were produced by submarine lithification. They are more frequent in areas of low overall sedimentation rate such as swells and the margins of old upstanding massifs. A few hardgrounds have been traced over very large distances and suggest regional episodes of shallowing.

A complex history for these features, involving a series of stages of progressive lithification, has been worked out by Kennedy and Garrison (1975b). A pause in sedimentation led to the formation of an omission surface followed by growth of discrete nodules below the sediment-water interface to form a nodular chalk. Erosion of this nodular chalk produced an intraformational conglomerate. Further growth and fusion of the nodules into continuous or semicontinuous layers gave rise to an incipient hardground, which became a true hardground if exposed by scour.

At this stage the exposed lithified chalk bottom was subject to boring and encrustation by a variety of organisms, and frequently glauconitized and phosphatized as well (Fig. 5.13). The hardground–nodular chalk fauna includes fauna rare in normal chalk, such as gastropods, corals, scaphopods and various species of bivalves requiring a firm surface for attachment.

The nodular chalk, with its characteristic flaser structure, is strikingly similar to the nodular limestones of the Albian *Red Chalk* of north-eastern England, the Jurassic *Ammonitico Rosso* of southern Europe and various Palaeozoic *griottes* more widely distributed in the continent. A similar mode or origin is probable (Jenkyns 1974). The chalk of the Normandy coast exhibits a number of prominent banks or buildups up to 50 m high and 1500 m across, with bedding being picked out by hardgrounds, nodular chalks and horizons of flints which represent the silicified burrow-fills of *Thalassinoides* (Fig. 5.14). Frame-building, sediment-trapping and stabilizing organisms are absent, and bank development and stabilization is thought by Kennedy and Juignet (1974) to be the consequence of an algal or sea grass covering. Bearing in mind, however, the deep-water buildups recently discovered in the Florida Straits (see chapter 3), it may be unnecessary to invoke plants, though this increases the difficulty of making a satisfactory interpretation. Similar structures occur in the Maastrichtian and Danian chalk of Denmark, which were originally interpreted as giant dunes. A striking resemblance with the Waulsortian-type buildups of the Palaeozoic is evident.

Although most of the chalk is almost pure $CaCO_3$ the Cenomanian chalk is more marly. This, and the occurrence of glauconitic sands (on the margins of old massifs) which disappear above the Cenomanian everywhere northwest of Central Europe, indicates shallower-water conditions. By late Campanian times, corresponding to the maximum inundative phase, the land was almost completely submerged. It has been generally assumed that the late

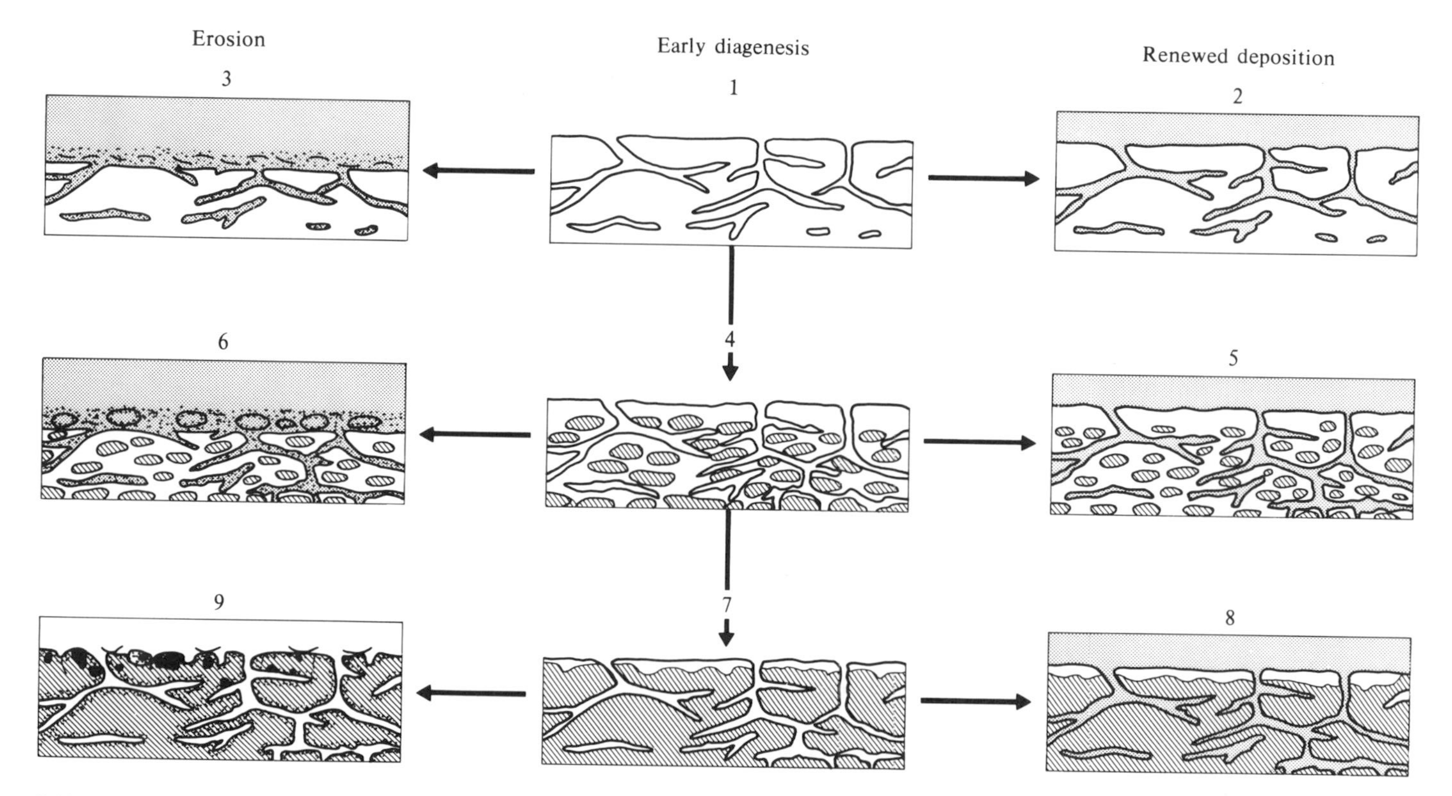

5.13 The relationship between diagenesis, erosion, and the formation of nodular chalks and hardgrounds. Note the *Thalassinoides* burrow systems. After Kennedy and Garrison (1975b).

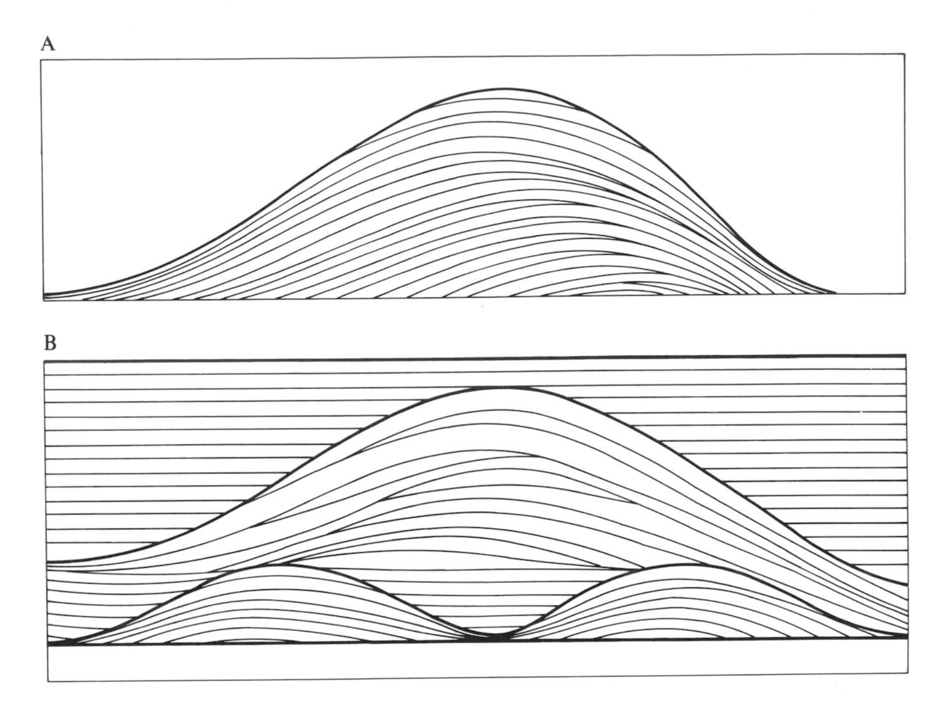

5.14 Simplified sketches of simple (A) and composite (B) buildups in the Chalk of the Normandy coast. After Kennedy and Juignet (1974).

Cretaceous land masses were peneplaned and hence of very subdued relief, but Hancock (1975) points out that this fails to account for the high proportion of sand in the siliciclastic sediments when present. His preferred explanation is that the climate was either arid or that of a tropical rain forest, because a non-seasonal climate nowadays is associated with negligible rates of erosion.

6. Eustatic changes of sea level

The rock sequence in epicontinental and marginal marine deposits rarely persists for long without some change of facies suggestive of changing depth of sea; indeed there is often a systematic variation usually referred to as cyclic sedimentation, with each cyclic unit being referred to as a *cyclothem* (Duff *et al.* 1967). A typical cyclothem in the Upper Cretaceous of the United States Western Interior provides a good example (Fig. 6.1). Taking in account also the regional facies distribution such a cyclothem can be interpreted in terms of a marine transgression followed by a regression, with the deepest-water phase in the depositional area corresponding to the time of maximum transgression at the sea margins.

Now such a cyclothem can be a response *either* to a regional tectonic event, such as alternating subsidence and uplift of the sedimentary basin, *or*

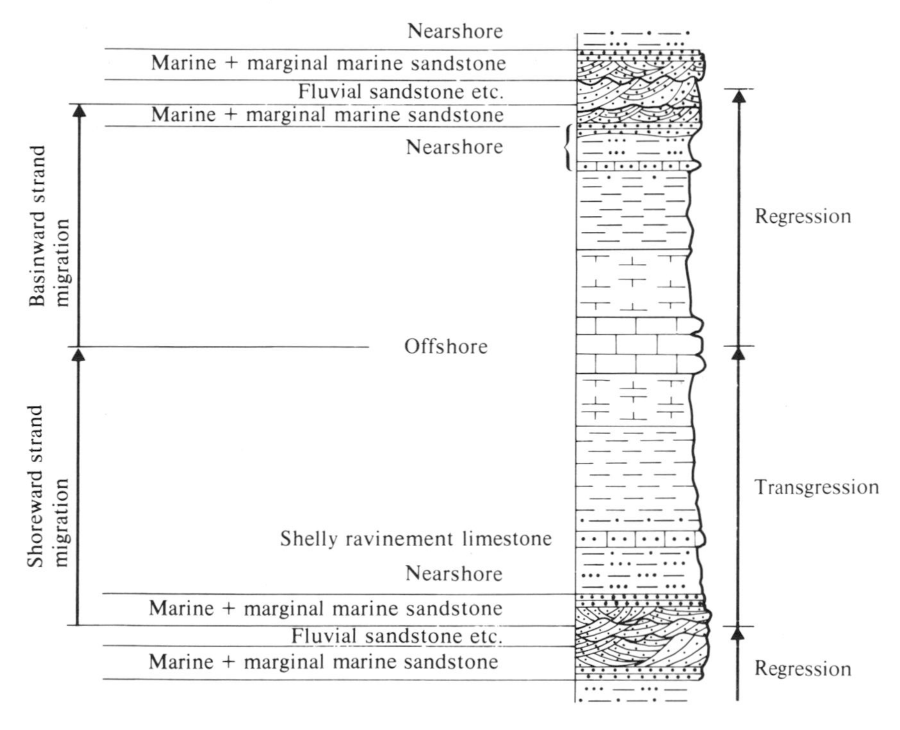

6.1 Interpretation of a typical cyclothem in the Upper Cretaceous of the Western Interior of the United States in terms of transgression and regression. After Hancock and Kauffman (1979).

a worldwide rise and fall of sea level. The term *eustatic* for global events of this sort was proposed by the great Austrian geologist Edward Suess (1906), who suggested three types of approach to their study. The first involves examination of existing coastlines and is therefore only relevant to very recent geological events. The second and third, respectively involving analysing the character of sedimentary formations and tracing the extension of ancient seas, can be applied to the whole of the Phanerozoic.

The notion of glacially-controlled eustatic changes during the Quaternary has flourished since the time of Suess, and a wealth of detailed information has been gathered (Lisitzin 1974). Geologists have been particularly impressed by the striking physiographic and sedimentational effects of the rapid Holocene 'Flandrian' transgression following the low sea-level stand of the last glaciation. For earlier times, however, Suess's idea received little explicit support or even discussion during most of this century, with the notable exception of Grabau (1936), who argued that a series of major Palaeozoic transgressions and regressions were the consequence of eustasy rather than regional epeirogeny. The situation has changed considerably in recent years as a result of improved stratigraphic correlation across the world and geophysical investigations of the ocean floor, which suggest a possible mechanism for sea-level change for times when there is no evidence of polar ice caps. In consequence the exploration of possible eustatic events in the stratigraphic record has become an increasingly active field of research.

The obvious difficulty in this type of work is the disentangling of the influence of local tectonics, of separating the eustatic *signal* from the regional *noise.* Additional problems concern the determination as precisely as possible of the relative rates and amount of sea-level rise and fall, and in finding a sufficiently plausible cause.

Methods of analysis

THE CHARACTER OF ROCK SEQUENCES

It is clearly desirable to avoid as far as possible those parts of the world where intensive tectonic activity has manifestly taken place, as outlined in chapter 4, and concentrate on the shallow marine sequences of relatively stable cratons. If the depth of sea were sufficiently shallow, even a slight change of sea level should have exerted a considerable effect on sedimentation patterns. The existence of good zone fossils allowing refined stratigraphic subdivisions is a necessary requirement, though perhaps not for the grosser changes.

Different facies cannot be expected to respond in a similar way to eustasy, as is made evident in Fig. 6.2. Successive rises and falls of sea level may be well expressed in a sequence on the margins of a sedimentary basin by

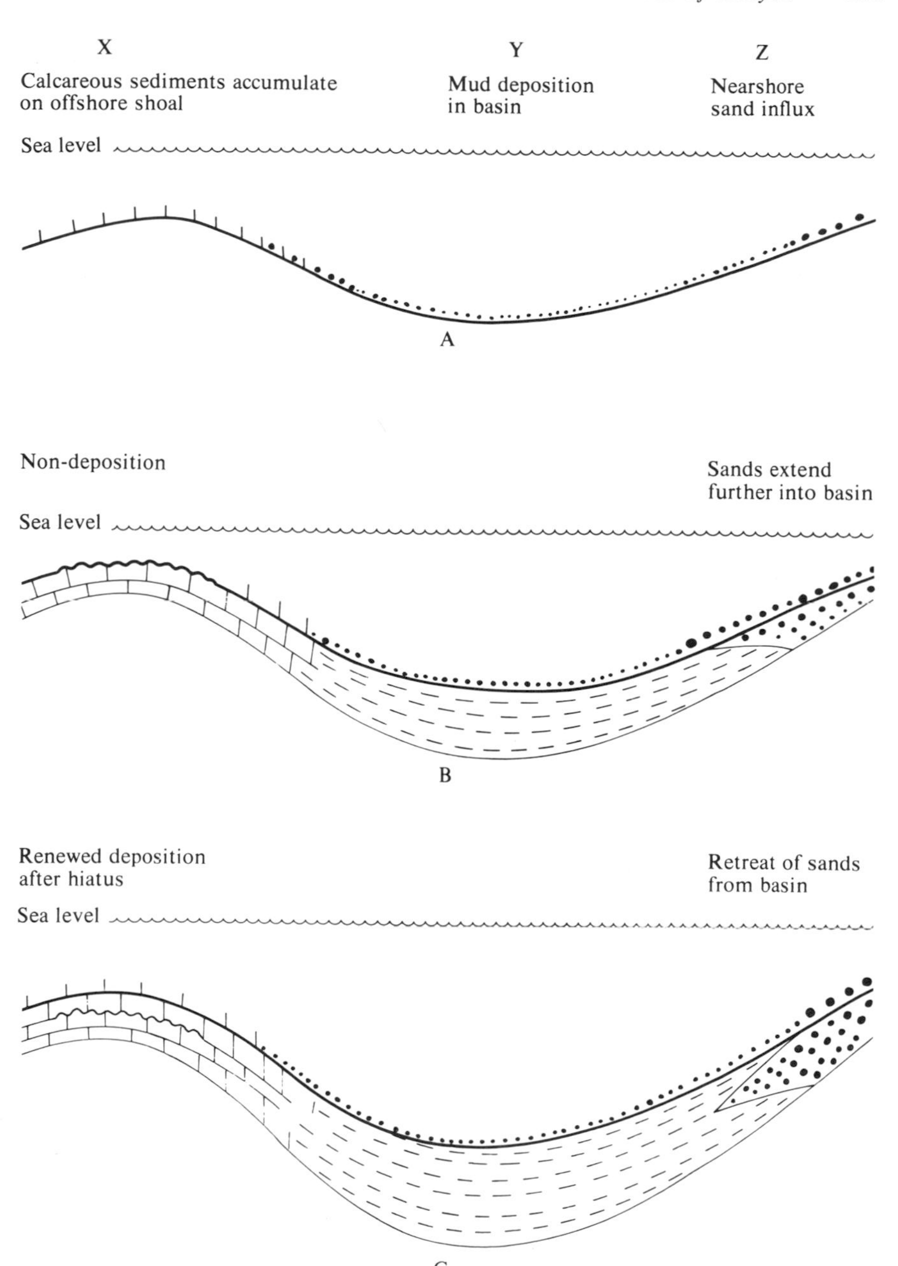

6.2 Stages in the development of a marine basin and offshore shale, with sea level falling from A to B and rising from B to C. In zone X the sea-level fall is expressed by a hiatus and in zone Z by a spread of terrigenous sand at the expense of shale. There is no facies expression of the change in zone Y.

changing facies whereas the deeper-water deposits of the basin centre might remain little affected, especially if the subsidence and depositional rate is considerable. Elsewhere, for instance on an offshore shoal free from siliciclastic influx, regressive events in a carbonate sequence may be manifested merely as pauses in sedimentation or hardgrounds. It should also be appreciated that rise of sea level may only correlate with transgression if the influx of terrigenous sediment is low (Fig. 6.3).

Nevertheless, if changes in the marine depositional regime indicative of varying water depth can be traced over a vast area of continental or subcontinental dimensions, *independent of* local sedimentational and tectonic features such as basins and swells (as expressed by varying facies and thickness),

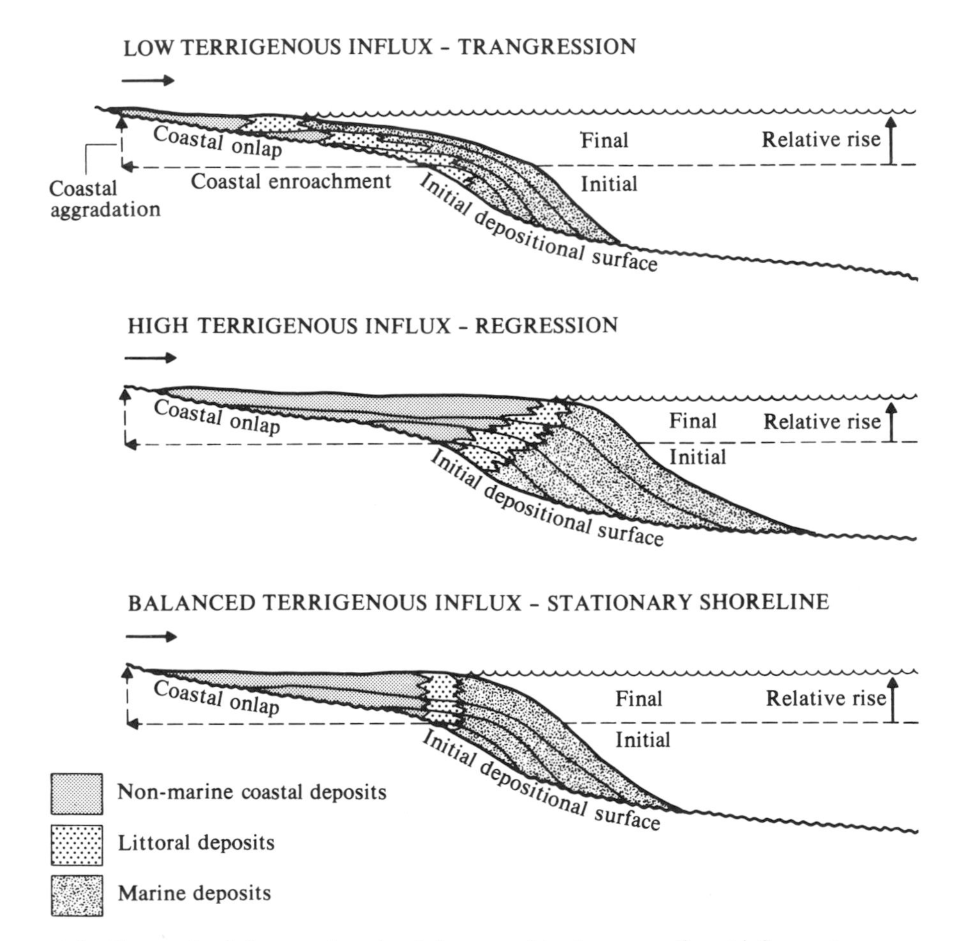

6.3 The varying influence of sea-level change and terrigenous sediment influx on transgression and regression. After Vail *et al.* (1977).

then the case for invoking eustatic control becomes plausible. It is strengthened further if times of water deepening can be closely matched with times of transgression on the land margins, and times of shallowing with regression. If intercontinental correlation of such events proves possible, then eustasy would seem to be conclusively established.

The only alternatives if the facies interpretation is substantially unambiguous are to deny the validity of the stratigraphic correlation or to argue that continents have moved up and down in concert. The idea of invoking simultaneous vertical movements of the continents seems far less plausible than sea-level change for various reasons. Sloss and Speed (1974) put forward a model whereby episodic changes in the the proportion of melt in the asthenosphere occur beneath the continents, causing uplift and subsidence. Not only is this speculation unsupported by independent data and effectively begs the question of why *all* continents should be affected at the same time, it ignores the fact that oceanographic work has demonstrated that oceanic ridges increase and diminish in volume through time, and hence *must* cause displacements of sea water over parts of the continents. The question then arises as to whether such displacements can be recognized in the stratigraphic record; there is rapidly accumulating evidence that they can.

Let us consider for instance the Jurassic, which is an especially well-favoured system for this type of analysis because of the refined stratigraphic subdivision made possible by ammonites. Jurassic ammonite zones had on average a duration of about one million years and most are divisible into several subzones. Correlation across the European continent, where the best-studied sections occur, is often possible at the level of subzone and European zones can be recognized as far away as South America (Hallam 1975). This is a level of precision unmatched in the stratigraphic record.

I have endeavoured to establish sedimentary and faunal criteria for recognizing upward-deepening and shallowing sequences in a wide range of facies in the epicontinental regime of north-west Europe, and have been able to demonstrate a correlation of events over areas of hundreds of thousands of square kilometres independent of local tectonic and facies patterns. Thus erosional intervals can be followed at precisely the same horizon from England to Germany. A marine interruption of a deltaic sequence may be correlative in other regions with the replacement of shallow-marine sandstones by a more 'offshore' limestone, or by a passage of shallow-water limestones and marls into deeper-water shales. The more striking of these changes can often be correlated with marine transgressions or deepening events in other continents and are therefore interpreted as eustatically-controlled events. Other regional shallowing events can be correlated with more extensive regressions. Thus I recognize major phases of sea-level rise in the early Hettangian, early to mid-Toarcian, early Bajocian, late Bathonian to early Callovian and mid-Oxfordian, and major phases of sea-level fall in the late Toarcian to Aalenian, early Bathonian, late Callovian and late Tithonian–Volgian (Hallam 1978a).

It is important to enquire into the relative rates of sea-level rise and fall, in the Jurassic or any other system. For my Jurassic analysis I considered the range of possibilities as portrayed in Fig. 6.4. These are: A, short phase of rapid sea-level rise interrupted by longer phase of stillstand; B, moderate rise followed by moderate fall without intervening phase of stillstand; C, slow rise followed immediately by rapid fall; D, rapid rise followed immediately by slow fall; E, rapid rise and fall interrupted by longer phase of stillstand. Deciding between these various models is not an easy matter and obviously depends on the best possible environmental interpretations of the different facies, but my present view is that most, if not all, of the Jurassic cycles in Europe can be accounted for best by model E or by a combination of D and E.

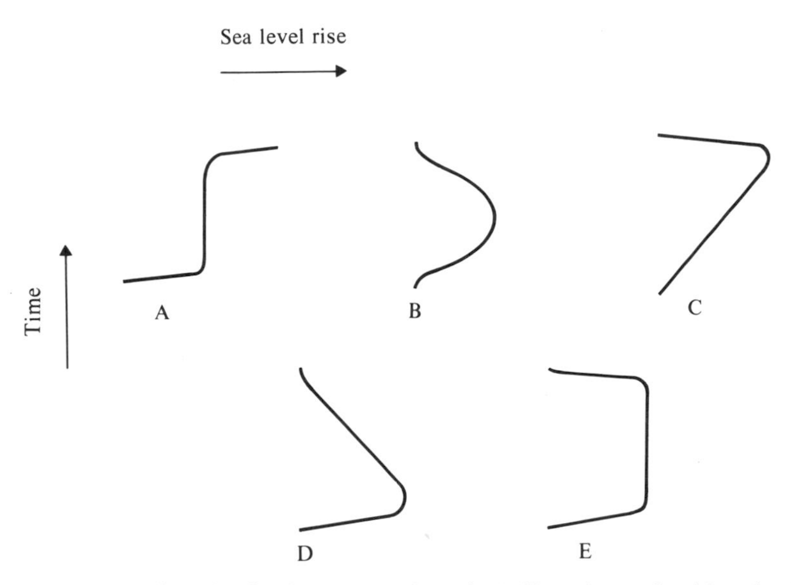

6.4 Possible eustatic models for the European Jurassic. A, Short phases of rapid sea-level rise interrupted by longer phase of stillstand; B, moderate rise followed by moderate fall without intervening phase of stillstand; C, slow rise followed immediately by rapid fall; D, rapid rise followed immediately by slow fall; E, rapid rise and fall interrupted by longer phase of stillstand. After Hallam (1978a).

AREAL DISTRIBUTION OF MARINE DEPOSITS

The third of Suess's methods involves plotting on an equal area map the distribution of marine facies on the continents for rocks of a given age, based on the fossil content, and determining the area by planimeter or by using a fine-mesh grid and counting the squares, and then determining how the area changes through time. So described, this seems to be essentially objective but because of subsequent erosion, and inaccessibility as a consequence of deep burial of many older deposits, one must in practice make best estimates by means of palaeogeographic maps, which of course involves a degree of interpretation. Furthermore, published information on the rocks underlying the present continental shelves is usually inadequate and therefore it is safer to exclude these areas from consideration. Provided, however, that the imprecision of the method is admitted, it can yield very informative results.

Persisting with the example of the Jurassic, a stage-by-stage areal plot reveals that the sea progressively encroached over the continents from a minimum of less than 5% cover in the Hettangian to a maximum of nearly 25% in the Oxfordian (Hallam 1975 and Fig. 6.5).

A remarkable method of independently checking this type of analysis has been demonstrated by Spooner (1976) using strontium-isotope ratios. The principal factor controlling variations of $^{87}Sr/^{86}Sr$ through time is probably variation in continental-runoff flux produced by variation in land area. The ratio in sea water is less than that of continental-runoff water because of isotopic exchange with the ocean crust, bound up with hydrothermal convective systems within spreading ridges. There turns out to be excellent agreement both with the Oxfordian result cited above and those determined by similar areal plots for the Cenozoic.

The results obtained from the two methods of studying sequential change of the strata and determining changing areas covered by sea can be combined to produce a composite *eustatic curve* for the Jurassic (Fig. 6.6). This is of necessity only tentative, and the relationship of the relative extents of shorter-term transgressions and regressions compared with the longer-term secular trend is especially uncertain. Furthermore regressions are more difficult to analyse than transgressions both because the stratigraphic data tend to be less precise and because there is a greater risk of confusing eustatic lowerings of sea level with events involving regional uplift. Accordingly the curve can at best be only a working model, an 'Aunt Sally' to be shied at, and may well require considerable amendment in the future.

Nevertheless such an approach points the way to disentangling the complicating effects of local epeirogeny. If the secular trend can be regarded as a reasonable approximation to the world picture, it can be used as a guiding framework for interpreting regional phenomena. Thus the Bajocian and Bathonian stages are widely developed in a 'regressive' or shallower-water facies compared with the underlying Lower Jurassic over most of the British area, especially in the North Sea region, which is counter to the world trend. It appears consequently to be the result of regional uplift centred on the North Sea (Hallam and Sellwood 1976).

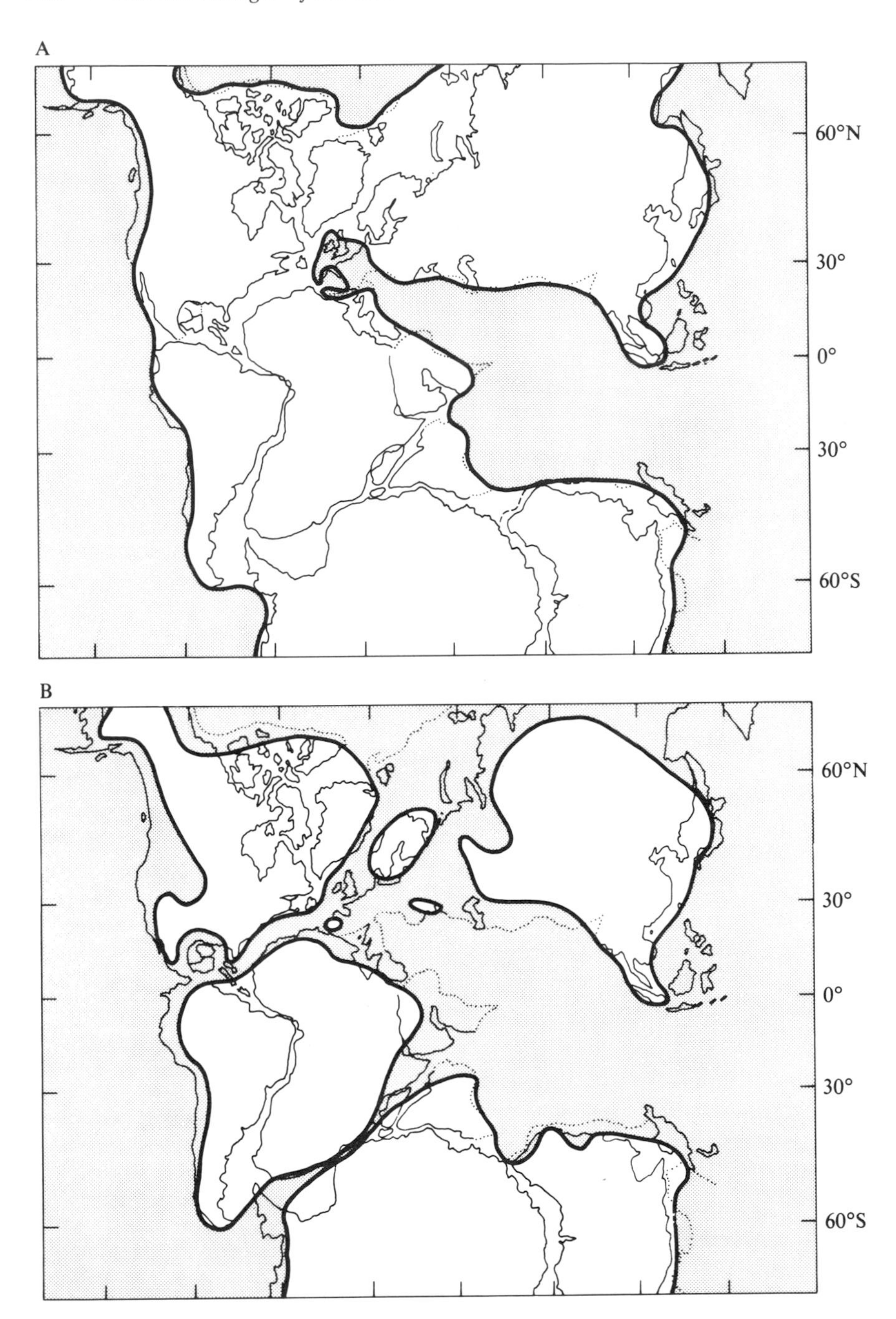

6.5 Approximate distribution of land and sea (stippled) in (A) the Hettangian and (B) the Oxfordian; small islands excluded. After Hallam (1975).

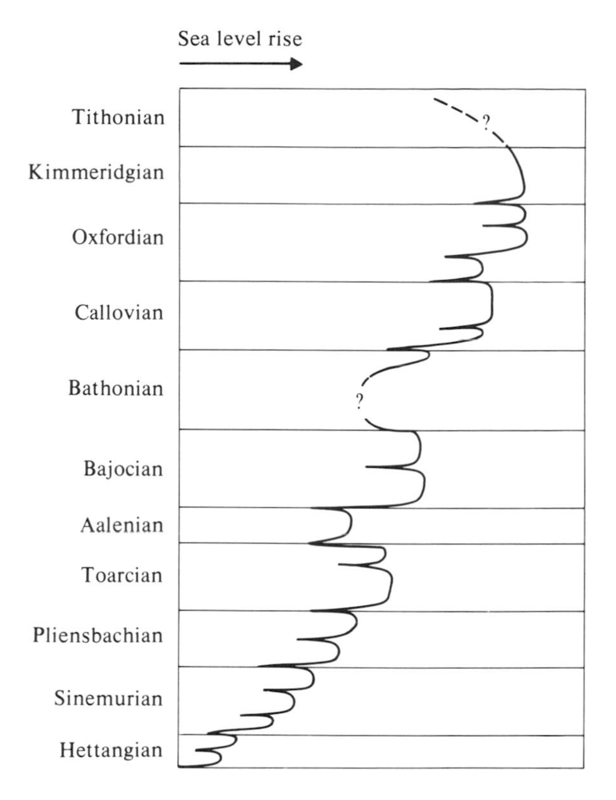

6.6 Tentative eustatic curve for the Jurassic, indicating the more important short-term phases of rise and fall of sea level and the longer-term secular trend. The relative extent of sea-level changes in the short- and long-term events are highly speculative. Broken lines signify uncertainty because of inadequate stratigraphic correlation. After Hallam (1978a).

SEISMIC STRATIGRAPHY

The development in recent years by oil companies of the technique of seismic reflection profiling has introduced a sophisticated and powerful new tool in the subsurface exploration of sedimentary rock sequences. In conjunction with drilling operations it is beginning to revolutionize our knowledge of what lies beneath the continental shelves and has promoted the new discipline of *seismic stratigraphy* (Payton 1977).

Seismic stratigraphy is basically a geological approach to the stratigraphic interpretation of seismic data. Primary seismic reflections are generated by physical surfaces in the rocks, consisting mainly of bedding surfaces and unconformities with velocity–density contrasts. These evidently follow chronostratigraphic rather than lithostratigraphic boundaries. Major stratigraphic units composed of a relatively conformable succession of strata, with upper and lower boundaries defined by unconformities, are termed *depositional sequences.* These resemble the *sythems* of Chang (1975) in being unconformity-bound units but sythems do not necessarily have chronostratigraphic boundaries, and it is important to stress that seismic reflections tend to parallel stratification surfaces rather than the gross boundaries of lithological units that may cut across stratification surfaces. The scale of depositional sequences may roughly correspond to stages or series and their age is determined by tracing the bounding unconformities laterally into conformable successions.

Stratal discordance is the principal criterion for distinguishing sequence boundaries (Fig. 6.7). *Onlap* and *downlap* indicate non-depositional hiatuses rather than erosion. Stratal lapout at the upper boundary is known as *toplap* and also represents non-depositional hiatuses resulting from depositional base level being too low to permit strata to extend further updip.

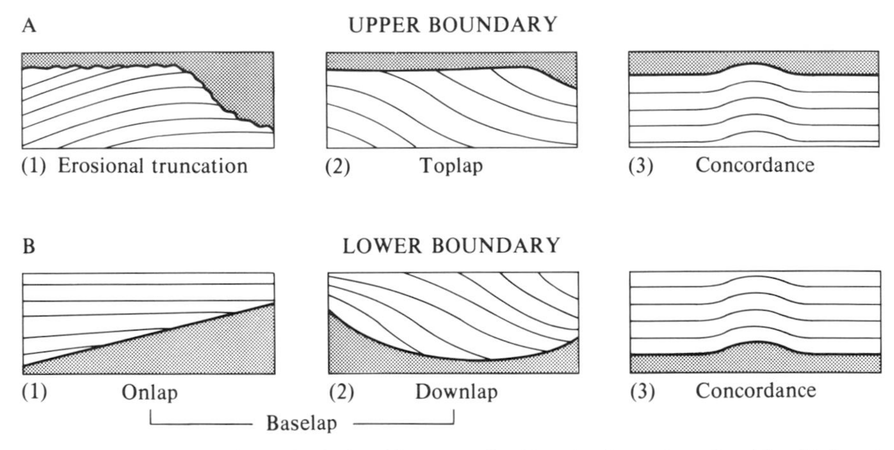

6.7 Different types of stratal relationships recognized in seismic stratigraphy. After Vail *et al.* (1977).

Taken together with lithofacies data from boreholes, these various features can be used to infer relative changes of sea level in a given region. Relative rise of sea level is indicated by *coastal onlap*, the progressive landward onlap of littoral and/or coastal non-marine deposits in a marine sequence. The vertical and horizontal components of coastal onlap are termed respectively *coastal aggradation* and *coastal encroachment* and can be used to measure the amount of sea-level rise.

Coastal toplap is characteristic of relative stillstand of sea level and can also be produced during a period of relative rise if there is a rapid excess deposition of siliciclastic sediment (Fig. 6.3). A relative fall is signified by a downward shift of coastal onlap and is usually abrupt.

Regional sea-level curves can be drawn on the basis of the analysis of marine sequences and construction made therefrom of chronostratigraphic correlation charts in the manner indicated in Fig. 6.8. If cycles of relative rise and fall can be correlated in three or more regions across the world than eustatic control is taken as firmly established. The results obtained by Vail and his co-workers using this technique will be discussed in the next section.

Major eustatic changes in the Phanerozoic

It is beyond the scope of this short book to attempt a comprehensive review of such an enormous subject and instead attention will be concentrated on the results of a number of recent workers.

The start of the Phanerozoic is marked by an early Cambrian transgression that spread progressively across the cratons. Stratigraphic control is poor, however, and it is difficult to estimate, even approximately, the rate at which it proceeded. Matthews and Cowie (1979) consider that the best comparison is the late Cretaceous transgression. McKerrow (1979) utilizes information from graptolite zonation and depth-controlled brachiopod communities to analyse Ordovician and Silurian sea-level changes. A rise in the Llandeilo and earliest Caradoc was followed by a fall in the late Ashgill which was in turn succeeded by a rise in the earliest Llandovery. There was a further rise in the earliest late Llandovery followed by a fall. The sea-level changes are thought to have occurred in comparatively rapid pulses, of the order of 1–2 million years. For the Devonian, House (1974) inferred that the major transgressions and regressions were eustatic. A widespread late Gedinnian to early Siegenian regression was followed by a late Siegenian to Emsian transgression. A major deepening event at the beginning of the mid-Devonian is marked in Europe by a widespread return to argillite deposition and correlates with an important transgression over the Russian Platform; an approximately similar story can be worked out for North America. The most important transgression, and most obviously worldwide, took place at the start of the late Devonian, in the early Frasnian.

Ramsbottom (1979) recognized a hierarchy of biostratigraphically-based stratigraphic units in the Carboniferous of north-west Europe termed cyclothems, mesothems and synthems, and interpreted them in terms of eustatic control, with the major transgressive events being traceable to the United States and Russia (Fig. 6.9). Relatively slow, multiphase transgressions were succeeded by relatively rapid regressions, marked in the siliciclastic Namurian sediments by the spread over marine shales of thick, rapidly deposited sandstones signifying prograding deltas. Regressions in the cal-

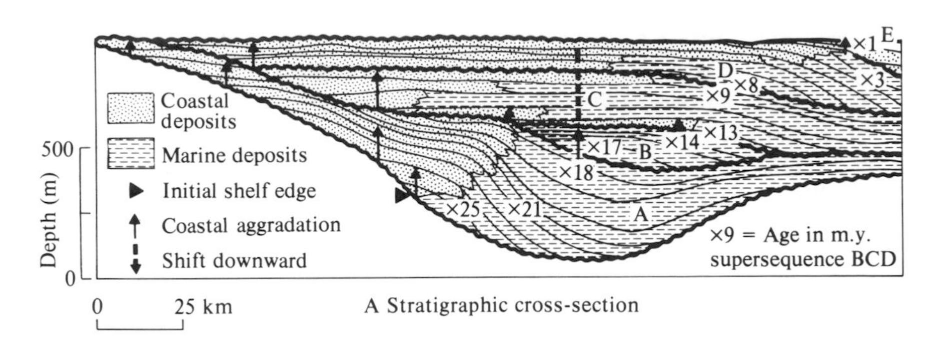

A Stratigraphic cross-section

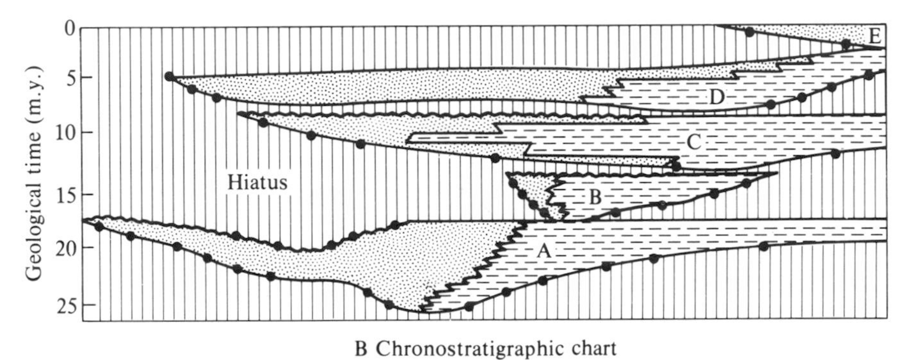

B Chronostratigraphic chart

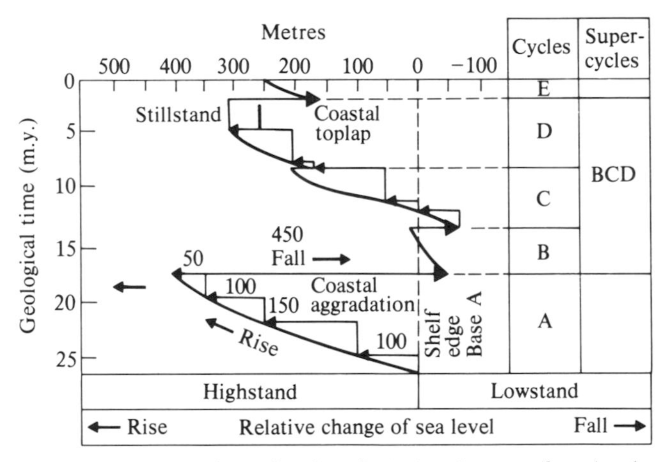

C Regional chart of cycles of relative changes of sea level

6.8 Conversion of stratigraphic sequence relationships to chronostratigraphic charts and regional charts of cycles of relative sea-level change. After Vail *et al.* (1977).

careous facies of the underlying Dinantian are indicated by karstic features and hiatuses in areas marginal to the basin depocentres. The eustatic interpretation for the Dinantian, put forward in an earlier paper, was strongly challenged by George (1978), who considered that eustatic effects were subordinate to regional tectonism. In the discussion to his paper Ramsbottom pointed out that his eustatic hypothesis had been successful in prediction, whereas the alternative could not be, by its very nature, as it had an *ad hoc* flavour.

This particular controversy illustrates well the problems of interpretation that arise when the biostratigraphic control is not very refined. Correlation in the European Namurian is much easier because of the abundant goniatites and hence this stage provides a better test case.

According to Heckel (1977) and others, eustatic changes must be invoked in conjunction with other factors to account for the Upper Carboniferous cyclothems of the American mid-continent region.

By plotting areal distributions of marine inundation of the continents Schopf (1974) estimated that eustatically-controlled regression caused a reduction of epicontinental seas in the Permian from about 40% cover early in the period to less than 15% at the end, with an accelerated drop in sea level in the late Permian; it was followed by an early Triassic rise.

No one has made a recent study of sea-level change in the rest of the Triassic, but it seems likely that there was a notable rise in the late Anisian to early Ladinian because the Muschelkalk transgression of western Europe is reflected in a comparable event in other continents. Though the pattern is regionally complex there appears also to have been a general tendency towards regression in the later Triassic.

The Jurassic has already been dealt with, so we can turn our attention immediately to to the Cretaceous, which is one of the most promising of all the periods for the study of eustasy. Cooper (1977) claimed no fewer than fifteen major eustatically-controlled transgressions, but his interpretation has been challenged by Hancock and Kauffman (1979) on the grounds that he ignored diachronism. Hancock and Kauffman base their work primarily on a detailed comparison of late Cretaceous sequences in two tectonically relatively-stable cratonic regimes, namely the United States Western Interior and north-west Europe. They point out that, to make the case for eustatic control, it is necessary to demonstrate synchroneity of major transgressions and repressions, and a more or less similar size of transgression in each stable region with similar topography.

It is acknowledged that regressive sequences may develop at times of rising sea level in areas of rapid sedimentation but this could hardly apply to the Upper Cretaceous of the two areas in question because there are no regional regressive sequences associated with a major rise of sea level. In the Western Interior there is a close correlation between cyclothems and strand-

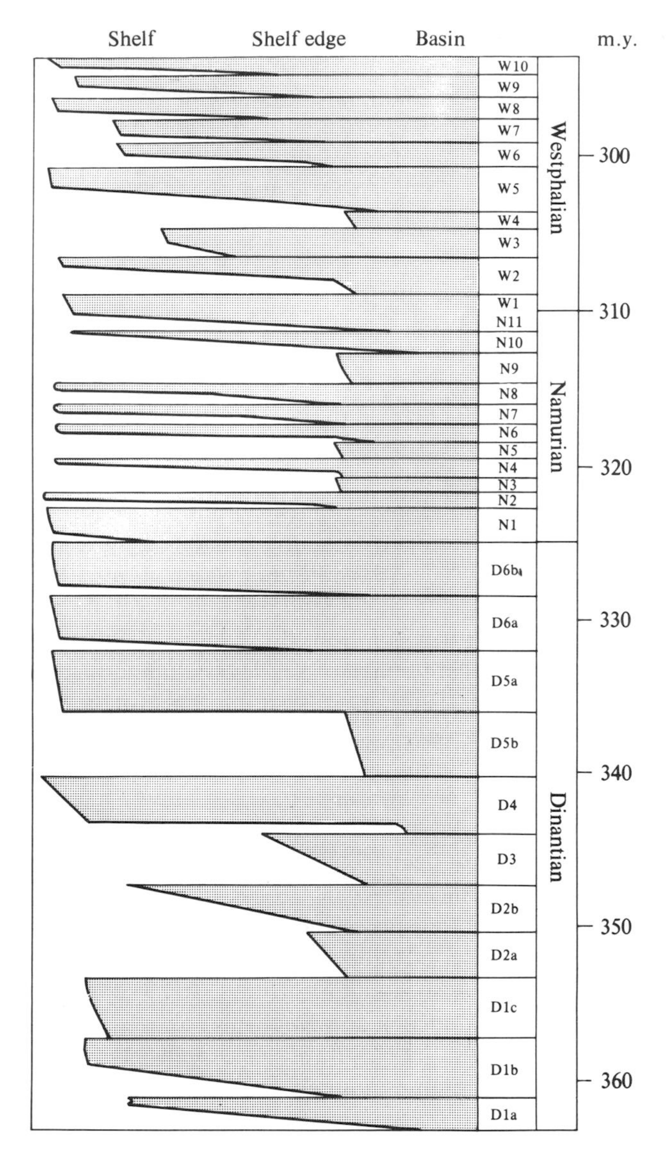

6.9 Mesothemic cycles in Carboniferous rocks of north-west Europe up to the middle of Westphalian C. After Ramsbottom (1979).

line migrations but in Europe one must often use the evidence of disconformities (hiatuses) in the Chalk.

Hancock and Kauffman's results are portrayed in Fig. 6.10, based both upon the distribution of facies and the extent of marine inundation. It will be seen that the intercontinental correlation is very close except for the late

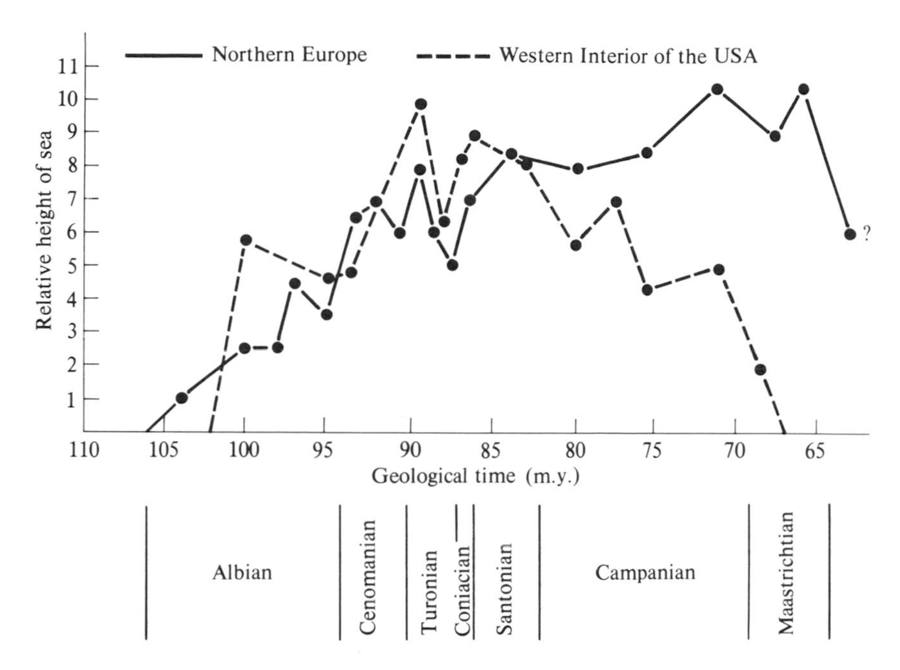

6.10 Relative highs and lows of sea level in northern Europe and the Western Interior of the United States. After Hancock and Kauffman (1979).

Campanian and Maastrichtian. This anomaly can be readily attributed to the tectonic overprint of the Laramide uplift in the western United States. By taking into account evidence from other continents five eustatically-controlled transgressive peaks are recognized: early late Albian, earliest Turonian, Coniacian, mid-Santonian and late Campanian to early Maastrichtian. The great late Maastrichtian regression at the end of the period is the most obviously world-wide event of all.

A general survey of the distribution of deposits on the continents indicates a more or less progressive regression of the seas through the Cenozoic indicative of sea-level fall, interrupted by a series of transgressions, notably in the early Palaeocene, the early and mid-Eocene, the early Miocene and the early Pliocene, of which the most important was probably the mid-Eocene. By far the most important regression was in the late Oligocene, with other notable ones at the end of the Palaeocene, Eocene and Miocene (Hallam 1963). Using the information from seismic stratigraphy and therefore taking into account also offshore data, Vail *et al.* (1977) confirm the overall regressive tendency through the era, especially the marked late Oligocene regression, but they infer a more complex stepwise pattern of transgressions, and the mid-Eocene event appears less significant in their scheme.

THE PHANEROZOIC AS A WHOLE

In one of the most significant contributions to the study of eustasy for many years, Vail *et al.* (1977) utilize an immense amount of data principally from seismic stratigraphy to put forward a eustatic curve for the whole of the Phanerozoic (Fig. 6.11). Sea level is inferred to have risen comparatively rapidly to a late Cambrian peak and then undergone a progressive decline through the Palaeozoic, followed by a rise through the Mesozoic to a late Cretaceous peak comparable with that of the early Palaeozoic, and a further decline through the Cenozoic to its present abnormally low level. Superimposed on the secular trend is a series of second-order cycles, themselves subdivisible into third-order cycles recognizable from Jurassic times onwards. The time scale of the second-order cycles is 10–80 million years and the third-order cycles 1–10 million years. Both exhibit a pronounced asymmetry, with a gradual rise being succeeded by an abrupt fall.

Impressive though this work is, it is frustrating that the data on which it is based are not available in the published domain, lodged as they are in confidential Exxon files. It is therefore impossible to make an independent check, and one wonders if other oil companies utilizing similar data would come up with a comparable curve. There are in fact a number of comments that can be made which suggest that the results of the Exxon geologists should not be accepted uncritically.

Consider, for instance, their spectacular late Oligocene lowering of sea level which they claim took place within the limit of resolution of one planktonic foraminiferal zone, approximately one million years in duration. This appears to be based on data from four regions—the North Sea, north-west Africa, the San Joaquin Basin of California and the Gippsland Basin of Australia (Vail *et al.* 1977: fig. 5). Of these, only the North Sea and north-west Africa show a pronounced drop and the Gippsland Basin shows only a minor hiccup in their curve, no greater than a number of others. Their global curve is in fact heavily weighted in favour of the North Sea data, but no reason is given for this.

Similarly, in their detailed curve for the Jurassic (Vail *et al.* 1977: fig. 2) the Sinemurian is shown to be markedly regressive with respect to the Hettangian, whereas I am convinced from an analysis of rocks I know well that the reverse is true. One wonders therefore whether other specialists might have similar reservations about other points of detail. Again, offshore information from seismic stratigraphy is not available below the Mesozoic and their Palaeozoic global curve is based solely on evidence from North America.

Despite these reservations the results of the Exxon group are an invaluable basis for further research and are likely to be a reasonable estimate of the general picture.

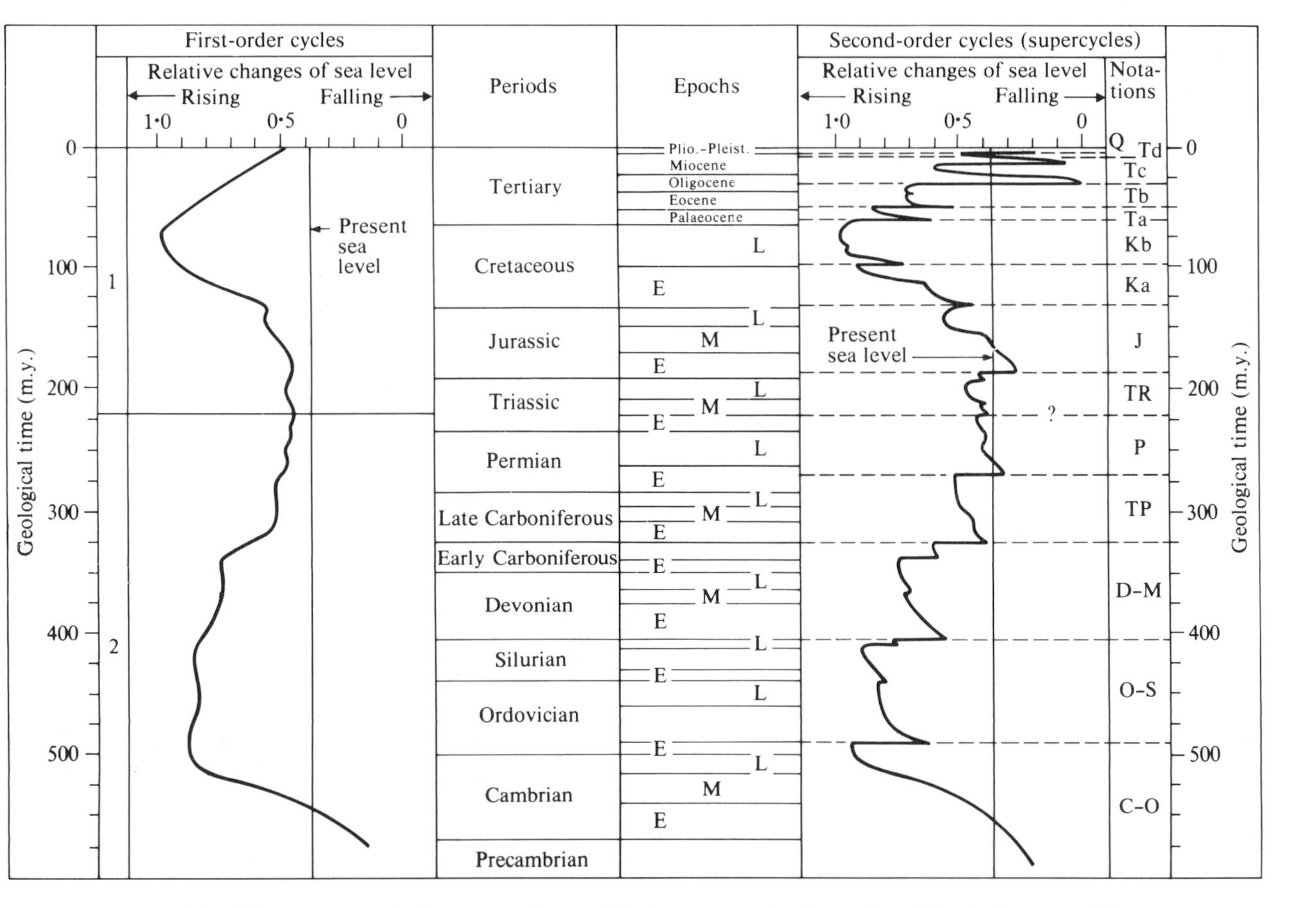

6.11 Phanerozoic eustatic curves in terms of first- and second-order cycles. After Vail *et al*. (1977).

Secular trends through the whole Phanerozoic have also been studied by means of plots of areas of continent flooded by sea for successive time intervals. The Soviet Union and North America together occupy about one-third of the total continental area and should therefore between them reflect fairly accurately any global trend; good palaeogeographic data are available for both.

The series of palaeogeographic and facies maps of the Soviet Union published by Vinogradov and colleagues (1967–69), as yet unmatched for any other extensive region, allow a detailed analysis (Hallam 1977b). The sea spread slowly in the early part of the Palaeozoic to cover approximately 60% of the USSR by mid-Ordovician times. Thereafter there was a faster withdrawal to reach a minimum value at the Silurian–Devonian boundary, followed by a relatively rapid restoration to mid-Ordovician values in the late Devonian and early Carboniferous. Subsequently a progressive withdrawal culminated in a minimum at the Permian–Triassic boundary. During the Mesozoic there were two transgressive peaks, in the late Jurassic and late Cretaceous, interrupted by an early Cretaceous minimum and followed by a sharp Cenozoic decline interrupted by a small peak in the mid- to late Eocene. These trends are smoothed in Fig. 6.12, which shows clearly that the Mesozoic transgressions failed to match the extent of those in the early and mid-Palaeozoic.

The close spacing of the data points (Hallam 1977b: fig. 1), at least back to the Devonian, rules out a systematic temporal bias, with the length of time intervals chosen increasing the probability of overestimating the extent of sea the earlier the period, as argued by Wise (1974). Indeed, the probability of losing the stratigraphic record through subsequent deep burial, metamorphism and erosion must increase with time, and so the maps for older periods are more likely to underestimate than overestimate the former extent of marine cover. Furthermore, cratonic areas such as the Russian Platform and the margins of the Siberian Shield—in fact everywhere outside eugeosynclinal regions—contain much more substantial proportions of carbonates and evaporites to terrigenous siliciclastics in the Palaeozoic than Mesozoic or Cenozoic, implying more areally restricted and topographically subdued sediment sources. Mechanical denudation rates on the present continents show a tendency towards an exponential increase with increasing topographic elevation (Garrels and Mackenzie 1971). Hence the Russian data strongly support the notion of a general trend towards secular withdrawal of sea through the Phanerozoic.

The only comprehensive series of palaeogeographic maps available for North America is that published in the atlas of Schuchert (1955), which was used by Wise (1974) to demonstrate a condition of essentially constant freeboard throughout the Phanerozoic, interrupted by short-term oscillations of sea level, 80% of which remained within about 60 m of a normal

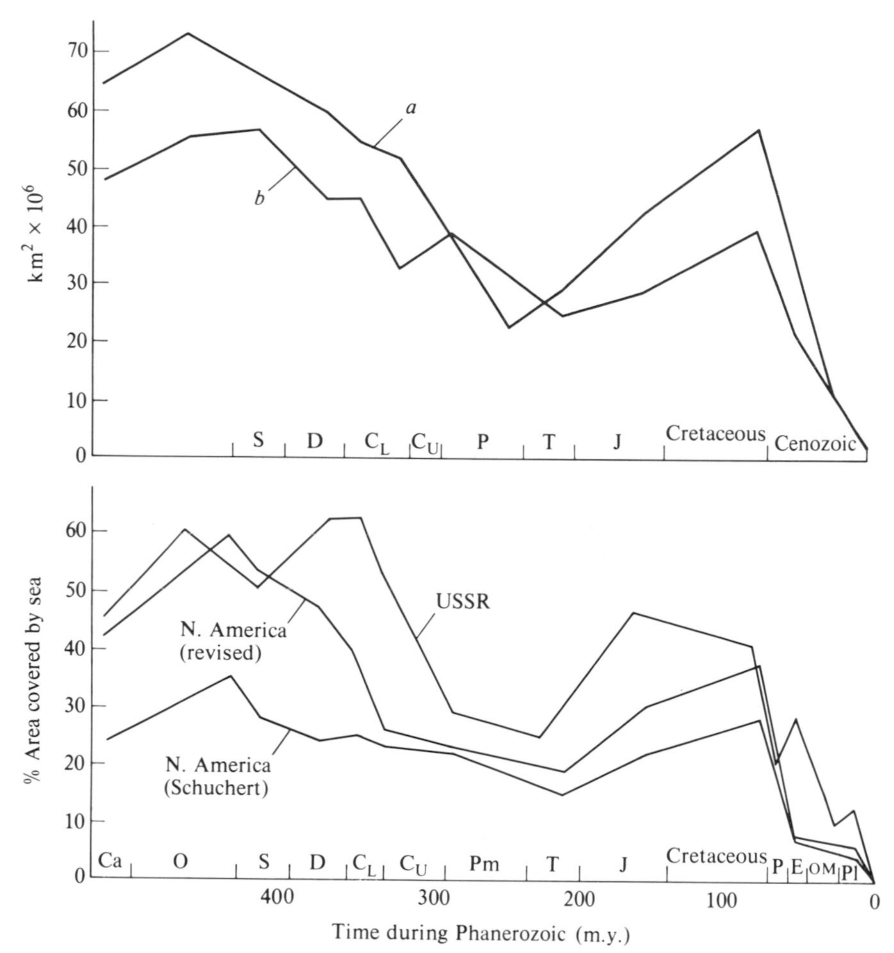

6.12 Maximum degree of marine inundation for each geological period for USSR and North America, and comparison with the world data of Strakhov (*a*) and the Termiers (*b*). After Hallam (1977b).

freeboard level about 20 m above the present level. These maps were drawn, however, several decades before the atlas was published, and therefore do not take into account extensive modern discoveries of strata, especially in the Arctic. Moreover, Schuchert left large areas of the cordilleran region blank for the Palaeozoic, but facies changes from areas further east clearly indicate a eugeosynclinal regime. As with the USSR, Palaeozoic sequences of the North American Central craton are much richer in carbonates than younger rocks, and testify to a correspondingly greater spread of sea, so that Schuchert's maps, based on erosional remnants, tend to be cautious and substantial underestimates for times earlier than the Mesozoic.

I have used data from a new atlas of facies distributions through time (Cook and Bally 1977) to propose a revised graphical interpretation of North American data (Fig. 6.12). In the absence of up-to-date palaeogeographic maps for individual stages, an attempt was made to determine the approximate maximum extent of sea for each period, using references cited by Cook and Bally and making fairly conservative inferences.

Schuchert considerably underestimated the former extent of seas from the Cambrian to the early Carboniferous. Subsequently the Schuchert and revised North America curves match quite closely with only a relatively slight but systematic increase in inferred extent of flooding in the revised version. The revised curve matches the similarly plotted Russian curve much more closely than the Schuchert curve. The main differences are that the North American seas attained their maximum extent in the late Ordovician rather than the mid-Ordovician and late Devonian–early Carboniferous, although the values are remarkably similar at around 60%, and the late Jurassic transgression was less extensive than the late Cretaceous.

That the Russian and North American curves reflect eustatic events is indicated by the close match with the more generalized curves based on worldwide data produced by H. and G. Termier and Strakhov (Fig. 6.12). The Phanerozoic curve of Vail *et al.* (Fig. 6.11) is remarkably similar, when smoothed, to the North American curve based on Schuchert's unrevised data. Since their Palaeozoic control comes only from North American onshore data I regard it as considerably underestimating the extent of Palaeozoic flooding.

The causes of eustasy

One has only to think of the long-delayed general acceptance of continental drift to appreciate that geological phenomena are rendered more plausible if an adequate controlling mechanism can be put forward.

For many years by far the most obvious cause of rises and falls of sea level was the successive melting and freezing of polar ice caps, with huge volumes of water being shed into the ocean system during intervals of climatic amelioration. No one seriously doubts that this was the cause of the well-documented Quaternary sea-level changes, but extrapolation back through time poses problems, because significant abstraction of water to form an extensive continental ice cap did not take place until the mid-Miocene. Before this time, until the late Palaeozoic, the world enjoyed a more equable climate, especially in the Mesozoic, and evidence of polar ice caps is lacking (see chapter 7). A case can be made for the early Silurian rise of sea level being the consequence of melting of the late Ordovician Saharan ice sheet (McKerrow 1979) and it has been argued that the late Carboniferous cyclothems in the northern hemisphere reflect minor sea-level changes resulting from melting and freezing of Gondwana ice. However, the late

Permian disappearance of the Gondwana ice cap correlates with a fall rather than a rise of sea level.

With the advent of a mass of new oceanographic data after the Second World War it seemed logical to propose that eustatic changes could also have been caused by uplift and subsidence of oceanic ridges (Hallam 1963). This idea has been widely taken up, most notably by Hays and Pitman (1973), who showed that the great late Cretaceous transgression correlates well with a time of accelerated seafloor spreading, which would have caused a significant expansion in volume of the oceanic ridge system. The subsequent fall of sea level could be attributed to slower spreading rates causing a volume contraction.

This interpretation is based on the age versus depth relationship of mid-oceanic ridges, which approximately follows a time-dependent exponential cooling curve (Fig. 6.13). In addition, part at least of the late Cretaceous eustatic rise component could have resulted from an extension in length of the ocean ridge system, as significant dispersion of the Pangaea components commenced (Hallam 1977b).

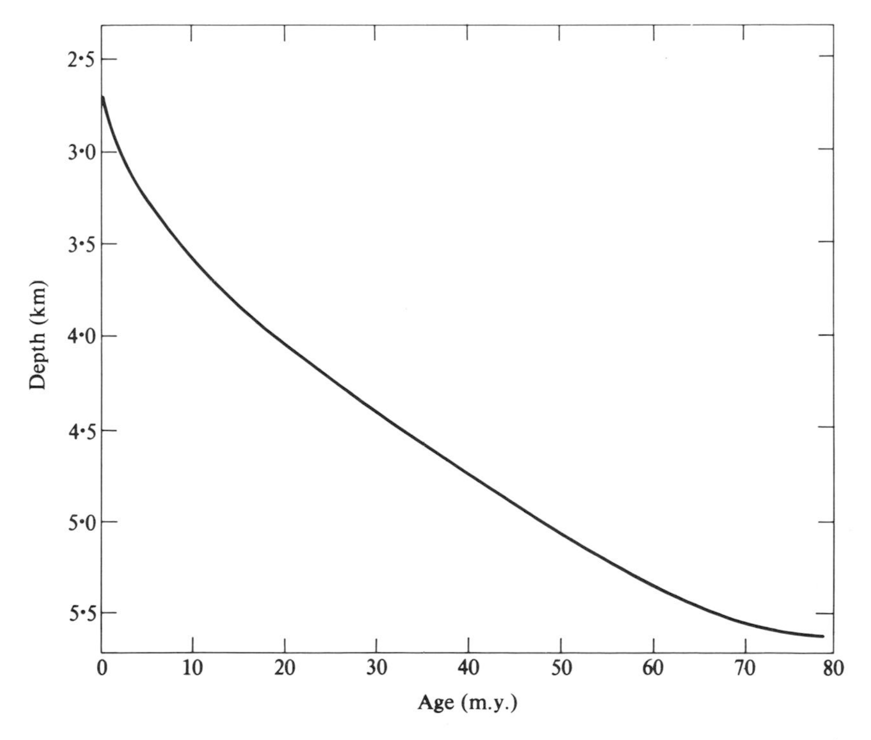

6.13 Age versus depth relationship for oceanic crust. Adapted from Pitman (1978).

In an extended analysis of the interaction of sea-level change with sedimentation and subsidence on Atlantic-type continental margins, Pitman (1978) has shown that transgressive or regressive events may not be simply indicative of sea-level rise and fall but of changes in the rate of sea-level change. Thus a decrease in the rate of sea-level rise and an increase in the rate of sea-level fall may result in regressions, while an increase in rate of sea-level rise and a decrease in rate of sea-level fall may produce transgressions. Calculations indicate that since the late Cretaceous, sea level may have been falling continuously, but slowly in the latest Cretaceous, rapidly in the Palaeocene, less rapidly in the late Eocene, more rapidly in the Oligocene and less rapidly in the early Miocene. The consequence has been a minor latest Cretaceous regression, a large Palaeocene regression, a large Eocene transgression, an Oligocene regression and an early Miocene transgression.

Other phenomena such as oceanic sedimentation, flooding of continental crust, ocean trench subsidence and continental collision are considered to have only minor effect on sea level (Donovan and Jones 1979; Pitman 1978). It has been argued that rapid changes in the shape of the geoid could cause significant rises and falls of sea level in different parts of the world (Fairbridge 1961; Mörner 1976) but not only is this phenomenon unproven, it is hardly relevant to biostratigraphically established global changes.

The long-term secular lowering of sea level through the Phanerozoic, with a major reversal only in the late Mesozoic, calls for a different kind of explanation. It has been used as an argument for slow Earth expansion during this time, but this poses various difficulties and other, more conservative explanations are available (Hallam 1977b). The best explanation may be one or more of the following factors: continental thickening by orogeny consequent on subduction or collision, variation in spreading rate and in the cumulative length of the ocean ridge system. Some form of continental underplating would also contribute, but this is more speculative and so far not strongly supported by independent evidence.

If changes in ridge volume have been the prime cause of sea-level change through the Phanerozoic, the ultimate controlling factor is likely to have been variations of heat flow from the mantle, with the long-term secular change perhaps signifying a gradual reduction in such heat flow (Turcotte and Burke 1978).

Quantitative estimates of sea-level change

The frequency distribution, or *hypsometric*, curve of continental elevations (Fig. 6.14) can be used to calculate the sea-level change that will flood varying portions of the continents. Forney (1975) has used this method to analyse Permo-Triassic sea-level changes using Schopf's (1974) data. On the assumption that the Permian hypsometric curve was similar to the present, a

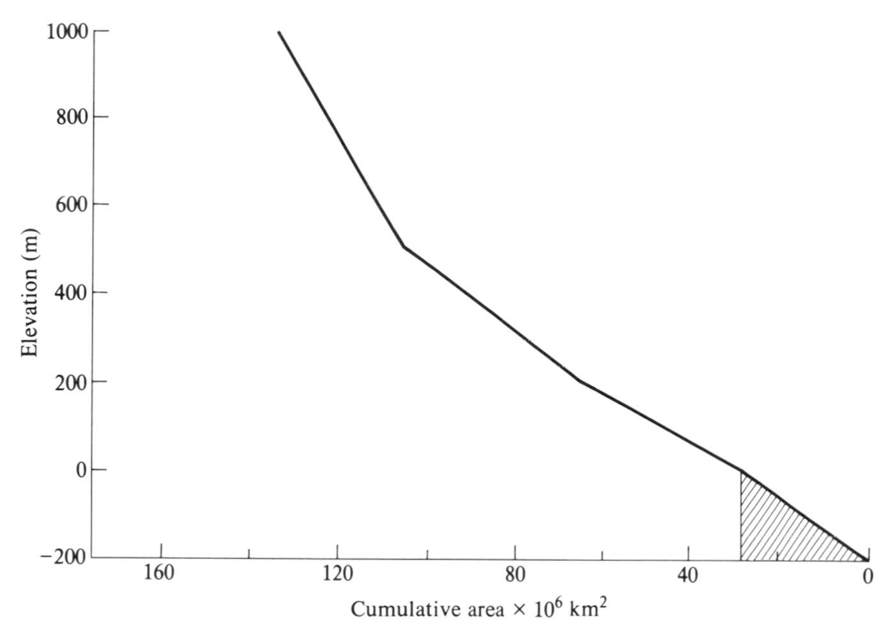

6.14 Hypsometric curve of the Earth from −0.2 km (approximate edge of the continental shelf) to 1.0 km. Shaded area indicates continental regions below present sea level. Adapted from Forney (1975).

late Permian fall and early Triassic rise of slightly over 200 m was estimated to have occurred (Fig. 6.15). As continental hypsometries differ, with Africa elevated about 200 m compared with other continents, this figure is likely to be an upper limit, and Forney's best estimate of the late Permian fall is between 125 and 225 m.

For younger rocks on cratons that are thought not to have undergone much subsequent epeirogenic movement, sea-level fall may be estimated more directly. Sleep (1976) has pointed out that the late Cretaceous coastline in Minnesota has a similar elevation over a wide area, implying only a modest amount of local tectonic disturbance. He estimated accordingly that in late Cenomanian times the sea level stood at 300 m above the present and in the late Turonian–early Coniacian at 375 m.

Hancock and Kauffman (1979) adopt a similar approach for the late Cretaceous. Their calculations are in three parts. The present height above sea level of a given stage in tectonically undisturbed areas is first determined. To this figure is added the thickness of Upper Cretaceous marine sediments up to that stage, and the assumed depth of the sea at the time of deposition (the least reliable). They arrive at a figure for the early Albian approximately the same as today—a late Albian figure of 250 m higher and a late Campanian–mid-Maastrichtian figure of 430–660 m higher.

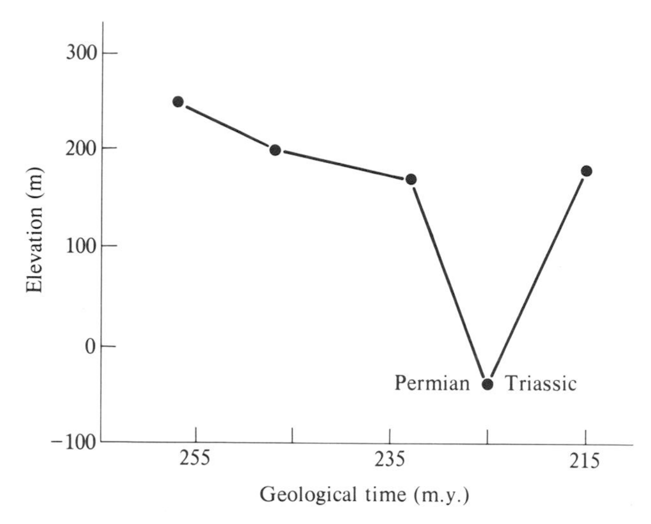

6.15 Hypothetical Permo-Triassic sea-level curve based on the shape of the modern hypsometric curve. After Forney (1975).

Bond (1978) develops Forney's method by using the scatter of data for different continental elevations to distinguish between sea-level changes and the vertical motions of large continental surfaces. By making corrections for post-Cretaceous continental uplifts he arrived at a lower figure of 150–200 m for the high stand of Cretaceous sea level.

Pitman (1978) has attempted to estimate the amount of sea-level fall since the late Cretaceous by measuring the changing volumes through time of the oceanic ridge system, utilizing the spreading model and taking into account the varying length of the ridge system (Fig. 6.16). His figure of 350 m above the present at 85 million years is in close agreement with that of Sleep, but Hancock and Kauffman's calculations would suggest a rise of sea level between this time and the end of the Cretaceous.

According to Pitman the maximum rate of change of sea level due to changing ridge geometry is of the order of 1 cm per 1000 years. This is three orders of magnitude lower than the rate of change due to glacial events. After allowing for isostatic adjustment, the complete melting of all present land ice should cause a rise in sea level of 40–50 m. Calculation of the maximum volume of Pleistocene ice sheets is less certain, but Donovan and Jones (1979) make an approximate estimate of 100 m.

Calculation of the smaller-scale cyclic changes is more difficult because we lack precise knowledge of the depth of deposition of the epicontinental marine strata in which they are expressed. By assuming a likely depth range of a few tens of metres, Hallam (1978a) and Ramsbottom (1979) obtain figures of up to a few tens of millimetres per 1000 years. This seems a

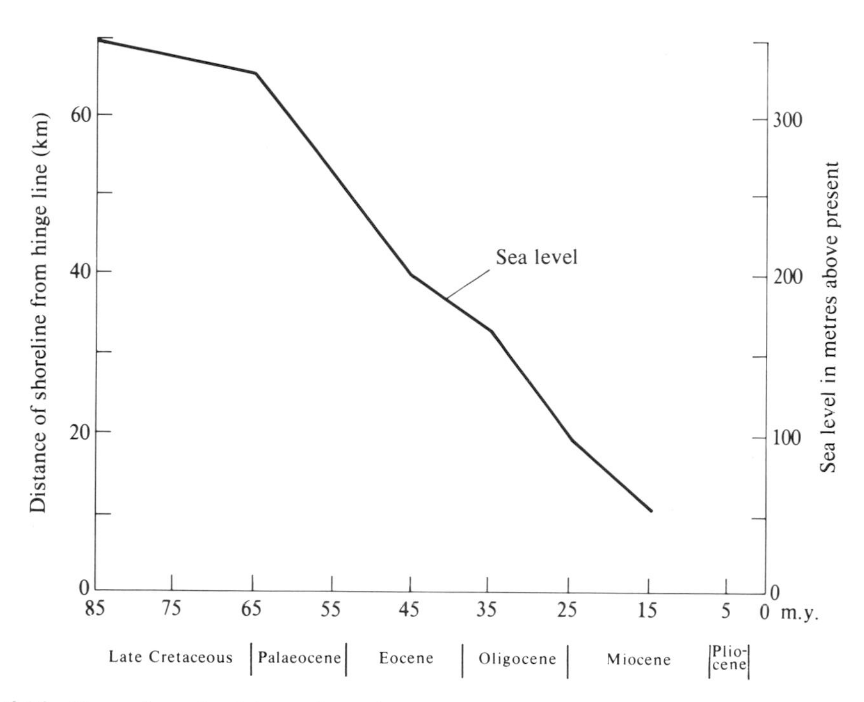

6.16 Change in sea level due to change in ocean ridge volume for period from 85 to 15 million years ago. Adapted from Pitman (1978).

reasonable figure according to Pitman's (1978) model, which would attribute small-scale fluctuations to an interaction of ridge spreading and sedimentation and subsidence on continental margins. By calibrating their eustatic curve with Pitman's data Vail *et al.* (1977) recognize a drastic late Oligocene fall of 350 m within a mere one million years. This rate of sea-level fall would seem plausible only with a glacio-eustatic interpretations, but there is no evidence of the necessary major glaciation event at this time. However, as noted earlier, their interpretation of the stratigraphic data on which their conclusions are based is open to question.

There is some agreement that regressions proceed more rapidly than transgressions (Hancock and Kauffman 1979; Ramsbottom 1979; Vail *et al.* 1977). Whether this reflects a corresponding asymmetry in the controlling mechanism is debatable.

Epilogue

Despite the high measure of uncertainty and disagreement that persists, considerable progress has been made in the study of eustasy in recent years and the subject is full of challenge, because it alters the whole perspective with which we view epicontinental marine sequences. It is probably reasonable to state that there is a fairly general measure of agreement that many of the grosser changes have been controlled by eustasy, but much less about the smaller changes. If they are less than the limit of biostratigraphic resolution the maintenance of a sceptical viewpoint is reasonable enough and probably the best we can do is to make as plausible a case as possible without hoping to eliminate all doubt.

In the case of many geological phenomena, such as earthquakes, the larger the event the fewer there are of them. In the words of the old rhyme:

Big fleas have little fleas
Upon their backs to bite 'em
And little fleas have lesser fleas
And so *ad infinitum*

Many geologists would dispute the truth of this for eustasy, but where, we must ask, is the cutoff point? In Ramsbottom's terminology, is it at the level of synthem, mesothem or cyclothem? The answer will only come with the detailed regional analysis of facies.

7. Phanerozoic climates

Climate is obviously one of the most important characteristics of the environment but as it features variations in air and water temperature, degree of humidity or aridity and wind strength and direction, there is no guarantee that it will be precisely expressed in the stratigraphic record of sedimentary deposits and their contained fossils. In fact before the Cenozoic quantitative information is hard to come by and we have to be content for the most part with broad qualitative generalizations with a considerable margin of uncertainty. The most up-to-date general review of palaeoclimates is by Frakes (1979).

Climatic criteria

Before going on to survey our present state of knowledge of changing climates through the Phanerozoic it is necessary to outline the principal criteria utilized.

Palaeomagnetism. Palaeolatitude determinations obtained from a variety of sedimentary and volcanic rocks provide data from which can be erected a geographic framework, vitally important in pre-Cenozoic times, with which to study the distribution of climatically significant rocks (McElhinny 1973).

Glacial deposits are the most important group of sediments because by establishing the former existence and extent of ice cover we can learn more about ancient climates than from any other source. Criteria for their recognition are dealt with briefly in chapter 2 and extensively by Frakes (1979).

Evaporites are probably the next most important, because substantial deposits are the best indication of both warmth and aridity. At the present day their principal zones of accumulation are approximately 15–35° north and south of the equator, though in some areas of extreme aridity such as central Asia they may occur at latitudes of as high as 50° (Drewry *et al.* 1974).

Coals. Generally speaking coals are good indicators of humidity though conditions of poor drainage can occur even in semi-arid regions given a sufficiently high water table. As regards temperature, the best indicators are the contained plant remains. Modern peat accumulations are largely confined to high latitudes because in low latitudes woody materials are more readily destroyed by oxidation and bacterial action.

Bauxites and *laterites* are, judging from their present distribution, produced exclusively in tropical and subtropical regions with high rainfall and temperature in conditions of intense chemical weathering. Unfortunately their preservation potential is not high and they are rare in pre-Mesozoic sequences. Furthermore the time of lateritization may long post-date the origin of a given soil. A comprehensive review of karst laterites and bauxites in relation to climatic and palaeogeographic changes through time is given by Nicolas and Bildgen (1979).

Several other types of sediment have been cited as climatic indicators but are less reliable. *Red beds* can form in both arid and humid conditions and the red coloration may develop long after deposition (see chapter 1). Some red bed sequences contain horizons of calcrete, which are generally taken as signifying a semi-arid or arid climate. Others contain aeolian sandstones, from which wind directions may be inferred, as noted in chapter 2. Nowadays aeolian sands are broadly confined to areas up to 30° north and south of the equator, but local data on wind directions are not particularly trustworthy because the trade winds are less regular over continents than oceans (Drewry *et al.* 1974). *Limestone* distributions by themselves indicate little more than a lack of influx of siliciclastic material, and are forming today in a wide range of latitudes, but they may contain climatically significant fossils. Phosphorites form today only in warm oceans in zones of upwelling (see chapter 5), while there is a modern equatorial deep-ocean belt characterized by radiolarian siliceous ooze. There is also, however, a high latitude belt of diatom ooze and a rich biogenic source for *chert* formation may signify a variety of environments having little to do with climate.

With regard to fossils, terrestrial plants are probably the most sensitive indicators of climate, especially in the Cenozoic, where the fossils often have close living relatives, but they are also very informative for the Mesozoic. In the marine realm, hermatypic corals, confined today to latitudes of less than 30°, with water temperatures generally above 21°C, would appear to be safe indicators of warm climate back at least to the early Mesozoic. As substantial organic reef developments are confined nowadays to warm waters those faunally-rich buildups that qualify as true reef limestones are probably a reasonable climatic guide. As with terrestrial plants, many Cenozoic marine invertebrate species have close living relatives whose climatic tolerances are well known, and study of their distribution may yield valuable information.

For times further back in the past, it is worth noting that organic diversities of virtually all groups at the present day increase towards the equator, and that gastropods and bivalves tend to have thicker and more strongly sculptured shells in shallow tropical seas, as a consequence probably of the greater ease of carbonate secretion in warm water and the increased predation pressures in low latitudes (Vermeij 1978).

Oxygen-isotope analysis

To determine temperature from a calcitic shell it is necessary to measure the $^{18}O/^{16}O$ ratio and to make two assumptions: that the calcite was precipitated in isotopic equilibrium with sea water (which of course can only be measured for living organisms), and that there has been no post-depositional alteration. Many fossilizable organisms fulfil the first condition and precautions can be taken to minimize the errors introduced by diagenesis and other post-depositional factors. Fossils from marginal marine environments should be avoided as far as possible because abnormal salinity can affect the ratio.

The temperature of calcite precipitation can be determined from the following palaeotemperature equation of Epstein *et al.* (1953), or one of its (slight) modifications.

$$T°C = 16.5 - 4.3\,(\delta_c - \delta_w) + 0.14\,(\delta_c - \delta_w)^2$$

where

$\delta_c = \delta^{18}O$ calcite relative to a standard (PDB carbonate)

$\delta_w = \delta^{18}O$ water relative to 'mean ocean water'

and

$$\delta^{18}O = \left[\frac{(^{18}O/^{16}O)\text{ sample} - (^{18}O/^{16}O)\text{ standard}}{(^{18}O/^{16}O)\text{ standard}}\right] \times 1000$$

Clearly δ_w cannot be known precisely for the geological past, which is a limitation of the method, but inferred variations in the isotopic composition of sea water have led to the development of a fascinating tool for studying the late Cenozoic, when polar ice caps developed. To understand why, we must first consider the factors that control isotope fractionation.

$H_2\,^{16}O$ is more volatile than $H_2\,^{18}O$ because it is lighter, so that marine vapour has a lower $^{18}O/^{16}O$ ratio than sea water. When the vapour condenses to form rain there is a preferential condensation of $H_2\,^{18}O$, so that the residual vapour becomes even more ^{18}O depleted. As high latitude snow is precipitated from much-depleted vapour far from the site of initial evaporation, polar ice has a very low $^{18}O/^{16}O$ ratio, with $\delta^{18}O$ as low as -60, compared with ~ 0 for mean ocean water. During a major phase of glaciation therefore, δ_w increases perceptibly (Fig. 7.1).

Now in the pioneer work on Pleistocene surface-water temperatures based on the isotopic composition of planktonic foraminiferal tests (Emiliani 1955), a rise of δ_c was taken solely to signify a fall in temperature. Shackleton (1967) was the first to point out that such a change could also be attributed to a rise in δ_w consequent on the abstraction of ice. Interestingly

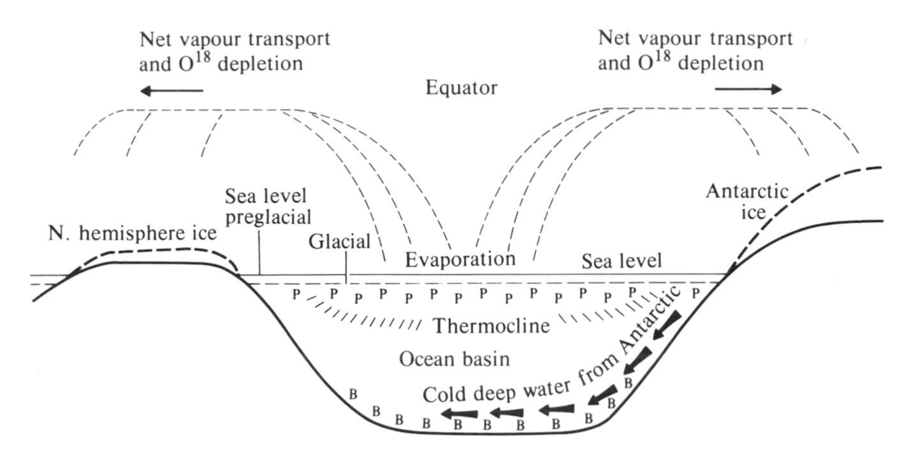

7.1 Schematic diagram to illustrate global pattern of oxygen isotope fractionation in the preglacial and glacial Cenozoic. P, Planktonic; B, benthic foraminifera. After Hudson (1977b).

enough, both effects should result from a climatic deterioration, so the validity of the method as a correlation tool is not necessarily impaired. As the water-mixing time is only a few hundred years the effect should be geologically 'instantaneous'.

An astonishingly good correlation of isotopic fluctuations in widely separated ocean cores has been achieved and since temperature fluctuations are unlikely to have been the same throughout the world the 'ice effect' was probably the dominant controlling factor. During the interglacial–glacial oscillations of the Pleistocene the temperature of the deep oceans is unlikely to have altered significantly, so that measurement of isotopic compositions of abyssal benthic foraminifera provides a method of disentangling the ice effect.

Unlike fossils collected from rocks exposed on the continents, diagenetic factors in deep-sea cores are comparatively insignificant, at least at low burial depths, but account has to be taken of other factors, such as mixing of sedimentary layers by bioturbation, and the differences in depth (and therefore temperature) habitat of different species of planktonic foraminifera. Shackleton and Kennett (1975) estimate δ_w of the pre-glacial ocean to have been -1.2 for the P D B and -1.0 for the SMOW (Standard Mean Ocean Water) standard.

Cambrian and Ordovician

Our knowledge of Cambrian and Ordovician climates is decidedly sketchy, especially as map reconstructions are of limited value because of the uncertain palaeogeography. Whereas there is scant evidence of Cambrian glaciation, the situation changes for the late Ordovician of the central Sahara, where numerous glacial features including striated pavements have been recorded (Beuf *et al.* 1971). There is also quite good evidence that the ice sheet extended to West Africa and to Arabia (McClure 1978) but claims for Newfoundland and Nova Scotia, where the tillites are associated with turbidites, appear to be more dubious. Palaeomagnetic data indicate that the Sahara was in a polar position in the Ordovician (Frakes 1979).

Evaporites occur in Australia, Siberia, Aden, India, Canada, the United States and the Spanish Sahara. On palaeomagnetically guided map reconstructions most lie within 45° of the equator but those of the Spanish Sahara are anomalous in having a nearly polar location. There is disagreement about the importance of evaporites in the Cambrian; according to Meyerhoff (1970) one of the Phanerozoic maxima occurred in the early Cambrian (notably the Siberian salt deposits) whereas Gordon (1975) considers both the Cambrian and Ordovician to be times of relatively low accumulation.

Both Cambrian archaeocyathid and Ordovician reefs occur within 30° of the equator but are not widespread. Whether this signifies that the climate was not unduly warm or that reef-building organisms were scarce at the time, is uncertain.

Silurian and Devonian

Whereas the late Ordovician glaciation might have persisted into the early Silurian there is no convincing evidence of any Devonian glaciation. On the other hand Siluro-Devonian reef limestones are more abundant than in the preceding two periods. Silurian reefs are confined to a narrow equatorial zone and include the famous examples of Gotland (Sweden) and Illinois. Devonian reefs occupy a wider zone extending to slightly more than 30° latitude. They are strikingly developed in Belgium, North Africa, western Canada and Western Australia (Fig. 7.2). Evaporites obtain their greatest Palaeozoic development in the Devonian (e.g. the Elk Point Basin, see chapter 5) but are uncommon in the early Silurian and latest Devonian. The distribution of Devonian coals poses a problem because they tend to occur in low-latitude areas and nowadays abundant accumulation of peats does not occur in such areas, as already noted.

The strongly endemic distribution of early Devonian brachiopods and other benthic invertebrates suggests a marked climatic zonation to Cocks and McKerrow (1973), Boucot (1974) and Copper (1977), with the South American–South African Malvinokaffric Realm being the coolest. Other factors can cause high endemism, however, as discussed in chapter 10.

According to Frakes, temperature and humidity patterns were not greatly different from today, although the climate was relatively arid except in the late Silurian and late Devonian. If climate was the dominant control on

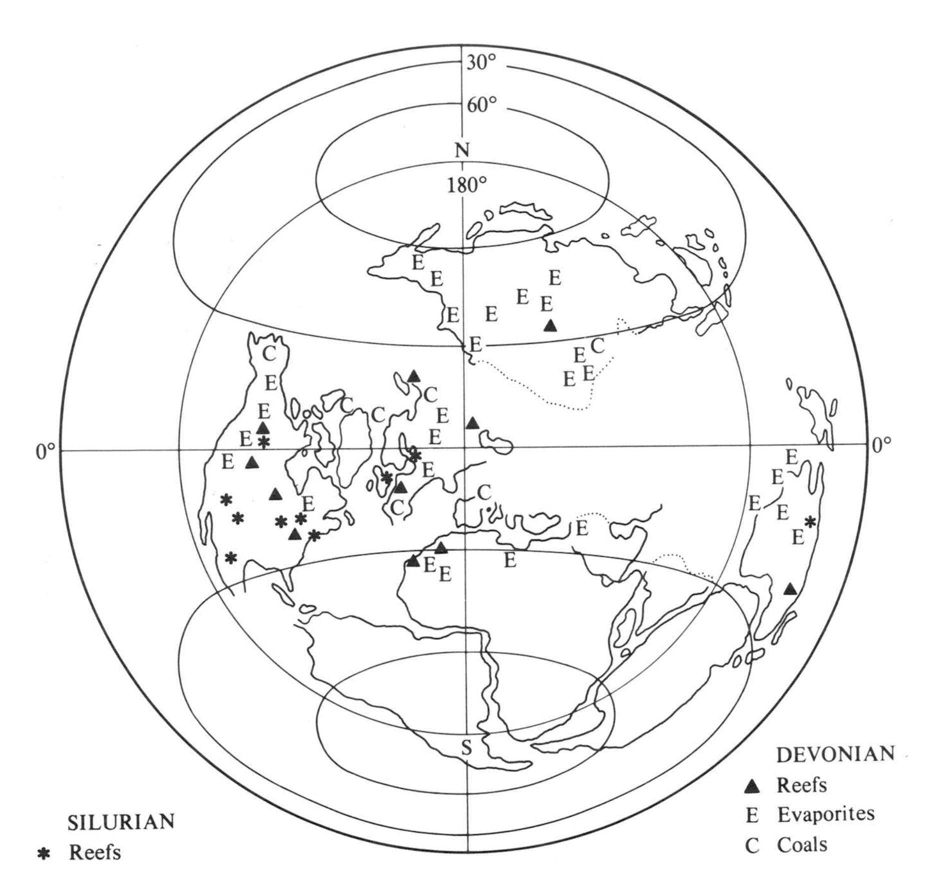

7.2 Siluro-Devonian global palaeogeography and climatic indicators. Adapted from Frakes (1979).

marine faunal provinces then the early Silurian and late Devonian were times of comparative equability while in the early Devonian there was a more pronounced latitudinal temperature gradient.

Carboniferous and Permian

The feature of outstanding interest is the abundant evidence of the most important ice age between the Precambrian and the Pleistocene, affecting all the southern continents and India (Crowell 1978; Crowell and Frakes 1975). It lasted some 90 million years, from the mid-Carboniferous to the mid-Permian, and was expressed as a series of ice centres that waxed and waned through time, rather than a single Gondwana super ice sheet. As the south pole moved from west to east, as originally predicted by Wegener and confirmed by palaeomagnetism, there was a shift in the principal locus of ice accumulation (Fig. 7.3). Thus the oldest glacial evidence comes from South America and the youngest from Australia, although the oldest Australian glacial deposits are only slightly younger than those in South America.

The fullest information comes from southern Africa and Australia. In South Africa the celebrated Dwyka Tillite and associated deposits occupy a vast extent exceeding 200 000 km^2 in the Karroo Basin and is more than 1000 m thick in the south; much of the sequence is glacio-marine. In the Zaire Basin to the north the glacial sequence is about 900 m thick. Dating is not precise but the age range is probably Stephanian to Sakmarian. This is broadly similar to the Paraná Basin of southern Brazil, Uruguay and Paraguay, except that the oldest deposits there are Namurian. In Australia the age range is Westphalian to Kazanian. The earliest, late Carboniferous event was an alpine glaciation on the borders of the Tasman Geosyncline. By Sakmarian times an ice sheet covered approximately half the subcontinent but waning took place towards the mid-Permian, with only glacio-marine deposits recorded.

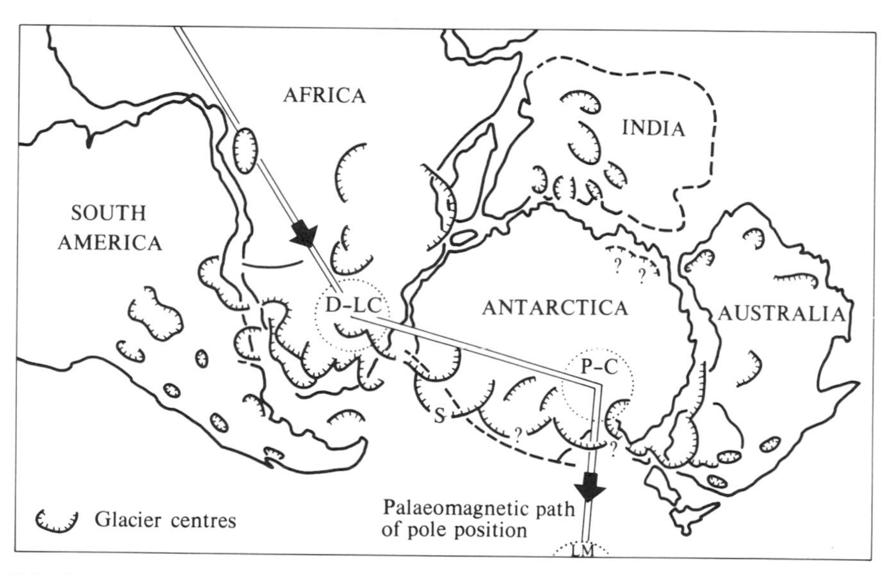

7.3 Late Palaeozoic ice centres in the Gondwana supercontinent. D, Devonian; LC, Lower Carboniferous; C, Carboniferous; P, Permian; LM, Lower Mesozoic. Adapted from Crowell (1978).

The Carboniferous was a time of very extensive coal formation (Fig. 7.4). Lower Carboniferous coals are most abundant in Asia but most of the European and North American coals are late Carboniferous in age. The coals occupy a wide range of latitude, up to 80° in Chile and such widespread

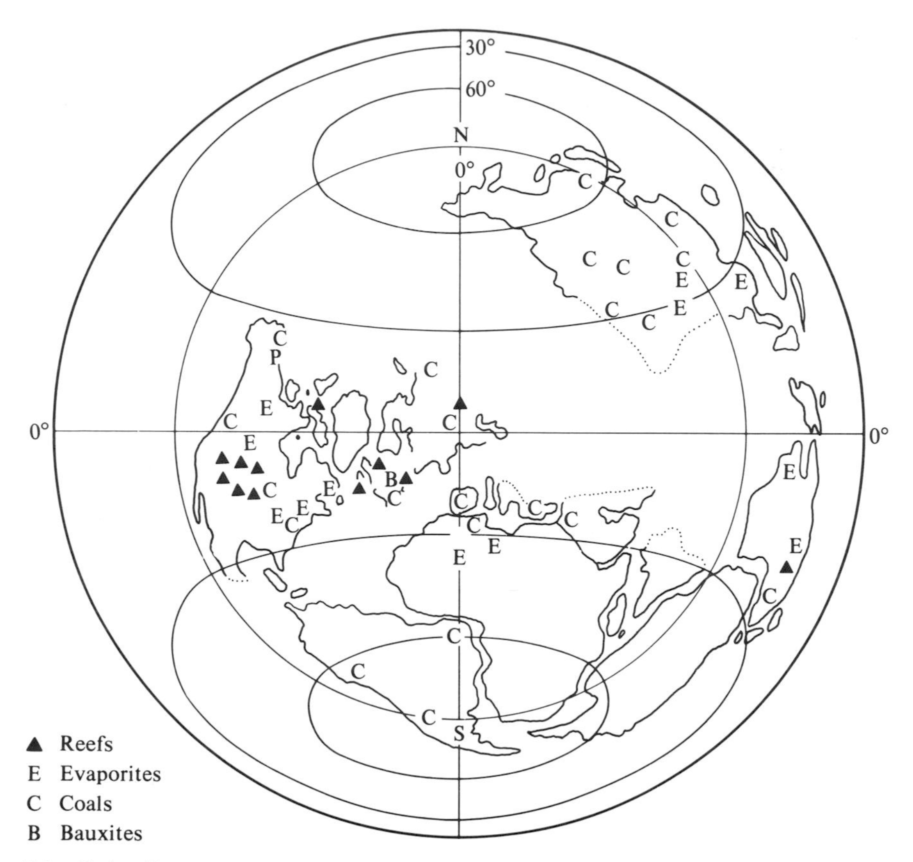

7.4 Carboniferous global palaeogeography and climatic indicators. Adapted from Frakes (1979).

deposits imply very humid conditions (Frakes 1979). Permian coals are abundant in the Gondwana continents and are also found in Asia; they occur mostly at latitudes above 40°. Palaeomagnetic determinations indicate that the European and eastern North American coals accumulated in an equatorial zone, which is confirmed by the virtual absence of growth rings in fossil logs. In contrast, trees associated with the Permian Gondwana coal deposits, which often occur only shortly above glacial deposits, have clearly defined growth rings, which is consistent with formation in temperate latitudes (Chaloner and Creber 1973). Because peats do not accumulate readily today in the equatorial zone, either very high precipitation rates or a combination of poor drainage and unusually good preservational conditions seem to be implied. As would be expected from the foregoing, Carboniferous evaporites are less abundant than in either the Devonian or Permian. Virtually all the Permian evaporites, which include the well-known deposits of Texas and New Mexico, the north-west European Zechstein and the Urals, occur in a zone of less than 40° latitude, like those of the present day. The orientation of dune bedding in aeolian sandstones of the European Permian gives a dominant easterly wind direction, indicative of a trade wind belt and consistent with palaeomagnetic determinations of 10–30°N (Glennie 1972).

The Lower Carboniferous Waulsortian buildups (not true reefs) of Belgium and the British Isles and Upper Carboniferous reefs of Texas and Utah occupy a low latitude zone. Permian reefs are likewise confined to a narrow equatorial belt (Fig. 7.5). Waterhouse and Bonham Carter (1972) have undertaken a diversity analysis on Permian brachiopods to determine the position of the Permian equator, about which there has been some dispute (Stehli 1970). Using a combination of statistical techniques, seven brachiopod groups of varying diversity were distinguished. The lowest

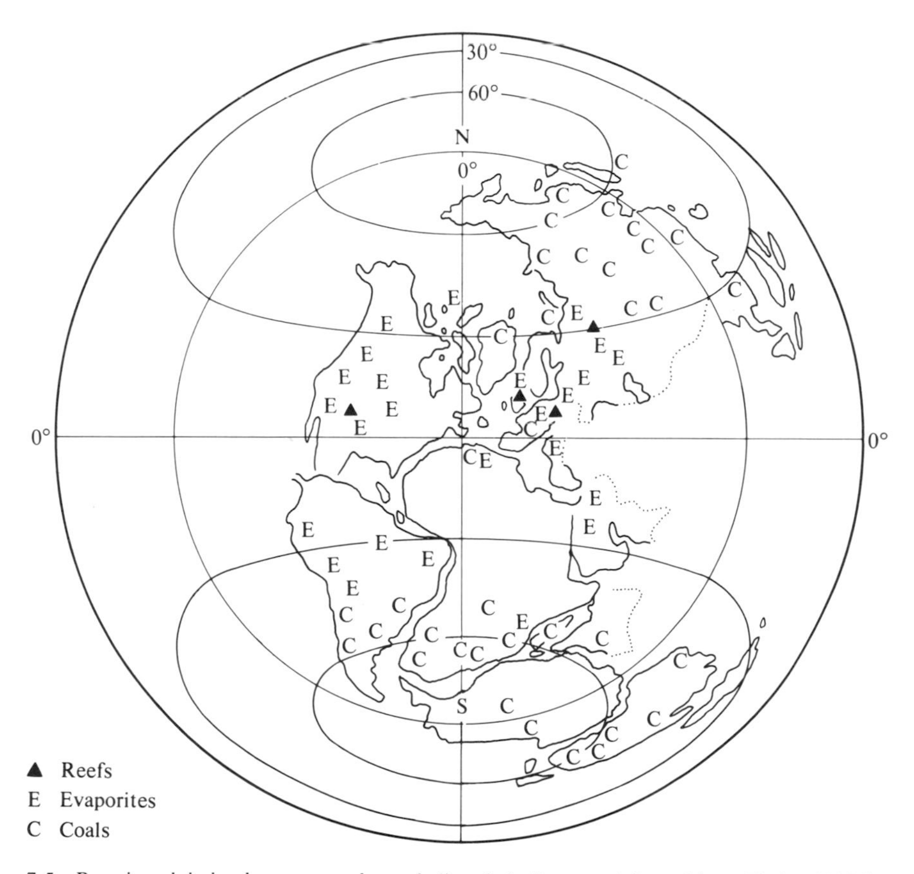

7.5 Permian global palaeogeography and climatic indicators. Adapted from Frakes (1979).

diversity, as might be expected, was found to occur in areas where glacial deposits are recorded, while the highest diversity group is associated with rugose corals, fusulinid foraminifera and richtofenid brachiopods, all groups widely regarded as having inhabited a tropical regime. The position of the Permian poles determined from rock magnetism accords well with the faunal data and with the generally accepted model for the reconstruction of Gondwanaland.

Several terrestrial floral provinces are distinguished in the Carboniferous and Permian, with the Glossopteris and Angara Floras occupying polar to subpolar regions (Chaloner and Lacey 1973). The existence of a large-leaved Glossopteris Flora poses a problem, because of the enforced six months' dormancy of growth.

Triassic and Jurassic

The keynote of Mesozoic climates is equability, with warm conditions extending far beyond their present latitudinal limits, and no well authenticated records of glacial deposits at the poles, so we must assume that ice caps were absent. The latitudinal distribution of evaporites suggests that a dry climate was characteristic of most of the Triassic and Jurassic. Both Meyerhoff (1970) and Gordon (1975) agree on a major peak in the Triassic, in their plots of the abundance of evaporites through the Phanerozoic. Extreme aridity characterized a zone in two bands between 10° and 40° latitude. Since modern evaporites occupy zones of 15 to 35° a poleward shift of at least 5° in air circulation patterns seems to be required. Jurassic evaporites are somewhat less abundant. Though they range in extent to 45° N and S they are concentrated more in lower latitudes. They are especially widespread in the late Jurassic, with large deposits in the southern USSR, the Middle East, the United States Western Interior and western South America. In contrast, early Jurassic evaporites are largely restricted to the Argo Salt of the Scotia Shelf. If the Navajo Sandstone of the western United States is a true aeolian sandstone then an early Jurassic desert must also be postulated for this area, but some doubt has been thrown on this interpretation (see chapter 2).

A change from an arid to a humid climate from the Triassic to the Jurassic in western Europe can be inferred from the sediments. The late Triassic Keuper consists of gypsum- and halite-bearing red beds with a distinctive suite of magnesium-rich clay minerals suggestive of hypersaline conditions; kaolinite is absent (Jeans 1978). Abundant kaolinite, indicative of weathering of a warm, humid land, enters in the uppermost Triassic 'Rhaetic' deposits and continues into the Jurassic (Hallam 1960; Will 1969). In addition, the Rhaeto-Hettangian deposits, where developed in a paralic-marginal marine siliciclastic facies, contain coals and there is abundant terrestrial plant debris in the associated sandstones. If the formation of oolitic ironstones demands a pre-concentration of iron on the land by lateritic weathering (see chapter 5) then we have a further indication of humidity, because such ironstones are most common in the early and early mid-Jurassic of this region.

Coal deposits are largely confined to the eastern parts of Laurasia and Gondwanaland in both the Triassic and Jurassic, with evaporites being concentrated in the western part of the supercontinents. This implies the existence of a western arid belt and two eastern humid belts at higher latitudes (Fig. 7.6).

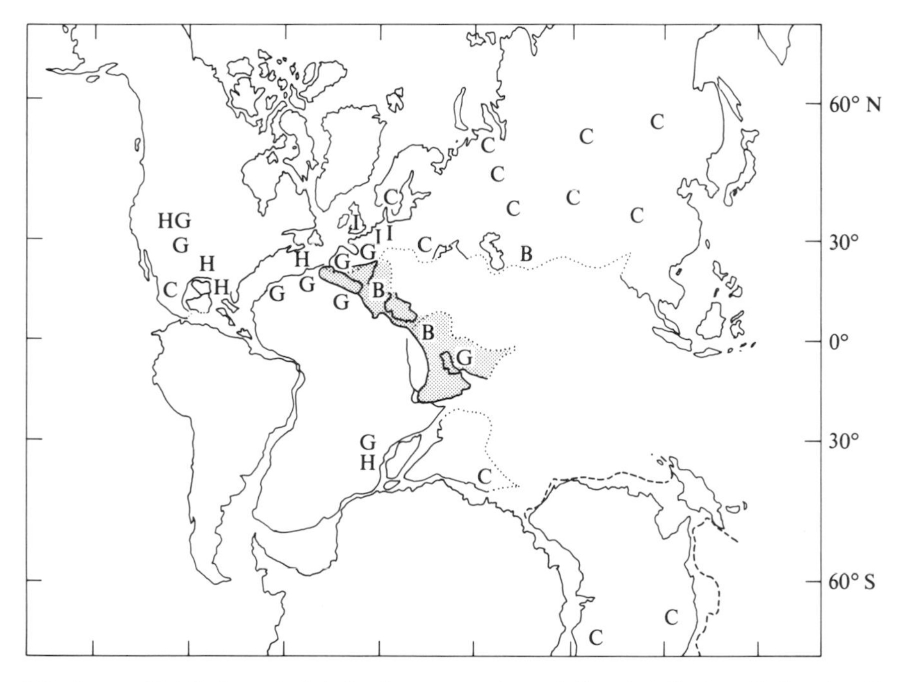

7.6 Lower–Middle Jurassic global palaeogeography and climatic indicators. B, Bauxite; C, coal; G, gypsum or anhydrite; H, halite; I, ironstone; stippled area, principal zone of carbonate sedimentation. After Hallam (1975).

Bearing in mind the climatic role of the Eurasian landmass at the present day in creating monsoonal conditions to the south, Robinson (1971), in her analysis of Triassic climates, considered that Laurasia and Gondwanaland might have acted likewise, so that winds reaching the eastern parts in middle and low latitudes brought monsoon-type summer rains, while a dry hot season occurred in winter as winds blew offshore. The central and western parts of the supercontinents would have tended to have an appreciably less humid climate because many of the dominant easterly winds would have travelled over land for a considerable distance, or, blowing towards the equator without the intervention of mountains, could not readily have jettisoned their moisture. Coals formed in the eastern, peninsular parts of the landmass in middle to high latitudes where the temperature was more moderate and the rainfall less strictly seasonal, being under the influence of westerly and polar easterly winds.

During the late Jurassic coal deposits became much more restricted in distribution, being largely confined to limited areas in central and eastern Asia. Taken together with the more extensive development of evaporites, occurring on the northern as well as the southern side of the Tethys, in a zone extending from Moldavia to Uzbekistan, this is a likely indication of increasing aridity through the period. It may also be significant that the areal extent and volume of ironstones diminishes from the Lower and Middle to the Upper Jurassic.

As already noted, laterites and bauxites have been widely considered to be good indicators of humidity, because of the intensity of chemical weathering required for their formation. However, in southern Israel there is a horizon of reworked laterites, including pisolitic conglomerates, sandwiched between evaporite-bearing Upper Triassic and Lower Jurassic deposits and attributed to the basal Jurassic. Following the development of a karstic surface after an episode of regression, leading to the subaerial exposure of hypersaline mud flats, so-called flint clays were generated by chemical weathering in the vadose zone (Goldbery 1979). Whereas swampy conditions might have occurred locally, no drastic regional increase of humidity after the late Triassic, followed by a return to aridity, apparently need be invoked. Other Lower Jurassic bauxites, which have conventionally been held to signify intervals of emergence and weathering in a warm climate with a humid season, occur in a zone extending eastwards from Yugoslavia to the Middle East and Uzbekistan.

Fossil evidence points unequivocally to an equable climate. The ferns are a particularly useful group to study because related living genera cannot tolerate frost. By plotting the relatively cosmopolitan Triassic and Jurassic fern distributions such as *Dictyophyllum*, Barnard (1973) has been able to demonstrate that climatic conditions characteristic of the present sub-tropical–warm temperate zone extended up to 60° N and S from the equator. The evidence from Jurassic cycadophytes suggested to Vakhrameev (1964) a latitudinal temperature gradient in Eurasia significantly less than today, with winter temperatures in Siberia never falling below 0°C. Seasonality in this region is shown by growth rings in coniferous wood.

A modest latitudinal temperature gradient is also indicated in the marine realm, perhaps most strikingly by the fact that the same endemic bivalve species of, for instance *Monotis* in the Norian and *Weyla* in the Lower Jurassic, extend all the way from Alaska to Chile. Other examples of such broad latitudinal distributions abound and stand in stark contrast to recent species distributions. Furthermore, Upper Jurassic reef corals occur in Sakhalin, nearly 70° N, and at over 50° N in western Europe. The more northerly occurrence in eastern Asia may relate to Coriolis and landmass deflections of a warm Pacific equatorial current. The main zone of reef limestones in both the Triassic and Jurassic follows the line of the Tethys and does not extend beyond 35° N. Associated with this Tethyan low latitude belt are a number of other groups suggestive of tropical conditions, notably the large, thick shelled megalodontid bivalves of the late Triassic and their Jurassic evolutionary descendents, the rudists.

The oldest fossils on which isotopic palaeotemperature determinations were undertaken were Jurassic belemnites. Subsequently many further determinations were made on Jurassic fossils, principally belemnites, but by and large the earlier results have been so contradictory, unreasonable or manifestly subject to error as to be of little value in palaeoclimatology. The quality of analysis and selection of suitable material has improved considerably in recent years, but doubt persists about whether this kind of analysis can amplify what we can learn from other sources (Hallam 1975).

Cretaceous

Both Gordon (1975) and Meyerhoff (1970) note a diminution in evaporite abundance from the Jurassic to the Cretaceous. Change from an arid to a humid climate in the United States Western Interior is clearly shown by the fact that Middle and Upper Jurassic evaporite-bearing red beds are replaced by Lower Cretaceous coal measures. Nevertheless the overall latitudinal distribution of evaporites is considerable, and the boundary between evaporite and coal deposition is generally between 45 and 55°. A west-coast desert is signified by the evaporites of western South America and the early Cretaceous salt basins of eastern Brazil and Angola indicate an arid zone in the region where the South Atlantic was about to be created.

Extensive mid-Cretaceous bauxite formation in southern Europe may be taken as pointing to a humid climate, but the reservations expressed concerning the formation of basal Jurassic bauxites in Israel should be noted. Perhaps a better indication of a change from arid to humid conditions across the Jurassic–Cretaceous boundary over an extensive region of western and southern Europe is the replacement of carbonate- and evaporite-bearing 'Purbeck' by coarse siliciclastic paralic or fluvio-deltaic 'Wealden' facies, locally containing thin coals. Taken together with the evidence from the United States Western Interior, the data appear to contradict Frakes' (1979) claim for continuation of a late Jurassic global trend towards greater aridity into the Cretaceous. Conceivably we are dealing with a regional exception bound up with the establishment from early late Jurassic times onwards of an ever widening ocean between North America and

Africa–South America with a continuous marine connection being established from the western Tethys to the Pacific. Such a major palaeogeographic change, marking the start of the breakup of the Pangaean supercontinent, might well have had significant climatic consequences, which deserve to be more fully explored.

Reef limestones with abundant rudists, often of huge size, occur in Mexico, the Caribbean, southern Europe and the Middle East, which can therefore be taken as signifying a tropical belt.

The terrestrial flora also signifies warm and equable conditions persisting to 60° N and S, as in the earlier Mesozoic, though the presence of higher latitude floras in Alaska poses the problem of continuous darkness in the winter months (Barnard 1973). An especially interesting study of Eurasian floral provinces was undertaken by Vakhrameev (1964). By determining the northward shift through time of the boundary between the Indo-European and Siberian provinces he argued for a progressive global warming from the early Jurassic to the mid-Cretaceous (Fig. 7.7).

With regard to the younger Cretaceous, Krassilov (1975) inferred from a study of east Asian floras that there was a temperature optimum in the

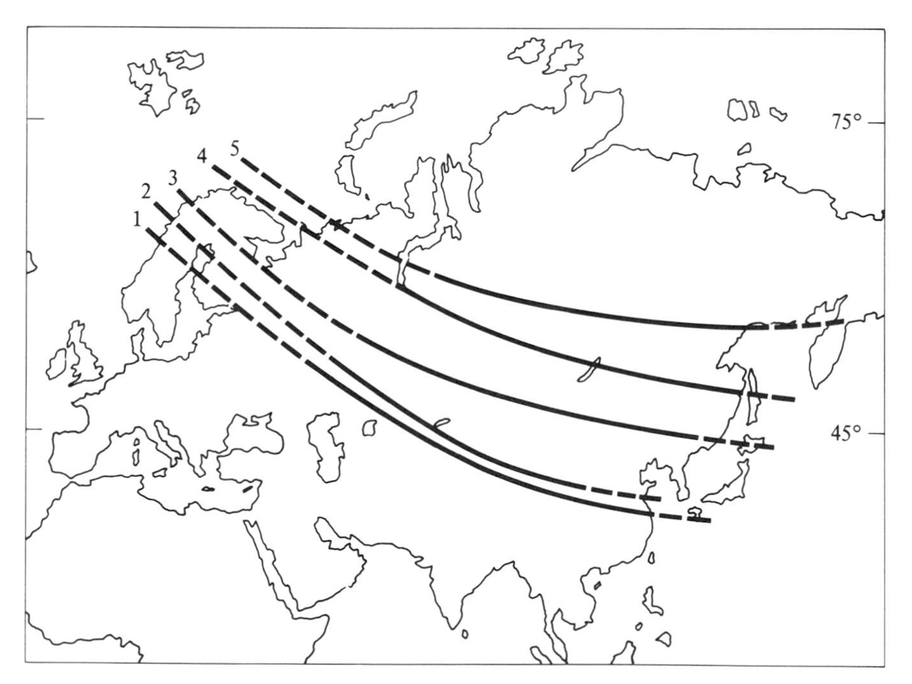

7.7 Shifts in the boundary separating the Indo–European and Siberian palaeophytographic provinces in Eurasia. 1, Early Jurassic; 2, mid-Jurassic; 3, late Jurassic; 4, early Cretaceous; 5, mid-Cretaceous (Aptian–Albian). Adapted from Vakhrameev (1964).

Campanian and cooler periods in the Coniacian and Maastrichtian, with climatic deterioration continuing across the Cretaceous–Tertiary boundary to reach a culmination in the late Danian. This was based on the percentage of angiosperm leaves with entire margins, the percentage of platanoid leaves and the frequency of temperate elements such as *Gingkoites*. He also cited evidence that comparable changes took place in Alaska and Canada.

Oxygen-isotope data from the continents, derived mainly from analysis of belemnites, is summarized by Stevens (1971). There are many differences between the results of different workers, which is not surprising if the post-depositional complications mentioned earlier are taken into account, but there is a consensus that the Campanian–Maastrichtian was a time of cooling and a weaker indication, notably from the Soviet data, that a climatic optimum was achieved in the mid-Cretaceous.

It is interesting to compare these results with the data obtained from benthic and planktonic foraminifera from north-west Pacific cores, which ought to be more reliable. Douglas and Savin (1975) infer a slight warming phase from the Valanginian to an Albian optimum and thereafter a cooling, accentuated in the Campanian–Maastrichtian and continuing into the Palaeocene. Vertical temperature gradients were less than half those at present. The temperature of the bottom waters achieved an Albian maximum of ~17°C, cooling to ~10° at the end of the period; the corresponding figures for surface waters are ~28° and ~19°C. A similar analysis for South Atlantic and Indian Ocean deep-sea cores indicates a marked cooling in the late Maastrichtian (Saito and Van Donk 1974).

In summary, the Cretaceous was, except at the end, a period when tropical–subtropical climates extended to at least 45° and warm to cool temperate climates extending to the poles. Mean annual temperatures were some 10–15° higher and the latitudinal temperature gradient only about half those of today (Frakes 1979). While there is general agreement that the Maastrichtian was a time of cooling, Krassilov's warm Campanian phase appears anomalous and warrants further investigation.

Tertiary

With the advent of the Tertiary we enter an era in which the major features of present world geography had become blocked out and the organisms bear a close resemblance to those of which we have direct experience. Palaeoclimatological interpretation is thereby enormously faciliated and a fascinating history of progressive climatic deterioration has been worked out, with the onset of widespread glaciation after over 200 million years of equability. The general outlines of this story have long been discerned from the more traditional methods involving comparison with living relatives of Tertiary floras and faunas found on the continents, but in recent years extensive drilling operations on the deep-ocean floor have yielded an immense amount of new information which greatly adds to the precision of our knowledge.

It is appropriate to deal firstly with the information obtained from the study of rocks and fossils presently exposed on the continents.

Evaporites are relatively restricted in distribution in the Palaeogene, with west coast deserts restricted to parts of South America. The widespread occurrence of lignites, laterites and bauxites confirms the existence of a generally humid climate. Consider the situation in western Europe, where all these deposits occur.

Palaeogene lignites are abundant in Germany and laterites are widespread, from northern Ireland to southern France, as well as lacustrine limestones and bituminous shales, some with a rich insect fauna. The Upper Eocene of the Paris Basin includes, however, a celebrated evaporite deposit, the Montmartre Gypsum, and halite and potash salts such as sylvite occur in the Oligocene of the Rhine Graben in Alsace. These evaporites cannot indicate widespread aridity, however, because of the other types of deposits cited, and because the Alsace potash deposits contain clay intercalations that have yielded a rich insect fauna signifying warm temperate but not desert conditions (Gignoux 1955). There are also palaeobotanical indications of high humidity (see below). On the other hand, there are thick Miocene evaporites in the Red Sea and the Mediterranean region, especially in the deep basins, as discovered during the Deep Sea Drilling Project. These deposits, which will be discussed in the next chapter, must surely signify an environment of great warmth and aridity.

The *leitmotif* of the Tertiary is, however, temperature decline, which is recorded in both neritic invertebrates and terrestrial plants. Palaeocene and Eocene foraminifera, molluscs and other groups indicate tropical or subtropical conditions extending in Europe to 50° N, while on the west coast of North America Durham (1950) has deduced an Eocene to Recent cooling trend from the progressive southward displacement of molluscan provinces, with 'tropical' conditions (surface water temperatures in excess of 20°) extending to 45° N in the Eocene.

A fuller picture emerges from study of the climatically more sensitive terrestrial plants. Climatic interpretation of the Lower Eocene London Clay flora of southern England, made classic by the monograph of Reid and Chandler (1933), is based mainly on fruits and seeds. Daley (1972) challenges their comparison with the tropical rain forest of the Indonesian–Malaysian region, pointing out that the implied climate would be impossible at 40° N, as determined by palaeomagnetism. There must have been significant seasonal variations at this latitude, and Reid and Chandler's assumption that the 11%of non-tropical plants were derived from contemporary uplands must be wrong, because there is no palaeogeographic evidence for such uplands. Nor is it likely that they are relics from a cooler

Palaeocene climate. Daley argues that an absence of frost would have allowed tropical plants to spread northward. Furthermore, higher rainfall due to increased warming of ocean surface waters would also favour such a spread. Away from rivers and lakes, more temperate plants could grow.

In North America, extremely informative results have been obtained from analysis of fossil leaves (Wolfe 1978). An excellent correlation exists between the type of leaf margin and climate, with the percentage of species with entire-margined leaves, i.e. lacking lobes and teeth, increasing systematically with temperature (Fig. 7.8). In areas of high mean annual temperature and precipitation the leaves tend to be entire-margined, evergreen, large and with a leather-like (coriaceous) texture, a high proportion of 'drip tips' and tendency towards palmate venation. The mean annual range of temperature is more difficult to infer than the mean annual temperature but can be done accurately in some cases, for instance by the proportion of microphyllous to notophyllous leaves.

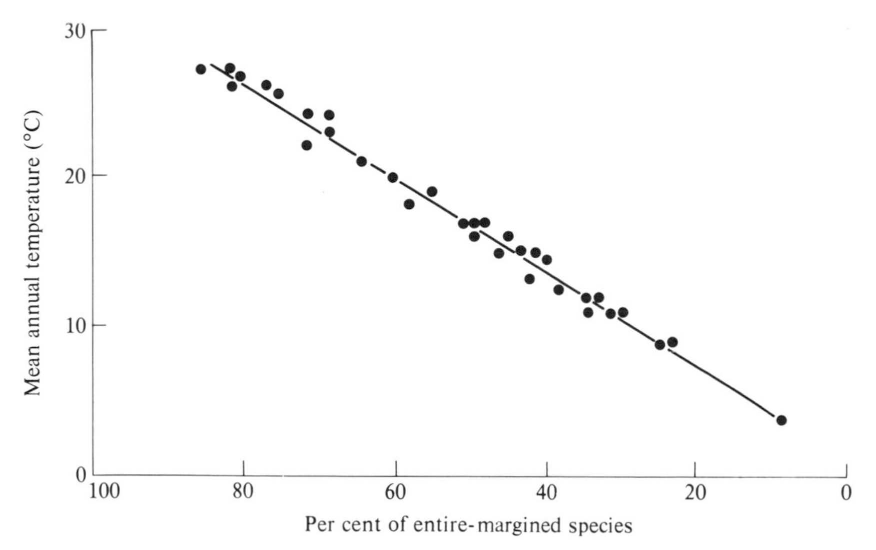

7.8 Correlation of mean annual temperature with the percentage of tree species with entire-margined leaves in the humid to mesic broad-leaved forests of eastern Asia. After Wolfe (1978).

Wolfe applies these various criteria to the interpretation of climatic change through the Tertiary in the northern hemisphere, the most comprehensive data coming from the north-western United States (Fig. 7.9). In Palaeocene and Eocene times tropical rain forest, and the 25° isotherm, extended 20–30° poleward of the present northern limit. There is general

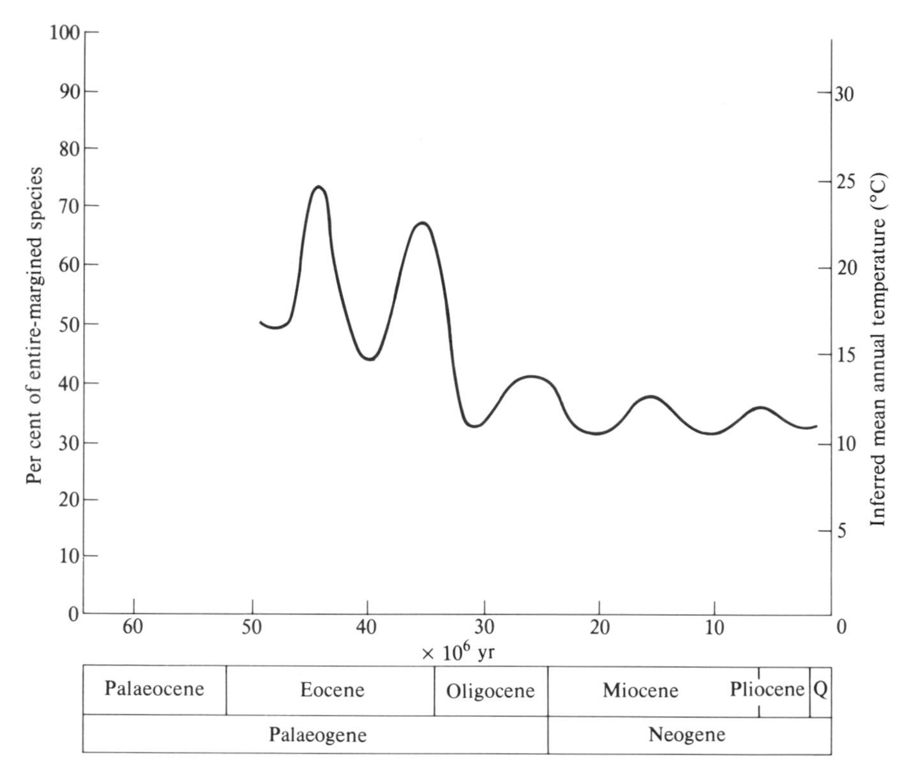

7.9 The Tertiary decline of temperature as determined from terrestrial plants in the north-western coterminous United States. Adapted from Wolfe (1978).

agreement that western North America shifted southwards in the Tertiary so that palaeolatitudes are probably higher than the present latitudes. During the Oligocene and Neogene, in mid-to-high latitudes of the northern hemisphere, areas of broad-leaved evergreen forests changed to temperate broad-leaved forests. There was a major drop from the Eocene to the Oligocene in mean annual temperature of about 12–13°C in Alaska and Washington State. The mean annual range of temperature increased from 3–5°C in the mid-Eocene to 21–25°C in the Oligocene. Both the European and Pacific north-west Neogene shows an overall change from broad-leaved deciduous to coniferous forest, with a corresponding decrease of both mean annual temperature and mean annual range of temperature.

An increase of the latitudinal temperature gradient should cause an increase in the intensity of subtropical high-pressure cells which in turn would effect an increase in west coast summer droughts. This is well documented from the plant record of western North America (Chaney 1944).

The best southern hemisphere data come from Australasia. Kemp's (1978) palynological analysis shows that in Eocene times Australia was covered by rain forest with a mixture of 'tropical' and cool temperate elements comparable to that in the London Clay, which perhaps relates to seasonal changes in high latitudes. The high fern diversity, the types of tree and the presence of epiphytic fungi point overwhelmingly to high humidity. Coastal vegetation existed in Antarctica, so that any ice must have been confined to alpine-type glaciers. Rain forest persisted in Australia into the Miocene.

The study of South Australian and New Zealand shallow-marine invertebrate faunas indicates an anomalous condition with respect to the rest of the world in that a climatic optimum occurred in Oligocene–early Miocene times (Fleming 1962; Gill 1961). This is apparently confirmed by oxygen-isotope analysis of fossil shells (Dorman and Gill 1959). Such an anomaly can be explained as a consequence of the northward movement of Australia (and presumably New Zealand) after severance from Antarctica in the Eocene.

Amplification of our knowledge of Tertiary climatic history from the results of deep-ocean drilling comes from study of the sediments, organic distributions and oxygen-isotope analysis of foraminifera.

The relevant sedimentary record has been reviewed by Kennett (1977). The first appearance of ice-rafted sediments in the form of dropstones and non-turbiditic sands containing quartz grains with characteristic surfaces revealed by scanning electron microscopy, signifies the initiation of extensive glaciation at sea level and the production of icebergs. A progressive spread in the Southern Ocean to lower latitudes through the Tertiary indicates a corresponding growth of Antarctic ice (Fig. 7.10).

Plotting the distribution of siliceous and calcareous oozes also yields valuable information. In the present ocean the position of the Antarctic Convergence, corresponding to the polar front, coincides precisely with the siliceous–calcareous ooze boundary. Siliceous oozes have approximated the position of surface waters with the temperature characteristics of the present Antarctic water mass. Diatomaceous sediments were first deposited close to Antarctica in the early Oligocene and have spread diachronously northwards during the Neogene. A considerable increase in the sedimentation rate of siliceous ooze, signifying increased biogenic productivity, began in about mid- to late Pliocene times, and probably indicates an intensification of upwelling associated with the Antarctic Convergence, related to increased glaciation.

The changing Palaeogene palaeobiogeography of calcareous nannoplankton and planktonic foraminifera has been related by Haq *et al.* (1977) to temperature fluctuations in the Atlantic. High-, mid- and low-latitude assemblages were recognized using factor analysis and the migrations of

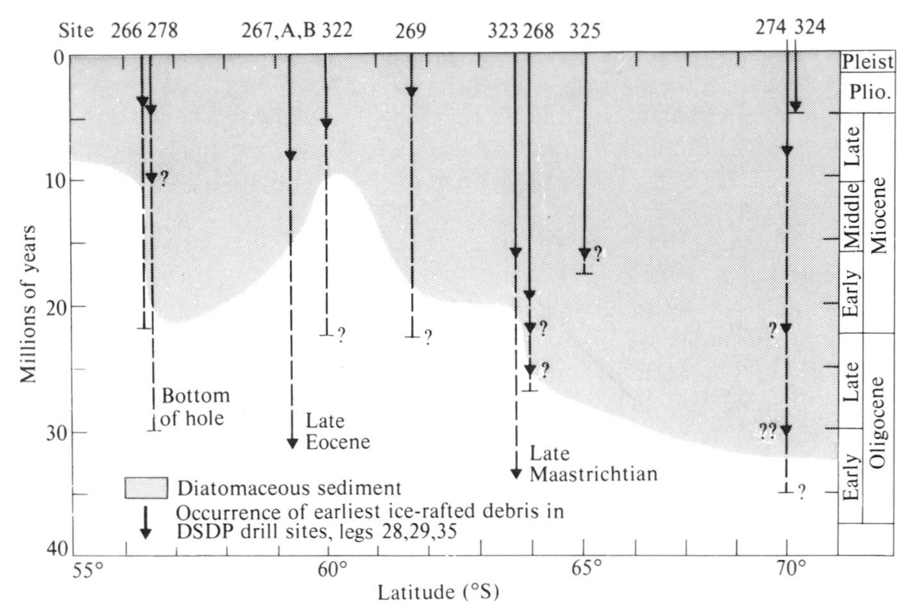

7.10 Temporal and geographic distribution of diatomaceous sediment and ice-rafted debris in the Southern Ocean. After Kennett (1977).

these assemblages interpreted as recording climatic change. No latitudinal provincialism was recognized in the earliest Palaeocene. Subsequently four marked cooling episodes were recorded, in the mid-Palaeocene, mid-Eocene, earliest Oligocene and mid-Oligocene. A marked warming event took place in the late Palaeocene–early Eocene and a less pronounced one in the late Oligocene.

Study of silicoflagellates confirms earlier work on radiolarians in indicating a warm early Pliocene phase in the Antarctic seas, with surface water temperatures 10°C higher than today. Following later cooling the Antarctic ice sheet could have reached its present size about 3.8 million years ago (Ciesielski and Weaver 1974).

The most striking result of oxygen-isotope analysis of subantarctic planktonic foraminifera is that there was a sudden drop of about 5° in surface water temperature at or close to the Eocene–Oligocene boundary after a more gradual Eocene decline (Shackleton and Kennett 1975; and Fig. 7.11). This pronounced change relates to the production of cold Antarctic bottom water, which had a temperature of less than 10°, comparable to modern values, and forming the so-called *psychrosphere.* After a slight early Miocene warming there was a further sharp change in the mid-Miocene reflecting a sudden increase in the heavy isotope. This is thought by Shackleton and Kennett to indicate the initiation of an Antarctic ice cap, so

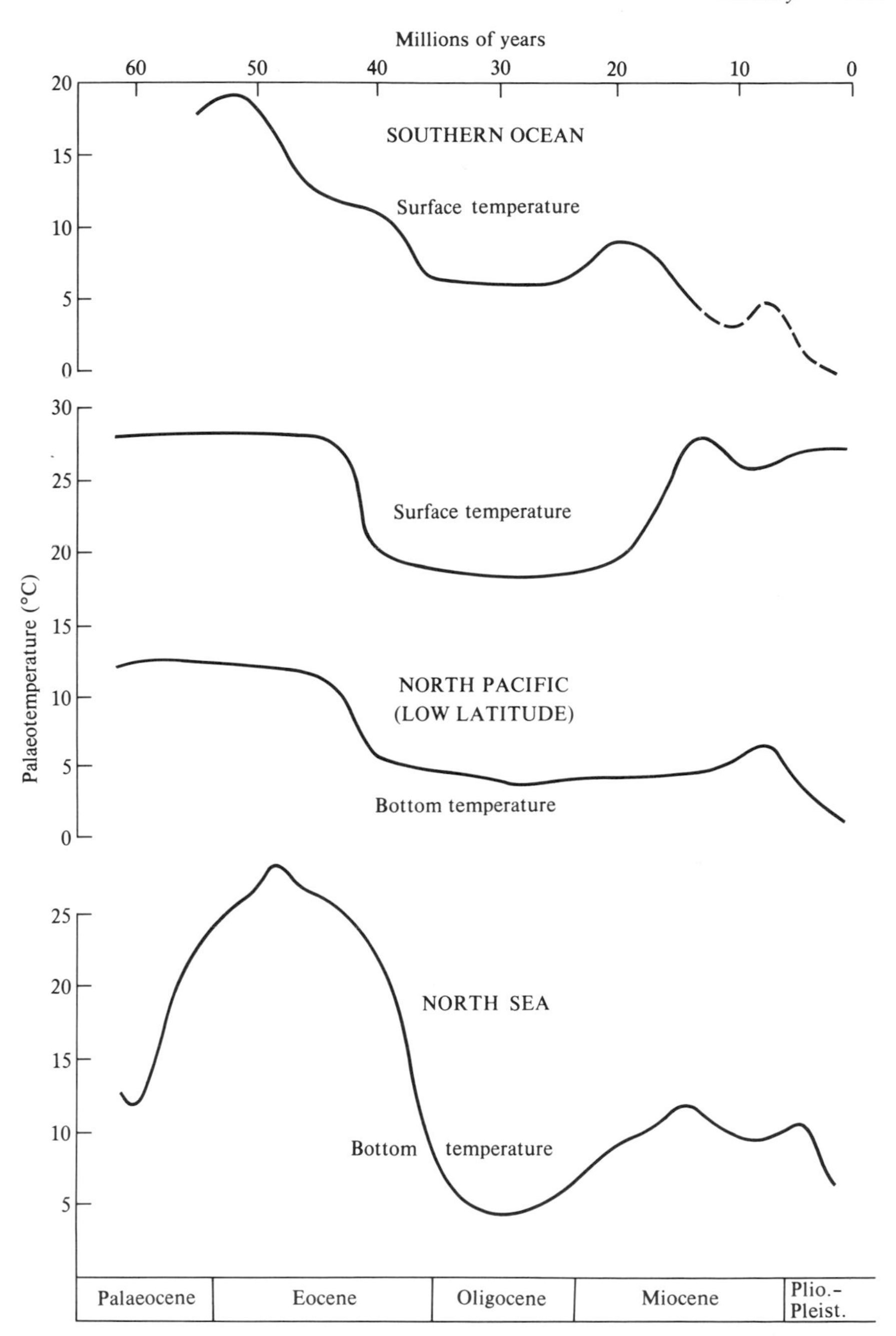

7.11 Smoothed Cenozoic temperature curves for different parts of the ocean system, based on oxygen-isotope determinations on foraminifera. Based on data of Shackleton and Kennett (1975), Savin *et al*. (1975) and Buchardt (1978).

that from mid-Miocene times onwards precise temperature estimation is not possible.

Savin *et al.* (1975) present data for both planktonic and benthic foraminifera from North Pacific sampling sites. Their surface-water temperature curve (corresponding to a tropical position in the Palaeogene) shows a pronounced drop in the late Eocene rather than at the Eocene–Oligocene boundary. A greater difference is that in mid-Miocene times there was apparently a sharp rise to restore the early Palaeogene values, so that Antarctic ice-cap formation had evidently no effect. Their bottom-water curve, on the other hand, shows sudden drops both in the late Eocene and mid-Miocene, implying the northward spread of cold Antarctic bottom waters (Fig. 7.11).

Data from planktonic foraminifera in another Pacific core indicate that glacial–interglacial fluctuations have characterized the world climate for the last 3.2 million years, before which there was a period of stable 'interglacial' or 'preglacial' climate (Shackleton and Opdyke 1977). This is taken to indicate the initiation of glaciation in the northern hemisphere.

These results from the ocean raise the question of whether comparable events can be recognized using the same technique for epicontinental marine regimes. Buchardt (1978) analysed benthic shells from deposits surrounding the southern North Sea. His isotopic-temperature curve shows a pronounced drop in the early Oligocene and a rise in the mid-Miocene, but in addition there is a mid-Eocene optimum following a mid-Palaeocene cool phase, not recognizable in the oceanic results.

It is intriguing to compare the various geochemical data with those obtained from more traditional palaeontological methods. Wolfe's (1978) North American plant data also show very clearly a sudden and pronounced fall of temperature at about the Eocene–Oligocene boundary (Fig. 7.9). The results of Haq *et al.* for Atlantic plankton agree with those of Buchardt in showing a marked mid-Palaeocene cooling episode followed by marked warming in the late Palaeocene and early Eocene. Similarly they confirm a pronounced cooling episode in the earliest Oligocene but a notable difference is their recognition of a mid-Eocene cool phase.

Ciesielski and Weaver's early Pliocene warm phase in the Antarctic seas might be reflected in Buchardt's minor peak, but the problem of accurately determining from the isotope record water temperatures after the mid-Miocene has already been alluded to.

The differences between the various isotopic and palaeontological analyses indicate a complex system of varying water and air temperatures, with no simple worldwide response, and further work is clearly required to disentangle 'signals' from 'noise'. The most notable paradox concerning the isotope data is that formation of the Antarctic Ice Cap in the mid-Miocene coincides with a phase of climatic amelioration in the North Sea and in low-latitude surface waters of the Pacific. On one point at least there is general agreement, and is confirmed by study of subantarctic deep-sea sediments. The most important phase of climatic deterioration before the initiation of polar ice caps took place at or near the Eocene–Oligocene boundary, with the development of the psychrosphere. This marks the end of a very long interval of benign climate.

Quaternary

This is not the place to present a comprehensive review of climatic change of the period in which we live and in which the interests of geologists overlap with those of geographers, glaciologists, archaeologists, anthropologists and indeed the lay public, because we are all concerned about what lies in store for us. A vast amount of information has been gathered over the last century and whole treatises have been devoted exclusively to the subject. All I can attempt is to outline some of the more exciting modern advances and indicate the range of research techniques now available.

Precision of stratigraphic correlation has improved enormously in recent years. Pollen analysis in conjunction with radiometric dating has enabled European and North American glacial and interglacial episodes to be correlated, while even better results are obtained from deep-ocean sediments, where the methods of isotope and magnetic stratigraphy supplement the more traditional approach of correlation by planktonic organisms. While problems persist in effecting accurate correlation of continental and marine deposits sufficient information has been gathered to indicate that gross changes of Quaternary climate were globally synchronous (Kukla 1977).

For many years our knowledge of Quaternary climates was based on glacial and interglacial deposits on land, locally with intercalated shallow marine beds (Flint 1971). Pollen grains of living species from lacustrine and peat deposits have provided the most sensitive climatic indicators, with the technique of analysis pioneered for the Holocene of Scandinavia being extended back in time to earlier interglacials (West 1979).

This method is now being supplemented and cross-checked by analysis of beetle elytra, which have a high preservation potential and hence are abundant fossils in terrestrial and freshwater deposits (Coope 1977). The beetle species exhibit remarkable evolutionary stability and their response to pronounced climatic change, like that of most other Quaternary species, has been to migrate rather than suffer extinction. Thus certain species common in British deposits are found today only in parts of Asia. Their value as climatic indicators is best for warmer interstadials, i.e. intervals of mild climate in otherwise cold periods, because colder conditions are inimical to insect life.

Fig. 7.12 indicates some of the most important results obtained so far by this method, which appears to rival pollen analysis in the quality of its

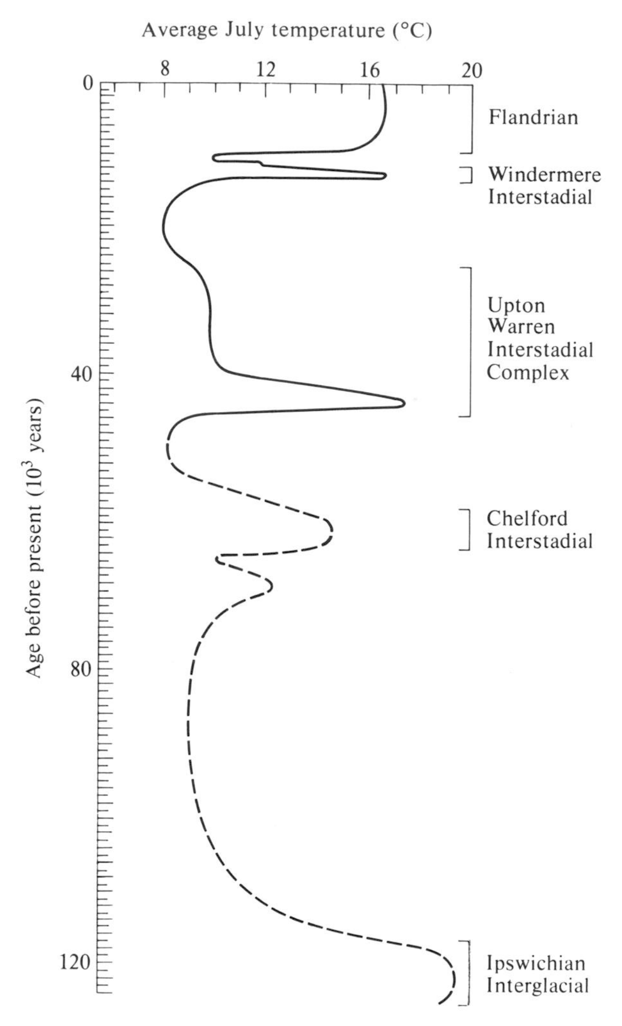

7.12 Variations in the average July temperatures in lowland area of England in the last 120 000 years, as inferred from beetle elytra. After Coope (1977).

results. Perhaps the most significant feature is the great rapidity, in geological terms, of drastic temperature change.

On a global scale the major advance has come from palaeoecological and oxygen-isotope analysis of the planktonic foraminifera of deep-sea cores. Thus in subpolar waters changes in the coiling ratio and percentage abundance of the cold-tolerant *Globigerina pachyderma* are an excellent first-order indicator of oceanic sensitivity to climate through the last few hundred thousand years (Ruddiman *et al.* 1977). This simple quantitative method of estimating proportions is now being supplemented by a more sophisticated method of analysis (Imbrie and Kipp 1971; Imbrie *et al.* 1973).

Factor analysis has enabled five statistically independent and ecologically significant assemblages to be separated, namely tropical, subtropical, subpolar, polar and gyre margin. All but the last of these are related to surface-water temperature. An ecological response model has been developed and a linear equation deduced which allows estimation to be made of surface temperature from information on the relative abundance of the assemblages.

This method of *transfer function analysis* has been applied with impressive results by a team working for the so-called CLIMAP project (McIntyre *et al.* 1976). Their goal was to obtain a world map of water surface temperatures 18 000 years before the present (BP), at the last glacial maximum. Not all the information acquired was readily predictable. Thus while the areas beyond the polar fronts, about 40° N and S, featured summer temperatures much cooler than today, the central Atlantic and Caribbean evidently cooled only slightly. More generally, for the North Atlantic, it appears that the average rate of change is 1–5°C per 1000 years for entire peak-to-peak transitions and 2–13°C per 1000 years for shorter intervals of faster change (Ruddiman *et al.* 1977).

Obviously it is desirable to use where possible faunal in conjunction with isotope data. The general measure of accord is impressive, and numerous pronounced climatic fluctuations have been recognized, into which the classic four Alpine and three northern European glaciations cannot easily be fitted (Kukla 1977). Quite clearly the world standard must come from oceanic studies. That climatic change in the ocean may in detail be diachronous is shown by the ingenious use of a widespread 9300 years BP band of volcanic ash in the North Atlantic as a time marker (Ruddiman and McIntyre 1973). By using this together with sedimentary, faunal and isotope data it is convincingly demonstrated that deglacial retreat of polar water commenced 13 500 years BP in the south-east, near the British Isles, but did not occur in the north-west, near Greenland, until 6500 years BP.

Even more fascinating is the apparently successful test by means of spectral analysis of the Milankovitch model of glaciations and deglaciations controlled by changes in the Earth's orbital geometry (Hays *et al.* 1976). According to this model three orbital parameters affect the amount of solar radiation received at the Earth's surface—the eccentricity of the elliptical orbit (period ~100 000 years), the obliquity of the ecliptic (~42 000) and the precession of the equinoxes (~23 000). Of these, the first is the dominant component. Fig. 7.13 shows a close accord of the faunal and isotope data and a good general agreement with the variations in eccentricity over the last half million years.

Demonstration of a correlation is not of course an explanation, and it

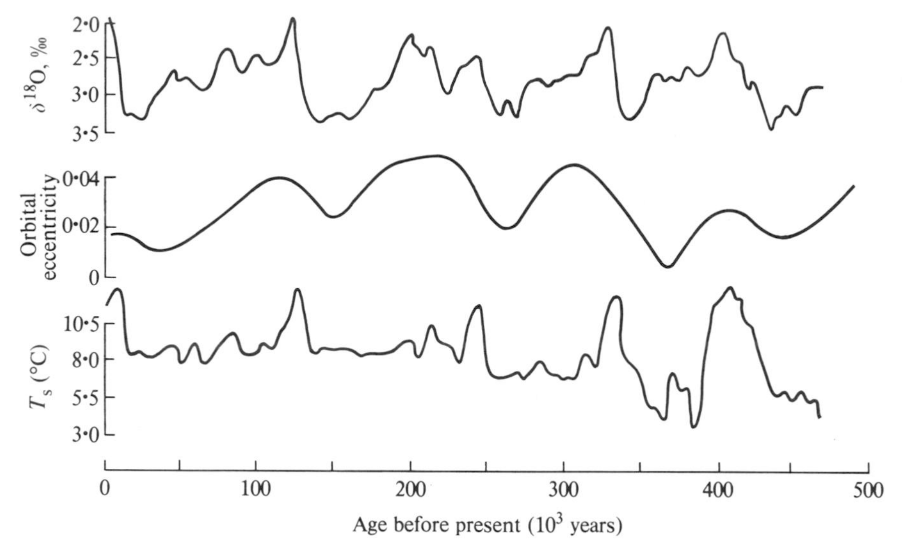

7.13 Late Quaternary variations in $\delta^{18}O$, estimated sea-surface temperature and orbital eccentricity. Adapted from Hays *et al.* (1976).

needs to be shown quantitatively how global variations in solar heat received are translated into climatic change. Schneider and Thompson (1979) have made a fairly encouraging start by developing a quantitative model involving the seasonal cycle of albedo and the zonal distribution of thermal inertia. While there is quite good agreement between their model and palaeontological and isotope data there are two major discrepancies: the temperature oscillations of 0.8°C between glacial and climatic optimum conditions are lower by a factor of seven than those inferred by McIntyre *et al.* (1976) and we should already have entered another glacial period!

General remarks

This chapter will have failed to make one of its major points if the reader fails to appreciate how drastically our knowledge of former climates diminishes back through time. It could be said that information for the Quaternary is excellent, for the Tertiary good, for the Mesozoic moderate and for the Palaeozoic poor, although we have some good documentation on glaciations. While the late Cenozoic researchers forge ahead at an impressive pace, those of us concerned with significantly earlier times usually have to be content with rather vague and qualitative generalizations.

Variations along the humidity–aridity spectrum are particularly difficult to investigate with precision. To what extent, it may be asked, is the latitudinal spread of evaporites a meaningful index of aridity, and is this more significant than the total volume of evaporites deposited? If the Palaeogene evaporites of western Europe do not signify extensive aridity how much can we rely on other evaporite deposits? How much more climatic information can be extracted from, say, coals? These and other matters pose challenging problems for the facies analyst.

No doubt the general problem that will continue to most exercise the minds of palaeoclimatologists will be the causes of ice ages. It may prove that the Milankovitch model is the best one for accounting for short-term glaciations and deglaciations, though as indicated above there are difficulties yet

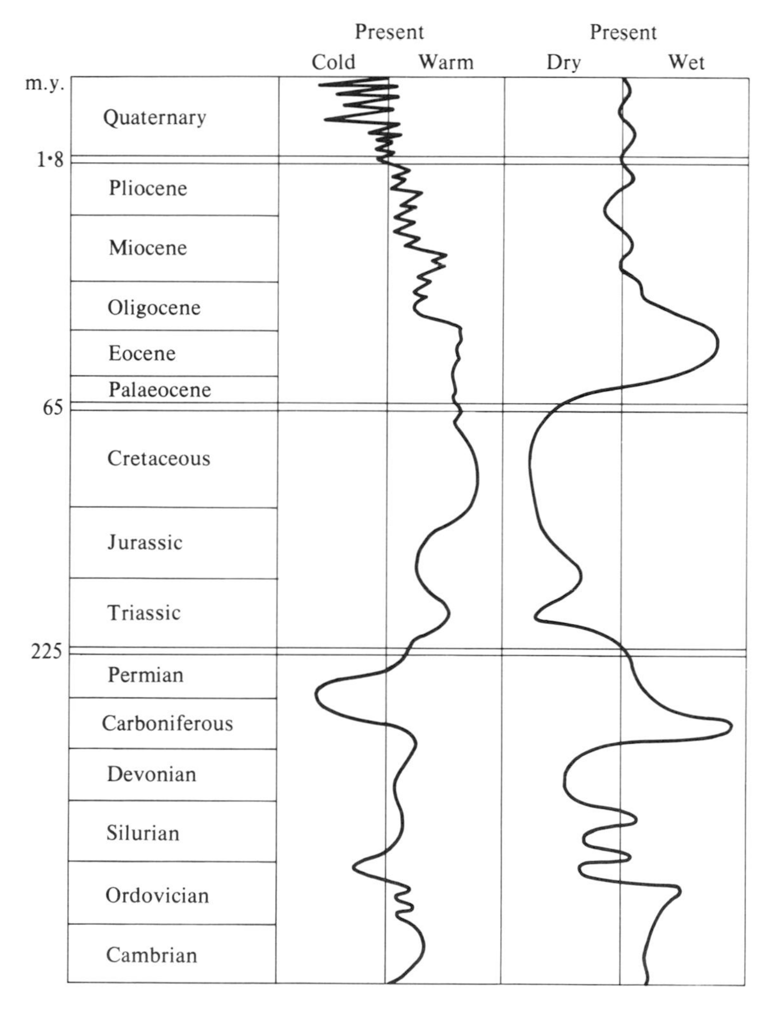

7.14 Generalized temperature and precipitation history through the Phanerozoic, according to Frakes (1979).

to be overcome, but we still have to explain the much longer-term appearance and disappearance of polar ice caps. The best presently available explanation invokes a concatenation of circumstances, such as polar positioning of major continents, ready access to moisture from the ocean and marginal seas, and a preceding period of high global precipitation (Crowell 1978; Frakes 1979; Fig. 7.14).

It is the geologist's task to explain the past rather than predict the future but those with a strong nerve might wish to ponder the following facts. The climate we experience at present has occurred for only a small fraction of the last 120 000 years. Interglacial episodes have rarely lasted longer than 10 000 years, approximately the time of the last significant glacial retreat, and significant temperature oscillations can happen with great rapidity, of the order of several human generations. Not that the longer-term prospect of a return to the benign early Palaeogene climate is much more comforting, because if all the polar ice melted many of the most densely peopled continental lowlands would be flooded.

8. Mesozoic and Cenozoic oceans

There are few more exciting developments in modern geology than deep-sea drilling, which has given us an opportunity for the first time to explore directly the history of the oceans back as far as the late Jurassic. We have already dealt with the record of temperature change but much can also be learned about bathymetry, ocean current systems, episodes of stagnation, controls on pelagic sedimentation and a number of other subjects, sufficient to justify the term *palaeoceanography* for what is in effect a new discipline.

Fluctuations of the carbonate line and palaeobathymetry

Approximately half the present ocean floor is covered by calcareous ooze composed mainly of coccoliths and planktonic foraminifera. We have seen in chapter 3 that its distribution is controlled by the *calcite compensation depth* (CCD), the level at which carbonate input from surface waters is balanced by dissolution. Related to this is the *lysocline*, the zone that separates well-preserved and poorly-preserved calcareous plankton assemblages, and thus held to mark a sharp increase in dissolution rate. The *carbonate line* is the intersection between the CCD and ocean floor topography. At the present day the CCD varies between about 4 and 5 km depth. It is depressed in the high fertility zone of the equatorial Pacific and tends to rise towards high latitudes because of increased rates of carbonate dissolution in cold, corrosive bottom waters. It also rises towards the continents, probably because of increased CO_2 production by benthic organisms and because of greater quantities of organic matter in the sediment (Berger and Winterer 1974).

Because the average elevation of mid-ocean ridges is between 2.5 and 3 km carbonate sediment is deposited on their upper and clay on their lower flanks. Therefore, in the course of spreading, clay comes to overlie carbonate and the boundary between the two types of sediment represents a fossil trace of the ancient carbonate line. Where the spreading direction is oblique to lines of latitude, as in the Pacific, equatorial depression of the CCD may cause further carbonates to overlie the clay in turn, which will subsequently be succeeded by more clay as the ocean floor moves away from the equatorial zone (Fig. 8.1). In order to determine the palaeobathymetry of the carbonate line from deep-sea drilling cores a method known as *vertical backtracking* is used (Berger and Winterer 1974). It is necessary to know the coordinates and depth of deposition of sediments adjacent to the carbonate–clay boundary. The basic tool for this purpose is the *age-depth curve* for spreading sea floor (Sclater *et al.* 1977), the constancy of which through time must be assumed. Taking 2700 m as the likely depth for newly created sea floor, one can obtain an estimate of the depth of deposition if the age of the

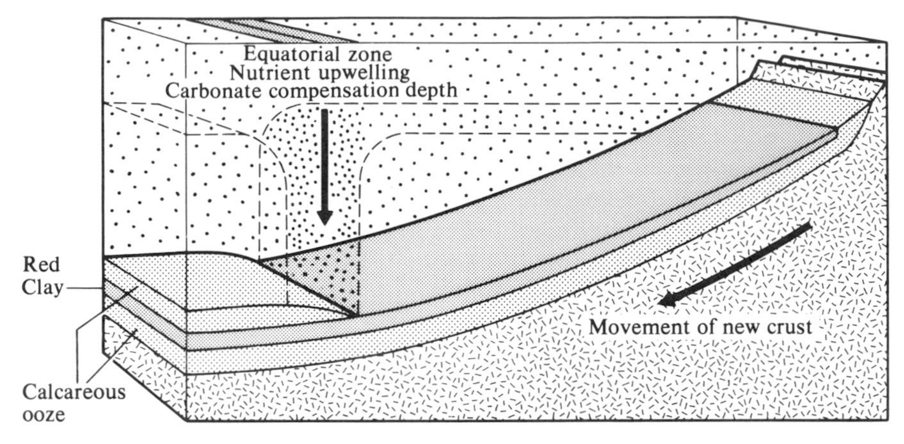

8.1 Model of deposition of sediments on the flanks of the East Pacific Rise, and changing patterns of sedimentation as new crust moves under the equatorial zone. From *The evolution of the Pacific* by B. C. Heezen and I. D. McGregor. Copyright © (1973) by Scientific American, Inc. All rights reserved.

basement and overlying sediment is known. It is also necessary to make an isostatic correction for the sediment and water load; this is approximately one-half (Fig. 8.2).

Consider as an example deep-sea drilling site 137 (Fig. 8.3). The sediment at 5361 m water depth is known to be 105 million years old. The amount of subsidence for 105 million years is estimated from the age-depth curve to be 3161–3160 + 199 m (i.e. $\frac{1}{2} \times 397$ m, which is the sediment thickness) = 2400 m palaeodepth. The subsidence track shows that calcareous sediments accumulated down to about 3500 m.

During the Quaternary, warm phases were characterized by increased dissolution of carbonates in the tropics, so that the CCD was shallower during interglacials. Further back in time marked fluctuations have been recognized (Fig. 8.4). The CCD descended sharply in the Oligocene from a high level of between 3 and 4 km in Cretaceous to Eocene time, and rose to about 4 km in the mid-Miocene, with a further fall subsequently. Berger and Winterer relate these changes to transgressions and regressions. At times of high sea-level stand such as the late Cretaceous and early Palaeogene there should have been substantial $CaCO_3$ precipitation on the wide shelves and this would have caused the CCD to be raised. The Oligocene drop correlates with a major regression and the Miocene rise to a further transgression.

There is no evidence from deep-sea drilling cores to determine the depth of the CCD before the Cretaceous, but if the ophiolite complexes of the Mediterranean region indeed represent fragments of ancient spreading ridges it is noteworthy that the Jurassic sediment directly overlying pillow

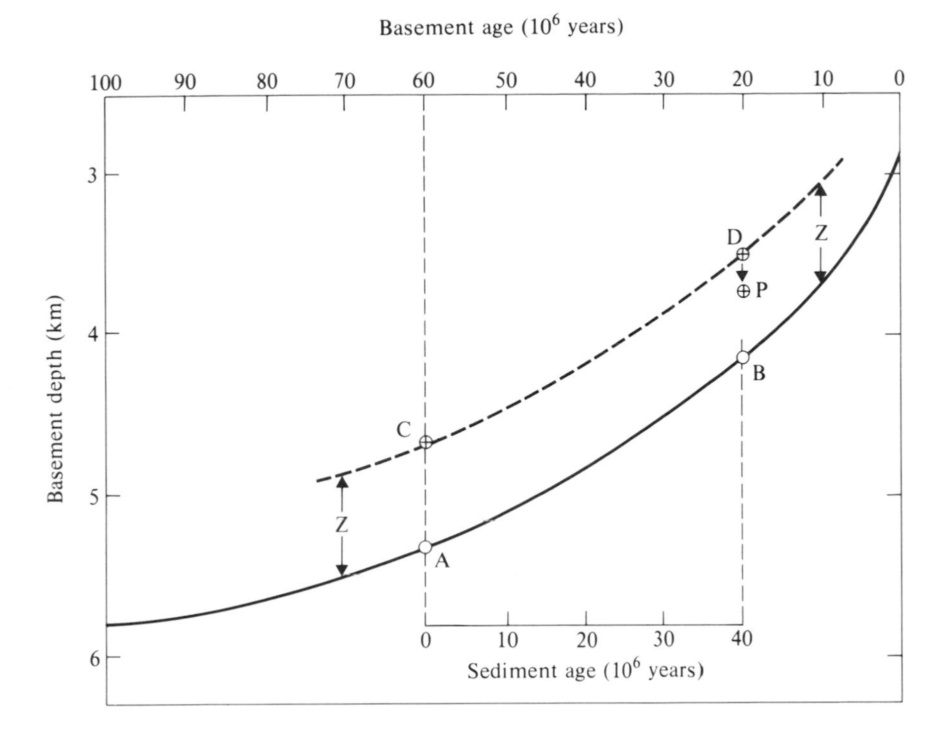

8.2 Palaeodepth determination by vertical backtracking parallel to idealized subsidence track. A and B, Present depth and palaeodepth (40 million years ago) on idealized curve; C, actual site; D, analogue to B on parallel curve; Z, distance between A and C; P, final palaeodepth after correction for isostatic loading. After Berger and Winterer (1974).

basalt is normally radiolarian chert rather than pelagic limestone. This could signify that the CCD was significantly depressed in the Cretaceous and Tertiary as a consequence of the explosive evolution of calcareous micro- and nanno-plankton and hence the vastly increased descent of calcite skeletons from surface waters.

The backtracking technique can also be used, in conjunction with plate tectonic data, to reconstruct the palaeobathymetric and depositional history of the oceans (Sclater *et al.* 1977; Van Andel *et al.* 1975, 1977). Reasonable inferences may also be made about circulation patterns. Thus in the case of the South Atlantic a free circulation of water between north and south was probably not established until Cenomanian–Turonian times, when the transverse Rio Grande–Walvis barrier sank below 1 km depth. By the end of the Cretaceous the establishment of 4 km-deep passages may have allowed free communication with the North Atlantic, and circulation patterns to a depth of 1 km may have differed little from those of today.

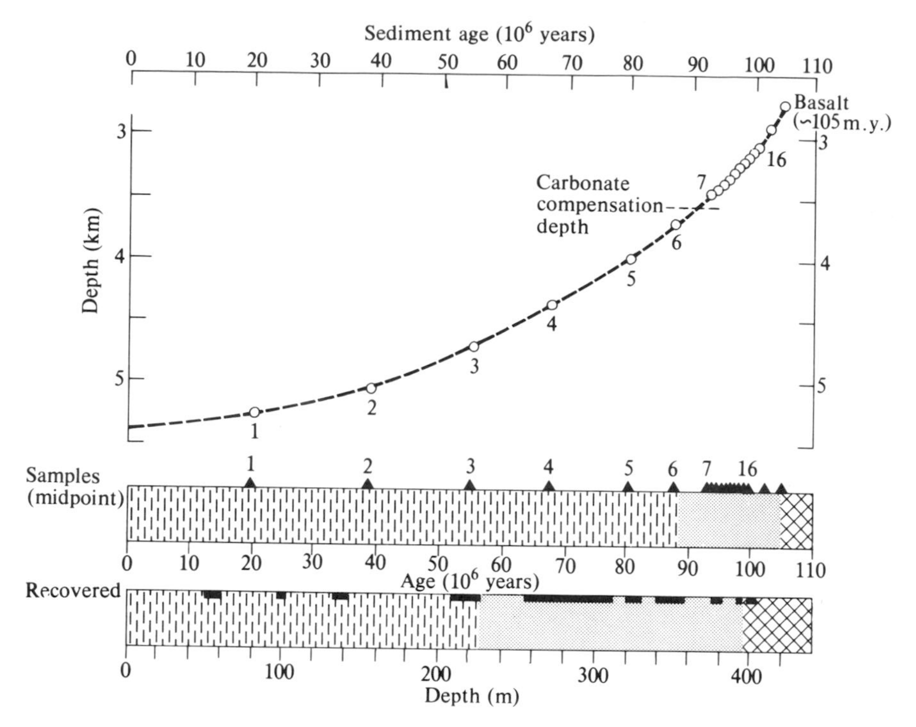

8.3 Backtracking of DSDP site 137, leg 14. Symbols from left to right: clay, calcareous ooze, basalt. Explanation in text. After Berger and Winterer (1974).

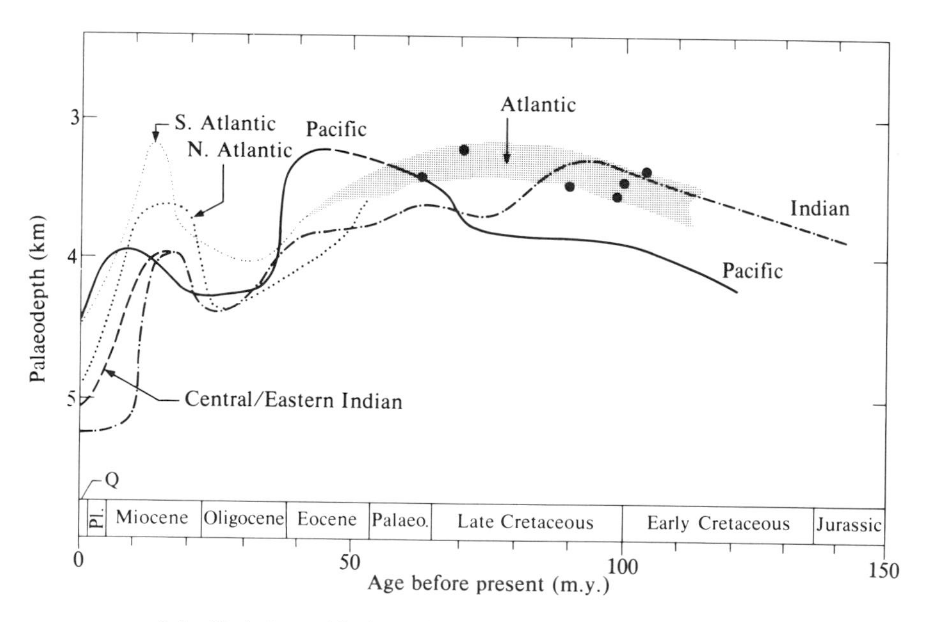

8.4 Variations with time of the CCD. After Van Andel (1975).

Sedimentation rates, hiatuses and ocean currents

A comparatively uniform pattern of Cenozoic pelagic sedimentation rate is inferred by Davies *et al.* (1977) for the Atlantic, Pacific and Indian Oceans, with high rates in the mid-Eocene and mid-Miocene to Recent, and low rates in the Palaeocene to early Eocene and late Eocene to early Miocene. These changes do not seem to relate to transgressions and regressions and the authors prefer to interpret them as being ultimately under climatic/weathering control. However, an overall low sedimentation rate may be inferred from a sequence of quite rapidly deposited sediment with large hiatuses representing substantial intervals of time.

One of the big surprises of the Deep Sea Drilling Project has been the discovery of numerous hiatuses in the Cenozoic record, because the deep-ocean basins have traditionally been regarded as the zones of ultimate sediment catchment. Erosion or non-deposition on the sea floor is determined by a dynamic balance between rates of supply and removal by erosion or corrosion. In pelagic regimes the supply rate is largely determined by productivity of surface waters. Therefore the areas most susceptible to hiatuses should be those beneath the relatively unproductive centres of gyres. Many studies have indicated the existence of deep-sea currents that can inhibit deposition, create sedimentary structures such as ripples and even cause erosion. Therefore if a pattern of hiatuses can be recognized it may be associated with the flow paths of bottom currents.

There is general agreement that the increased incidence of hiatuses in the Cenozoic is associated with more rapid circulation of bottom waters as a consequence of continental separation and climatic deterioration, what has been called 'commotion in the ocean' (Berggren and Hollister 1977); Moore *et al.* (1978) recognize three major episodes for which there is a worldwide concentration of hiatuses.

1. *Maastrichtian–Palaeocene.* This is thought to relate to the rapid opening of the Atlantic and Arctic Oceans.

2. *Eocene–Oligocene.* This was the time when a circum-Antarctic current system first became established, following the separation of Australia in the Eocene and opening of Drake Passage in the Oligocene. With the onset of localized glaciation in Antarctica cold, dense bottom waters began to circulate in the oceans. The opening of the Norwegian Sea–Arctic passage must also have had an important effect on circulation. Creation of the cold, deep psychrosphere in the late Eocene to early Oligocene had a drastic effect on deep-sea benthic ostracods and foraminifera, with many species extinctions (Benson 1975; Corliss 1979).

3. *Middle–Upper Miocene.* The formation of the Antarctic ice cap must have been responsible for substantial cooling of adjacent waters leading to the establishment of rapid currents of cold water moving northwards through the whole ocean system, with free communication being possible through the length of the Atlantic to the Arctic Ocean.

A less important hiatus episode occurred in the Pliocene, and is thought to be associated with the onset of Northern Hemisphere glaciation. In addition, elevation of the Central American isthmus about 3.5 million years ago not only had the effect of severing the Atlantic–Pacific connection, it also caused the deflection of water energy in the Atlantic, giving rise probably to a more vigorous Gulf Stream (Berggren and Hollister 1977). Back in the late Mesozoic, the relative sparsity of hiatuses indicates a less vigorous bottom circulation. Nevertheless there appears to be an important concentration of hiatuses in the Turonian (Fischer and Arthur 1977). The pattern of surface currents must have differed significantly from the present because of the different distribution of the continents. Luyendyk *et al.* (1972) have attempted an experimental simulation of the ice-free Northern Hemisphere circulation pattern for the mid-Cretaceous, using a planetary vortex model. With a continuous equatorial oceanic belt separating Laurasia and Gondwanaland a zonal east-to-west Tethyan current is the dominant feature of the circulation pattern up to 20° N. North of this clockwise-rotating gyres developed in both the North Pacific and North Atlantic.

Siliceous deposits

One of the earliest interesting discoveries of the Deep Sea Drilling Project was a major horizon of Middle Eocene chert in the Central Atlantic, which in the initial stages tended to play havoc with drilling bits. Other chert horizons were located in older deposits, back to the Upper Cretaceous. Cherts of corresponding age were also found in the equatorial Pacific. Although siliceous oozes occur in the present equatorial Pacific and Indian Oceans there is no corresponding zone across the Atlantic (Fig. 3.9). A different pattern of water circulation is the likeliest cause. Ramsay (1973) argues that nutrient-rich water entered the Atlantic from the highly productive equatorial Pacific (Fig. 8.5).

The source of the silica is at least predominantly biogenic, although the breakdown of volcanic debris could also have contributed, at least for the Atlantic cherts. A sequence of recrystallization phases has been recognized by study of core material, from the original biogenous opal via poorly ordered cristobelite to chalcedonic quartz. The Upper Cretaceous and Tertiary siliceous deposits are mostly cristobelite, and Calvert (1974) considers the term *porcellanite* more suitable than chert. Pre-Cretaceous rocks are mainly quartz.

The distribution of Oligocene and Neogene siliceous deposits resembles the Recent, and the sharp cut-off of Eocene cherts in the Atlantic probably relates to Oligocene depression of the CCD.

Anoxic deposits

It was indicated in chapter 3 that the bottom waters of the oceans are well

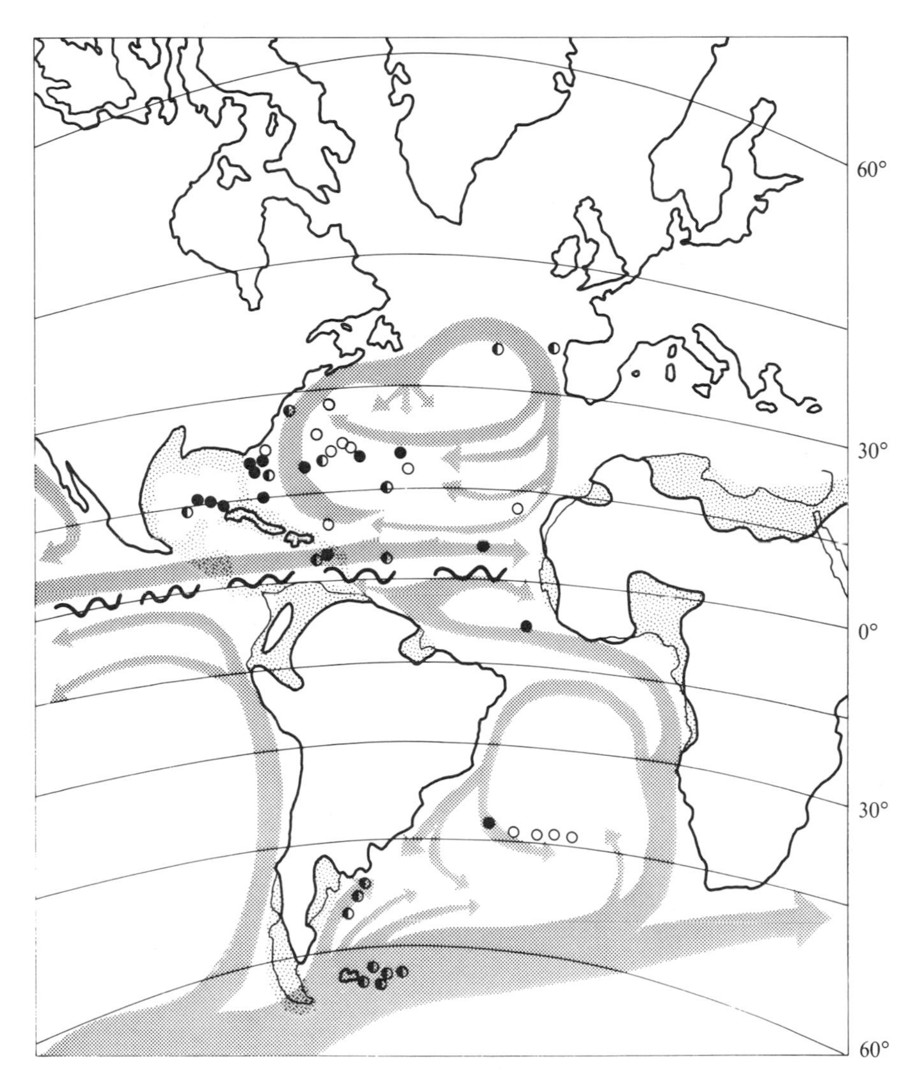

8.5 Chert distribution and proposed circulation patterns for the Eocene Atlantic. ○, Calcareous ooze; ◐, calcareous siliceous ooze; ●, cherts. Stippled regions on the continents represent shelf areas. After Ramsay (1973).

oxygenated except in a small number of land-locked or marginal 'barred basins' separated from the main ocean system by shallow sills. Here the water tends to become stratified and stagnant at depth, so that laminated bituminous muds accumulate below the zone where oxygen is replaced by bacterially-generated hydrogen sulphide. Scientists on some of the early legs of the Deep Sea Drilling Project were therefore intrigued to find thin layers of laminated black bituminous shales and mudstones of mid-Cretaceous age alternating with lighter coloured bioturbated mudstones, in the deep basins of the Central and South Atlantic. Much of the bituminous matter is plankton-derived, but in the eastern Central Atlantic a high proportion of terrestrial plant debris was found. Not surprisingly a barred-basin model was used to account for the formation of these deposits, which seemed plausible enough because at the time the Atlantic was a much narrower ocean, with the likelihood of numerous topographic restrictions to the free circulation of water.

Such an explanation completely fails to account, however, for comparable organic-rich deposits of similar age on the Hess and Shatsky Rises of the north-west Pacific and the Manihiki Plateau of the south Central Pacific. The volcaniclastic Manihiki deposits, of Barremian–Aptian age, are especially interesting because they contain as much as 29% organic carbon. Since Albian–Cenomanian bituminous deposits have also been recorded from the Indian Ocean some kind of worldwide phenomenon is suggested.

To establish this, it is necessary to examine the evidence of rocks of a similar age exposed on the continents, and indeed it turns out that in many cases, notably in North America, on the Tethyan continental margin and in north-west Europe, black bituminous deposits occur at the same horizons as in the ocean. Jenkyns (1980) singles out three major times of worldwide occurrence of such facies: late Barremian to Albian, the Cenomanian–Turonian boundary and, to a lesser extent, the Coniacian–Santonian.

The question can now be posed: what oceanographic situation is likely to promote anoxic conditions in water as shallow as about 300 m (a generous estimate for the Cenomanian–Turonian boundary Black Band in the Chalk of north-east England), and as deep as 2000–3000 m for the Pacific plateau? The answer favoured by Schlanger and Jenkyns (1976) and others is a much expanded oxygen-minimum layer compared with that at the present time (Fig. 8.6). The Schlanger and Jenkyns interpretation also takes into account the evidence of significant mid-to-late Cretaceous transgression, outlined in chapter 6. The vast increase in area of epicontinental seas is thought to have stimulated marine plankton production (see chapter 5), while an important by-product of the continental inundation was the seaward transport of plant detritus derived from richly vegetated land. Increased production of hydrogen sulphide by bacteria caused the oxygen-minimum layer to expand both upwards into the deeper epicontinental sea zone and downwards to envelop certain oceanic plateau.

That large areas of the ocean bottom were more or less anoxic is consistent with the much more sluggish deep-water circulation to be expected in a period of climatic equability, because the prime motor of the rapid circulation of today is polar ice, which cools and renders denser the water that

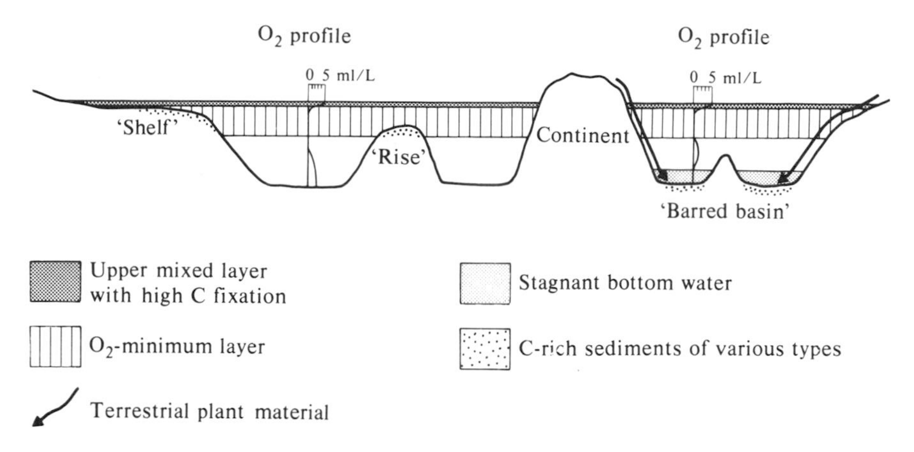

8.6 Stratification during an oceanic anoxic event. After Schlanger and Jenkyns (1976).

comes into contact with it. Furthermore, oxygen is less soluble in warmer water.

Although this kind of explanation seems much more satisfactory than application of the barred-basin model there remain problems to be accounted for. Although there is reasonable correspondence with the Hancock and Kauffman graphs (Fig. 6.10) indicating Albian, Cenomanian–Turonian and Coniacian–Santonian transgressive peaks, why are no anoxic deposits recorded for the major Campanian transgression? The Jurassic was also an equable period and there was an important Oxfordian transgression. Yet the Oxfordian deposits of the western North Atlantic, the oldest yet discovered by deep-sea drilling, include sediments closely comparable to the Ammonitico Rosso of the Mediterranean region, which undoubtedly signifies deposition in well-oxygenated water (Bernoulli 1972).

Further back in time one can only speculate about the occurrence of 'anoxic events' because of the absence of an oceanic record. The likeliest candidate in the Jurassic is the Lower Toarcian, which was a time of significant transgression when bituminous shales were deposited in several continents (see chapter 5). In the Palaeozoic the best example is probably the late Devonian (Famennian), with widespread bituminous shale facies in North America and Europe. Possible late Ordovician and early Silurian examples are discussed by Leggett (1980).

The discovery of the Cretaceous anoxic deposits in the deep ocean provoked Ryan and Cita (1977) to make an estimate of their volume and average carbon content and thereby calculate that they contain an order of magnitude more carbon than is present in all the known reserves of coal and petroleum. They must therefore form an enormous sink for carbon and sulphur (because of the abstraction of iron sulphide). Ryan and Cita argue for a significant increase, in consequence, in the amount of atmospheric oxygen, but it is difficult to conceive of an independent test for this interesting idea.

The tendency towards stagnation of a substantial part of the ocean system is thought by Jenkyns (1980) to have a bearing on the increased abundance during the Cretaceous of phosphorite and glauconite in epicontinental regimes compared with earlier times, because reducing conditions favour their formation (Fig. 8.7). Nevertheless the richest phosphorite deposits, in the belt extending from Morocco to Iraq, appear to be too young, ranging in age principally from the Maastrichtian to the Eocene. Furthermore it is arguable whether upwelling phenomena, to which phosphorite formation is usually attributed, might not be less pronounced at times of sluggish oceanic circulation. On the other hand the phosphorites in question are associated with black shales. Application of Heckel's (1977) model of phosphatic black shale deposition, outlined in chapter 5, might throw more light on the conditions of formation of these rocks.

The Fischer and Arthur model

With the advent of this wealth of new information from the oceans there is a natural temptation to synthesize and speculate on some general cause of change which can account for a wide diversity of observations. Even if such interpretations are found wanting in various respects they are a necessary part of any science because they serve as a guide to future research.

Quite the most general model so far proposed is that of Fischer and Arthur (1977). They suggest that the history of the last 200 million years or so can be understood in terms of cyclic changes between *polytaxic* and *oligotaxic* episodes. Polytaxic episodes are characterized by high organic diversity, higher and more uniform oceanic temperatures, with continuous pelagic deposition and widespread marine anoxicity, eustatic sea-level rises, and heavier carbon-isotope values in marine calcareous organisms and organic matter.

Oligotaxic episodes in contrast are characterized by lower marine temperatures with more pronounced latitudinal and vertical temperature gradients, interruptions of submarine sedimentation, marine regression, a lack of marine anoxicity and by lighter carbon-isotope values. Degradation of pelagic communities is reflected by loss of large predators and lowered diversity, with blooms of opportunistic species. During polytaxic episodes warm, globally equable climates result in reduced oceanic convection, causing expansion and intensification of the oxygen-minimum layer, while colder climatic intervals give rise to increased circulation rates and better oxygenation of ocean waters. No fewer than eight cyclic alterations, each lasting about 32 million years, are recognized back to the Triassic, with the present world being in an oligotaxic phase.

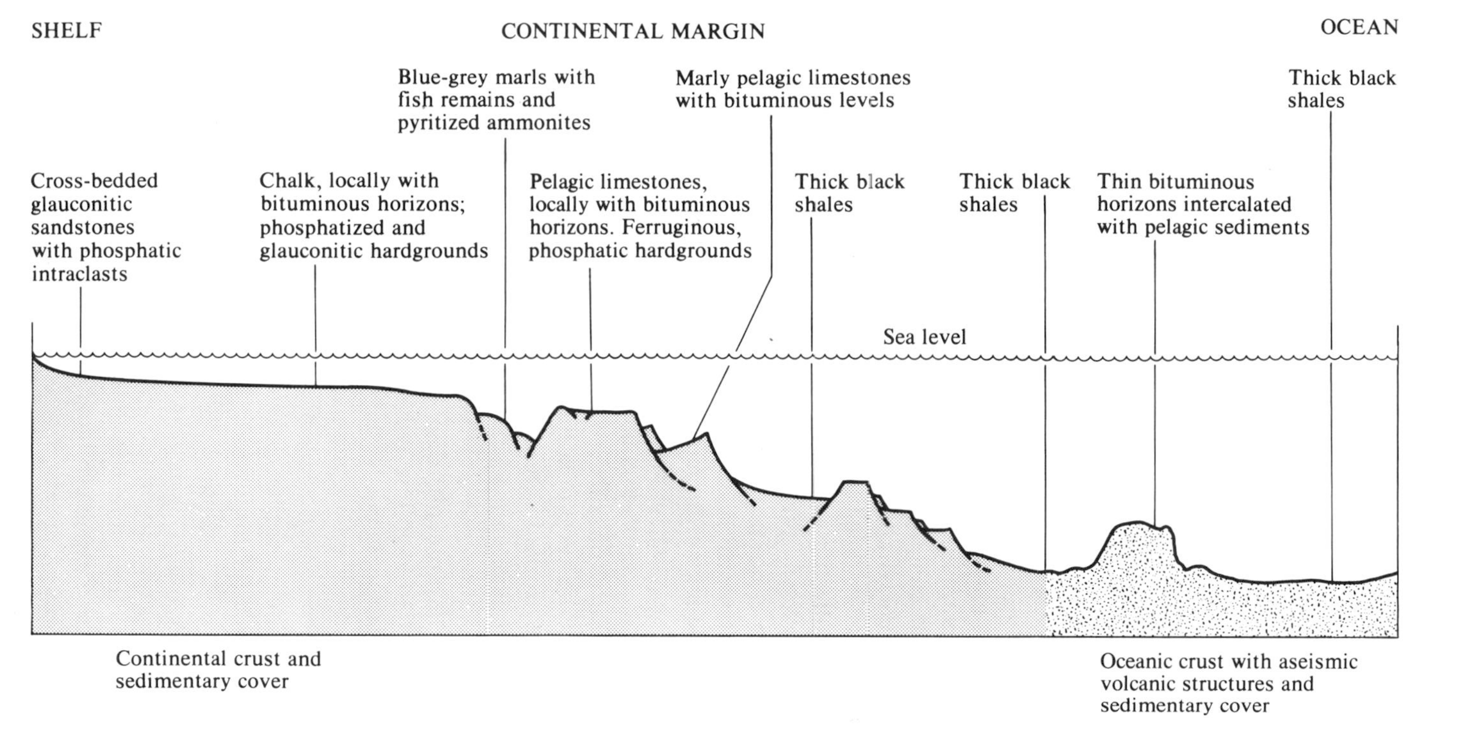

8.7 Model of Cretaceous sedimentation patterns from the deep ocean to the continental shelf during a transgressive episode. After Jenkyns (1980).

Let us consider in outline the evidence that Fischer and Arthur present in support of their model.

Organic diversity can be plotted as a simple count of fossil taxa through time. Fischer and Arthur record a series of broadly coincident peaks and troughs for a variety of plant and animal groups, of which perhaps the most interesting are pelagic invertebrates such as ammonites and globigerinid foraminifera (Fig. 8.8). For the ammonites the most striking features are pronounced diversity troughs at the end of the Triassic and Cretaceous and peaks in the late Triassic and mid-Cretaceous. For the globigerinids, a progressively increasing diversity is interrupted by significant troughs at the end of the Cretaceous and in the Oligocene. The troughs are prime evidence for oligotaxic episodes (which is after all what the word means), also marked in the Cretaceous and Tertiary by the occurrence of a few distinctive species occurring in enormous numbers, notably the coccolithophorid *Braarudosphaera*, which occurs today in certain bays, suggesting that it is tolerant of environments unfavourable to normal pelagic organisms. Such organisms, occurring sporadically in large numbers in high stress environments, indicating a high reproductive capacity, are termed *opportunistic.* Another feature of interest is the maximum size of predators, which is highest during polytaxic episodes; these 'superpredators' are thought to be an expression of community complexity.

Analysis of temperature change is based on oxygen isotopes and is held to display a corresponding cyclicity (Fig. 8.9.). Note in particular the Albian, Eocene and Miocene peaks and Maastrichtian–Palaeocene and Oligocene troughs. Their curve of oxidation state, on the same diagram, is based on the colour and other features of bottom sediment. Thus black sediments signify poorly oxidizing and red sediments highly oxidizing conditions. Aptian–Albian and Coniacian–Santonian low-oxidation alternate with Turonian and Cretaceous–Tertiary boundary high-oxidation states.

Fischer and Arthur also discern temporal fluctuations in the $^{13}C/^{12}C$ isotope ratios of foraminifera and organic matter in sediments. Most carbon in organisms and sediments is the result of photosynthetic fixation, which selectively favours the light isotope. Organic matter in live photosynthetic organisms is therefore enriched in ^{12}C and the atmosphere and hydrosphere are correspondingly depleted. Withdrawal of carbonaceous matter into sediments should therefore drive the oceanic reservoir to the heavier side but the converse effect should occur with carbonate deposition. In pelagic sediments strongly polytaxic episodes such as the Aptian–Albian commonly show a high organic carbon content combined with low carbonate values. The combined effect should be to increase the proportion of ^{13}C in the oceanic reservoir and shell carbonate should reflect this. The proportion of ^{13}C increases markedly in the Albian, which is singled out as a major anoxic event (Fig. 8.9). Marine transgressive pulses (high sea-level stands) are

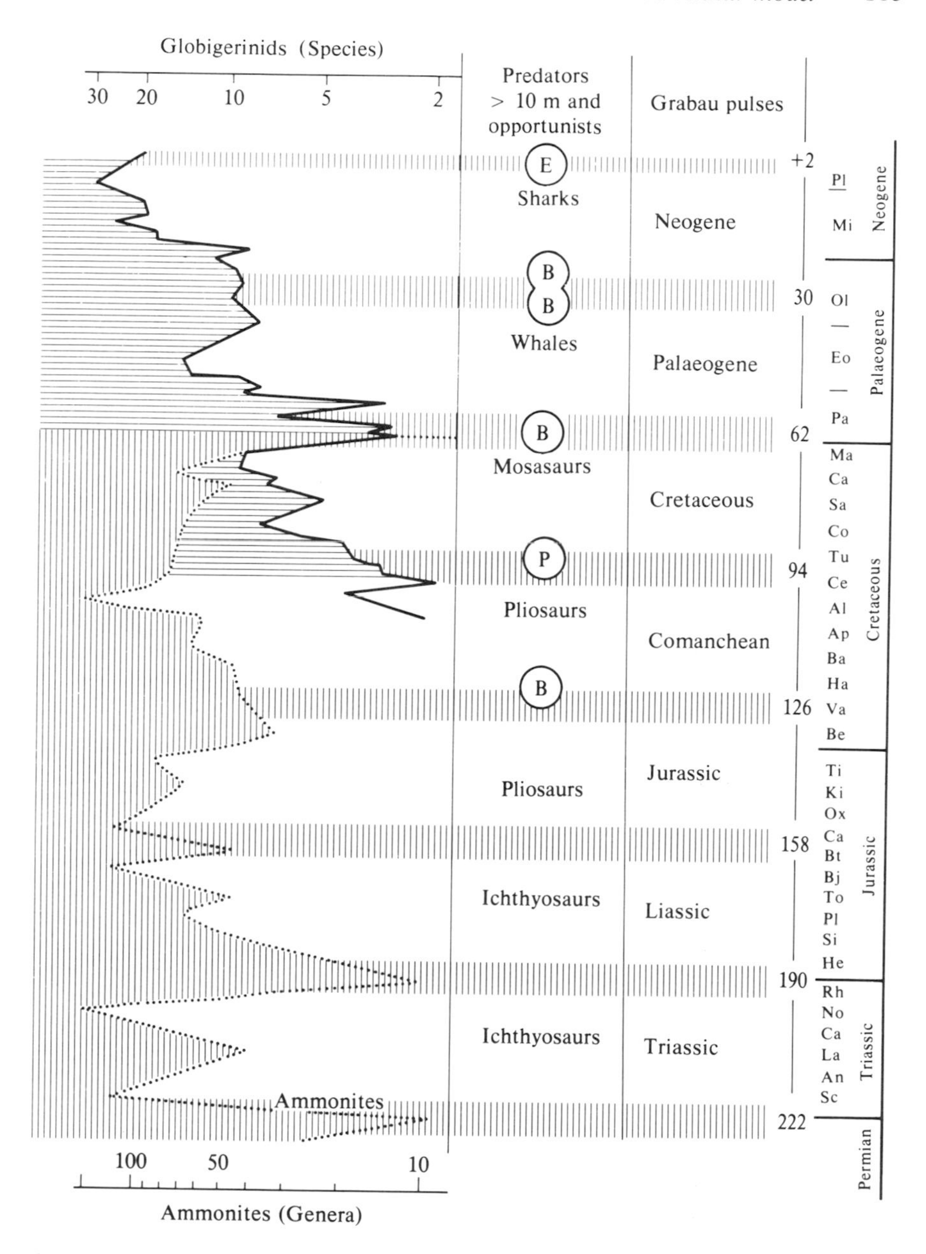

8.8 Fischer and Arthur's 32 m.y. cycle in diversity of important faunal groups since the Palaeozoic, and possible correlation with the occurrence of large predators and opportunists. B, *Braarudosphaera*; P, *Pithonella*; E, *Ethimodiscus*. After Fischer and Arthur (1977).

signified in Fig. 8.8 and correlated rises in the CCD and reduction in the incidence of deep-sea hiatuses indicated in Fig. 8.9.

Evaluation of this evidence is a difficult matter, because each of Fischer and Arthur's curves are generalizations of a mass of sometimes conflicting data and vary from precise quantitative estimates to subjective qualitative assessments, such as the oxidation state of sediments. It is rather easier to pick holes in the general thesis of regular cyclicity with a common cause, such as the following:

1. Variations through time of organic diversity are bound up with the complex subject of mass extinction, which will be discussed at some length in chapter 10. Mass extinction naturally causes a severe diversity reduction and a considerable time must elapse before adaptive radiation of the survivors can restore diversity levels to earlier values. Such extinction affects different groups of organisms in different ways and may have a number of causes, so that each case must be considered on its merits. Furthermore there was a marked increase in diversity of many groups through the late Cretaceous and Tertiary, probably associated with the increased endemism resulting from continental isolation and the greater niche partitioning as latitudinal climatic zonation became more pronounced (Valentine 1973). The effects of this secular 'overlay' must be carefully disentangled.
2. The palaeotemperature curve may be characterized as reasonably accurate for the Tertiary, less accurate for the Cretaceous and downright misleading for the Jurassic, based as it is for the most part on dubious results from belemnites collected on land, there being only an insignificant oceanic record.
3. The CCD curve extends back only to the Eocene. No cyclicity is evident from the, admittedly only generalized, curve(s) further back in time (Fig. 8.4).
4. The evidence for Eocene and Miocene deep-sea anoxic deposits is weak compared with the mid-Cretaceous, although one can presume more active oceanic circulation in the Miocene. The Albian is cited as a time when anoxic sediments were particularly widespread, which makes the well-known Albian red chalks of north-east England appear decidedly anomalous. Conversely, the Bathonian–Callovian is depicted as an oligotaxic episode, but Callovian bituminous shales are widespread in north-west Europe and the late Bathonian–early Callovian was characterized by significant transgression.
5. *Braarudosphaera*-rich sediments in the South Atlantic occur sporadically but not cyclically, as Fischer and Arthur maintain, and their origin is considered by Van Andel *et al.* (1977) to be a mystery.

Vulnerability to criticism on matters of detail is, however, the lot of anyone who attempts such a wide-ranging synthesis, and those who favour outright rejection of Fischer and Arthur's model risk throwing out the baby with the bathwater. My personal view is that this particular baby has as yet too much vitality to deserve such a fate, though I doubt very much that its behaviour exhibits such a simple cyclicity. The Tertiary data look impressive but we are still very much in the dark for earlier periods of Earth history. What can hardly be doubted is that if a prime task of the scientist is to stimulate thought, then Fischer and Arthur have succeeded admirably.

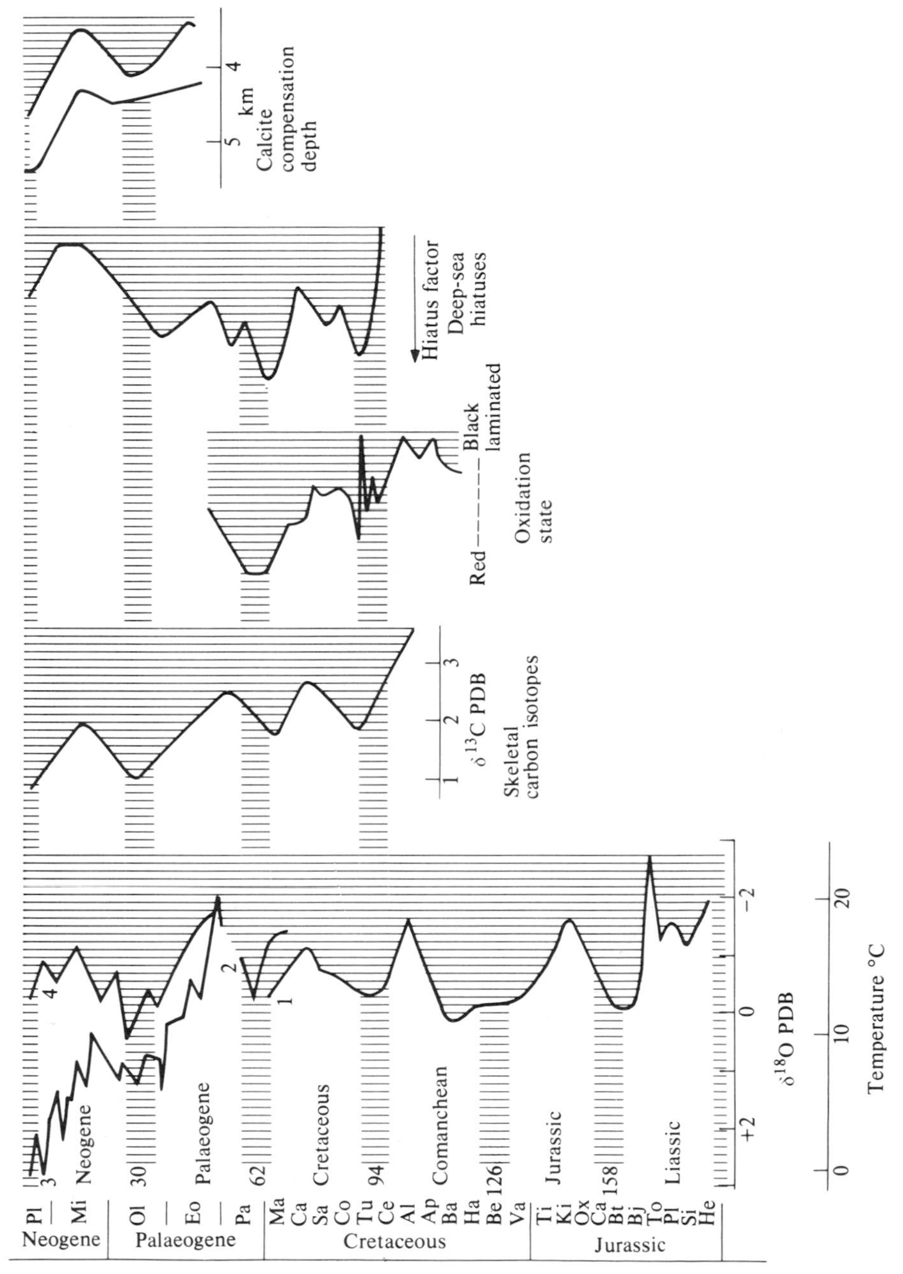

8.9 Synopsis of various geological and chemical parameters in relation to the Fischer–Arthur 32 m.y. cycle. After Fischer and Arthur (1977).

The Mediterranean evaporite basins

Being largely land-locked, the Mediterranean Sea is not normally considered to be part of the main ocean system but it nevertheless attains depths of several kilometres and fascinating information has been revealed in two legs of the Deep Sea Drilling Project, sufficient to justify consideration in this chapter.

One interesting finding has been layers of bituminous mud in the Quaternary deposits underlying the eastern Mediterranean. These indicate repeated episodes of stagnation, which appear to be the result of haline stratification following surface inflow of dilute glacial meltwaters from the Black Sea during terminal phases of deglaciation (Berggren and Hollister 1977).

Much more enigmatic, though, is the discovery that the deep basins of the Mediterranean are underlain by up to several kilometres of evaporites, beneath a cover of several hundred metres of clays. That a salt deposit underlies the Mediterranean had been suspected after a diapiric structure had been revealed by seismic reflection, but its age was controversial. In 1970 the results of deep-sea drilling showed the age of the evaporite deposit to be latest Miocene (Messinian), the same as much thinner evaporites exposed in the circum-Mediterranean countries, that had generally been thought to have been deposited in shallow seas or lagoons. Drastic faunal changes had been related to a so-called Messinian 'salinity crisis'.

The problem immediately posed itself: were the newly discovered evaporites formed from the evaporation of a deep-water Mediterranean Basin, which received constant inflow of water from the Atlantic and maintained its water level only slightly below the worldwide sea level? This would be consistent with the deep-water, deep-basin model of Schmalz (1969). Or were they deposited in shallow water, as presumed for the rocks exposed on land, with subsequent subsidence of several kilometres? Hsü *et al.* (1972) favoured a third, more spectacular, hypothesis, that of desiccation of the deep Mediterranean basins, which were at times completely isolated from the Atlantic; evaporites were precipitated in sabkhas or from desert salt lakes whose water levels were at several kilometres below world sea level.

The essential evidence of this radical view, which according to Hsü (1972a) may be relevant to other 'saline giants' in the stratigraphic record, is as follows. The Messinian evaporites are underlain and overlain by pelagic muds whose fauna indicates deposition in deep water. Yet many of the evaporites exhibit convincing evidence of shallow-water or marginal marine environments. Thus the common calcium sulphate mineral is mainly anhydrite, the high-temperature form found on sabkhas or in salinas, and structures thought to be characteristic of tidal flats or sabkhas such as nodular and 'chicken wire' anhydrite and stromatolites are common in the cores. Desiccation cracks and cross-laminated silt provide additional proof of subaerial exposure. Supporting evidence is drawn from buried gorges and channels beneath rivers flowing into the Mediterranean, such as the Nile, implying a much lower base level at some time in the recent past.

As regards the sudden replacement of the evaporites by Pliocene pelagic muds, Hsü (1972b) pictures the desiccated Mediterranean as a giant bathtub, with water cascading in from the Atlantic via the Straits of Gibraltar on a scale that makes the Niagara Falls seem like a mere trickle (Fig. 8.10).

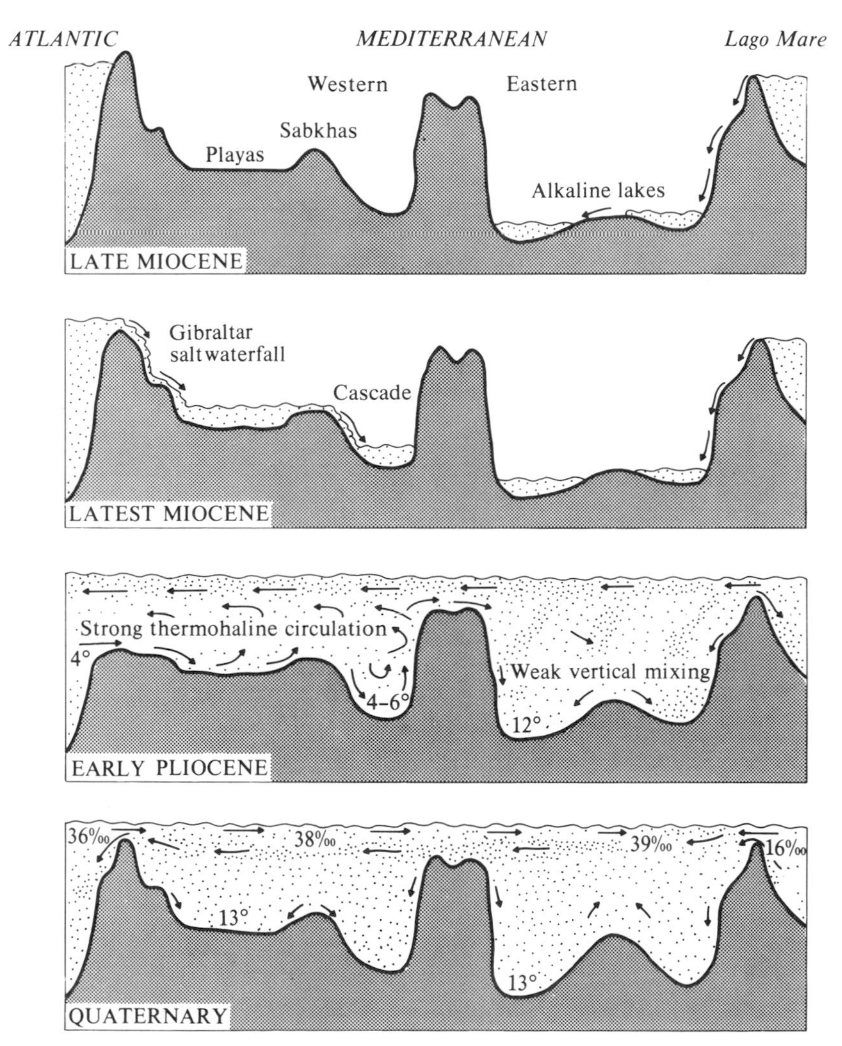

8.10 Idealized and highly speculative E–W sections across the western Mediterranean at different times during the late Neogene. After Cita (1973).

To say that this idea has proved somewhat controversial would be an English understatement. At a conference on the subject held in Utrecht (Drooger 1973) the participants tended to divide clearly into two camps, with the land-based majority favouring the shallow basin and the DSDP geologists the deep-basin desiccation model. There was general agreement, however, on a shallow-water origin for the evaporites and that the terrestrial and deep marine occurrences were intimately related to each other.

Before assessing the respective merits of the two hypotheses it is desirable to present a few more facts. In the comparatively well-studied Sicilian sequence, pre-Messinian Upper Miocene deposits suggestive of deposition in moderate water depths are sharply replaced by a Lower Evaporite unit containing halite, polyhalite and anhydrite unconformably overlain by an Upper Evaporite unit of gypsum and marls. A similar two-fold division of the deep-basin evaporites is suggested by drilling and seismic records, with the main salt sequence (Lower Evaporites) ranging up to several kilometres thick in the Levantine Basin. The overlying Upper Evaporites, consisting of anhydrite plus halite and carbonates, are thinner and more extensive.

In both the onshore and offshore sequences, which bear close resemblances in facies details, there is evidence of both sabhka/salina, already referred to, and marine deposition. Thus there are laterally extensive anhydrite–calcite laminae at some horizons, and a low-diversity flora of coccoliths has been found in some calcitic laminae from onshore sections (Rouchy 1976). Occasional intercalations of clay with marine pelagic microfossils, and evaporitic turbidite slump deposits, provide confirmation of deposition at an appreciable depth. Evidently a whole series of environments must be envisaged, and evaporites appear to have been deposited in a variety of situations with respect to water level.

At the end of the Messinian there was a sudden change in environment to a series of brackish and freshwater lakes, as revealed by ostracods, to which name *Lago Mare* has been given. This is thought to signify a rapid inundation of the Mediterranean from an eastern European inland sea, the Paratethys, which had been separated from the Mediterranean in mid-Miocene times so that the Mediterranean could no longer receive fresh water from many Eurasian rivers (Hsü *et al.* 1977).

A number of weaknesses of the deep-basin desiccation model have been pointed out. If such an enormous waterfall existed at the Gibraltar portal there should have been significant scour at the western end of the Mediterranean, for which no evidence has been forthcoming. The close similarity, except in thickness, of the onshore and offshore evaporites suggests lateral continuity of deposition, which implies several kilometres of Pliocene–Quaternary uplift of the onshore areas, so that large vertical tectonic movements cannot be discounted. On the other hand, if the onshore evaporites were deposited at a much higher level with respect to the ocean surface, why were these deposits not swept off the shelf into the deep basin? The evidence of buried river channels is indecisive. Nowhere, except in the Nile Delta which has surely subsided, do they exceed a few hundred metres in depth, whereas the floor of the desiccated Mediterranean is claimed to have lain at 2 km or more. Furthermore their age of incision has not been precisely established and they may well post-date the Messinian.

Clearly the bathymetric evidence provided by microfossils is crucial. It proved possible on DSDP leg 42A to drill through the Balearic Basin evaporites to the underlying Miocene pelagic clays, which contain benthic foraminifera and ostracods suggesting a depth of 1500 m in the mid-Miocene. This convinced the leg scientists (Hsü *et al.* 1977) that the basin was already in existence before the Messinian, having been created by rifting in latest Oligocene or earliest Miocene time, and was held to provide decisive support for the deep-basin model.

Determination of the precise depth of deposition of sediments is one of the thorniest problems facing the geologist and one wonders just how reliable a bathymetric guide the cited fossils provide. No one seriously doubts that both the sub- and supra-Messinian pelagic clays were deposited in an appreciable depth of water, but whether this depth was of the order of a few hundred or a few thousand metres is much less certain. Since considerable uplift of such clays in the circum-Mediterranean countries has manifestly taken place it is surely not unreasonable to propose that comparable subsidence might have occurred in the Mediterranean basins. This would still imply a water level for the evaporites well below that of the ocean, but much less than proposed in the extreme formulation of the deep-basin desiccation model, which denies the importance of epeirogenic movements. If the buried channels were indeed cut in the Messinian, they might give an indication of the true water level. As with so many scientific controversies which are initially polarized, the best solution may lie in some degree of reconciliation. The geotectonic implications of the alternative shallow- and deep-basin models are considerable, because the former implies that the Neogene sediments of the Balearic Basin are underlain by continental crust whereas the latter demands a basement of oceanic crust.

A broader perspective is brought to bear on the problem by Adams *et al.* (1977) who treat the Messinian salinity crisis in the context of eustatic changes, and the continental collisional plate movements that virtually sealed off the Mediterranean from the Atlantic. Entry of water from the Atlantic could have been effected through either the Gibraltar, Rif or Betic straits, of which the last is thought the most probable. The salinity crisis could not have taken place until the Andalusian Portal became shallow enough to prevent reflux of water from the Mediterranean and so allow salts to concentrate. Initially there was perhaps a wide shallow channel with a west–east current and subsequently a series of cataracts.

There are worldwide indications of late Miocene regression and other evidence, cited in chapter 7, of formation of an Antarctic ice cap in the mid-Miocene. A sea-level drop in consequence of perhaps 50–70 m is consistent with the magnitude of ice-volume increase as recorded in isotopic data. It is calculated that isolation of the Mediterranean resulted in the precipitation of more than a million cubic kilometres of gypsum and salts from a volume of water thirty times that of the present Mediterranean. This implies extraction of about 6% of the dissolved salts of the world ocean. Such an event could conceivably have induced Miocene glaciation by lowering the salinity in high latitudes sufficient to raise the freezing point of sea water. However, it is considered more probable that Antarctic glacial expansion was responsible for lowering of sea level.

The work of Adams *et al.* suggests the following thought. By analogy with the Quaternary, one might expect that climatic fluctuations promoted minor sea-level changes as a result of waxing and waning of the ice sheet. The evaporite record suggests a fine condition of balance between subaerial and subaqueous conditions. Given a western sill in the right level, small eustatic falls during times of colder climate could have promoted desiccation and rises at times of climatic amelioration could have caused renewed flooding from the Atlantic. Support for such an interpretation comes from the work of McKenzie *et al.* (1980) on the Tripoli Formation of Sicily, in a contribution to project 96 of the International Geological Correlation Programme, devoted to the Messinian. The Tripoli Formation directly underlies the Messinian evaporites and consists of cyclic alternations of diatomite and dolomitic clays. Isotopic data from the dolomites imply the former presence of evaporites in the clay bands. It is argued that diatomaceous sedimentation took place during high, and evaporitic sedimentation during low stands of sea level related to waxing and waning of Antarctic ice, with Atlantic water periodically entering the Mediterranean from the western sill. The final, Messinian, phase was one of total restriction which led to the desiccation of the deep basins and the onset of the salinity crisis.

In support of the deep-basin desiccation model, Hsü (1972b) has quoted Sherlock Holmes: 'When you have excluded the impossible, whatever remains, however improbable, must be the truth.' To that I am tempted to reply: 'My dear Watson, it's not so elementary.'

9. Precambrian environments

So far attention has been confined to that most recent eighth of Earth history known as the Phanerozoic, and with good reason. In the absence of biostratigraphic control comprehensive facies analysis is impossible and, as we are all aware, a large proportion of Precambrian rocks are igneous or metamorphic and the sedimentary record is often obscure. Nevertheless the situation is far from hopeless and enormous strides in knowledge have been made in recent years. Many Proterozoic sequences are now being studied with profit, and shown to preserve many features amenable to environmental diagnosis, while even Archaean sediments are beginning to yield their secrets. It has also become apparent that Precambrian sediments are far from being abiotic, though the fossils discovered are humble or sparse compared with their Phanerozoic descendents.

The kinds of questions that Precambrian sediments and fossils can help to solve are some of the most fascinating that we can pose about the Earth, concerning as they do matters of such general interest as the early world climate and the evolution of the continents, atmosphere, hydrosphere and biosphere. While our knowledge inevitably remains sketchy considerable advance has already been made in achieving a better understanding of these subjects. In particular the information acquired recently can be used to test a number of more or less speculative hypotheses that have been put forward about the early condition and evolution of the Earth.

Distinction of marine and non-marine environments

Whereas in most Phanerozoic successions there are usually some fossils to allow distinction between marine, lagoonal and lacustrine environments there is no such aid for the Precambrian, so that this is one of the biggest uncertainties in our knowledge. It allows some geologists to infer, for instance, that the banded iron formations, among the most important of Precambrian deposits, are marine in origin, while others argue that they are lacustrine. Geochemical criteria may prove helpful because, as noted in chapter 1, boron and other trace elements are consistently more enriched in marine than non-marine argillaceous sediments through a long time range back from the present (Potter *et al.* 1963). This field of research remains, however, to be explored for the Precambrian.

In the absence of diagnostic fossils, the best hope lies in a detailed examination of the sediments. For instance, Clemmey (1978) studied a 30-m unit of dolomitic and siliciclastic sediments in the late Precambrian of the Zambian Copperbelt and inferred a range of marine and lacustrine environments. The Orebody Member of the Roan Group exhibits indications of a calm-water body. Frequent subaerial exposure is signified by beautifully preserved desiccation cracks, rainprints and gas-bubble bursts. There are also *syneresis cracks*, which are randomly oriented incomplete cracks often having a characteristic birdsfoot shape. They are believed to form during episodes of subaqueous dewatering and have been recorded from undoubted lacustrine environments. The various cracks often have a filling of dolomite and anhydrite. Such evidence clearly indicates conditions of extremely shallow water, but a marginal marine environment cannot be conclusively discounted.

Glaciations

Hunt (1979) has produced a novel meteorological model for the Precambrian Earth, based on the belief that the rotation rate was significantly faster then. There is a well-known astronomical theory of lunar tidal friction, which holds that the rotation rate has gradually slowed down through time as the Moon has receded, and data on the growth of fossil corals and bivalves appear to support this for the Phanerozoic (Scrutton 1978). Hunt, however, cites the work of Mohr (1975) to support the idea that the Earth's rotation rate has slowed down since the Precambrian more than twice as rapidly as inferred from tidal friction theory.

Mohr inferred from his study of stromatolite growth layers in the Biwabik Formation in Minnesota (approximate age 2000, not 1500 million years (m.y.) as stated by Hunt) that there were only 25 days per month at that time, from which he calculated between 800 and 900 solar days per year, the figure used by Hunt. For reasons not explained, Hunt chose to ignore the work of Pannella (1975) on roughly contemporary stromatolites in Africa, suggesting a much lower rate of slowing down. Moreover, there is considerable doubt about the value of stromatolites as palaeontological clocks (Scrutton 1978).

Despite the problems of the data base it is worth considering Hunt's model, which may retain a general validity even if the quantitative assumptions are inaccurate. A more rapidly rotating Earth should induce a reduced poleward transfer of heat in the troposphere, reduced wind strength and incidence of oceanic gyres and upwelling. The net effect would be a general warming of the tropical oceans and atmosphere and a colder, somewhat arid polar region. There should in consequence have been substantial glaciation throughout the Precambrian.

Now inferences from the rock record about Precambrian climates are extremely difficult, with the notable exception of glaciation. There is in fact nothing like the continuous record demanded by Hunt's theory, but clear indications of two important episodes (Frakes 1979).

The earliest known glaciation episode took place around 2300 m.y. ago, with an error of several hundred million years, and is recorded in North America, South Africa and Australia. The best data come from the Gowganda Formation of Quebec and Ontario, with tillites, dropstones in laminites and striated floors. Much more striking, however, is the evidence of extensive glaciation in the late Precambrian, over a period from about 950 to 615 m.y. ago, more than half the time span of the whole Phanerozoic. Strata regarded as signifying glacial conditions have been recorded from every continent except Antarctica.

Glacial deposits are well developed and exposed in Australia and include the Sturt Tillite in the Adelaide Geosyncline. The various tillites are locally very thick, and dropstones also occur; glacial pavements are found in the Kimberley region. The European deposits have been studied in considerable detail. They include the well-known tillites and associated deposits, with at least one striated surface cut in unconsolidated sand, of the Varanger Subgroup of Finnmark in northern Norway, interpreted as ground moraine deposited on the margin of the sea, and dropstone laminites. Permafrost conditions are signified in the Port Askaig beds of Scotland by polygonal sand wedges.

One of the big problems is that, unlike in the case of Phanerozoic glacial deposits, the associated sediment is frequently dolomite or limestone, which are often held to signify warm conditions. In fact Schermerhorn (1976) considers that the presence of dolomite disproves glaciation, but it is difficult to see how extensive tillite and dropstone deposits in cratonic regimes can signify anything but cold conditions. Another problem is the almost global extent of the deposits, with palaeomagnetically determined low latitudes for the European, Australian and African examples. On the other hand intercontinental correlation is poor and imprecise, so that, while the evidence for glaciation is compelling, there is no good reason to think that it was worldwide in extent at any one time. It is highly likely, in view of the long time span, that there were a whole series of glaciations, with the major centres perhaps located in Central Australia, west Central Africa, the Baltic Shield, Greenland/Spitsbergen, Siberia and north-western North America.

It could be that Hunt's model could help to ease the problem but it is then necessary to account for the lack of a glacial record through most of the Proterozoic. On the contrary, the abundance of carbonates with stromatolites and ooids and the evidence of evaporite deposition suggests, by comparison with the present day, a relatively warm climate.

Tidal range

Another consequence of closer proximity of the Earth to the Moon in the Precambrian is that tidal forces should have been stronger. However, as observed earlier in this book, tides are strongly influenced by a variety of terrestrial factors including physiography of the marine basins and resonance, and tidal range is extremely difficult if not impossible to determine for the geological past. It has been suggested that the greater height of many

Precambrian stromatolite domes compared with Phanerozoic examples might be an indication of greater tidal range, but this interpretation is dubious because many other factors could have controlled the shape and size of stromatolites (Hoffman 1973). Von Brunn and Hobday (1976) claim to have recognized an example of late Archaean tidal sedimentation in the Pongola Supergroup of South Africa (age 3000 m.y.). Their interpretation calls for upward regressive sequences from low intertidal sandflats with herringbone cross bedding and reactivation surfaces to mid- and high-tidal mudflats. Application of Klein's (1971) model for determining palaeotidal range leads to figures between 12 and 25 m. The latter figure is in excess of any present-day tidal amplitudes, and is held to support ideas on the early Precambrian capture of the Moon. However, for the reasons given in chapter 2, I believe Klein's model to be invalid.

Evolution of the continents

Based essentially on a thermal model, Hargraves (1976) has proposed that the separation of the primordial shells of the Earth into continents and oceans began only about 3700 m.y. ago and continued through most of the Precambrian, and that the continents did not begin to emerge from the sea until after 1400 m.y. ago. This interpretation has been strongly criticized by Windley (1977b) and Knoll (1978) on a number of grounds.

Mineral assemblages in exposed Archaean rocks suggest formation at depths in excess of 30 km, where the depth to the Mohorovicic Discontinuity is of normal continental amount. The sedimentary facies of early Proterozoic rocks indicates the existence of extensive cratons that witnessed shallow marine deposition and terrestrial erosion and deposition. There is in fact abundant evidence that growth of the present-day continents in terms of surface area and thickness was almost complete 2500 m.y. ago.

A related problem concerns whether a primordial sialic crust has been continuously recycled through geological time or whether new continental crust has been generated by chemical differentiation of part of the upper mantle. Strontium-isotope data convincingly support the latter interpretation and Moorbath (1977a) considers it certain that typical continental crust was in existence as long as 3800–3700 m.y. ago.

Archaean and Proterozoic sediments typically have strikingly different facies and tectonic settings. The most characteristic Archaean sediments are greywackes, which occur in association with volcanic rocks in *greenstone belts* embedded in larger areas of granite-gneiss terrains (Fig. 9.1), whereas Proterozoic sediments are laterally more extensive, often little deformed and include a high proportion of dolomites and limestones. There are many reasons for regarding the Archaean–Proterozoic time boundary (2500 ± 200 m.y.) as the most important turning point in continental evolution. It appears to signify a changeover from a permobile regime to a platform-

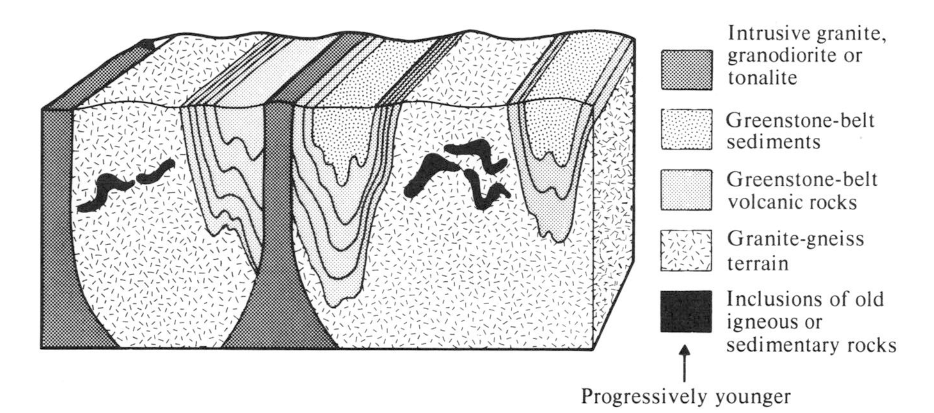

9.1 Highly schematic diagram of a typical Archaean shield area, to show the relationship of greenstone belts to other geological terrains. From *The oldest rocks and the growth of continents* by S. Moorbath. Copyright © (1977) by Scientific American, Inc. All rights reserved.

geosynclinal style of tectonics comparable to modern times (Windley 1977a). Thus a good case has been made for an early Proterozoic aulacogen in northern Canada (Hoffman *et al.* 1974 and Fig. 9.2). Most geologists would probably accept that plate-tectonic processes operated back to the early Proterozoic but whether they also operated in the Archaean is more controversial. Moorbath (1977a) argues that the differences were in degree rather than in kind, with probably more and smaller continents.

There are as yet only few detailed studies in which Archaean sediments have been interpreted in the light of modern sedimentological knowledge. There are daunting handicaps to such studies in greenstone belts, such as intense structural deformation and a general absence of long, continuously exposed stratigraphic sequences. Walker and Pettijohn (1971) interpret a thick series of graded quartz-rich greywackes and interbedded argillites, commonly 10–100 cm thick, in the Minnitaki Basin of Ontario as a turbidite sequence comparable to much younger examples. There are also graded conglomerates with abundant granite boulders which are thought to represent a proximal resedimented facies. As greenstone debris is very rare, the source area is presumed to have been devoid of a greenstone cover. Precise environmental diagnosis is, of course, impossible and there is no way of determining the original size of the depositional basin. There is no evidence, however, of shallow agitated-water deposits.

Resedimented deposits of the turbidite-debris flow association appear to be widespread in greenstone belt terrains but continental and shallow marine deposits are poorly defined. They may include arkoses and conglomerates of the Moodies Group of South Africa and the Kurrawang Conglom-

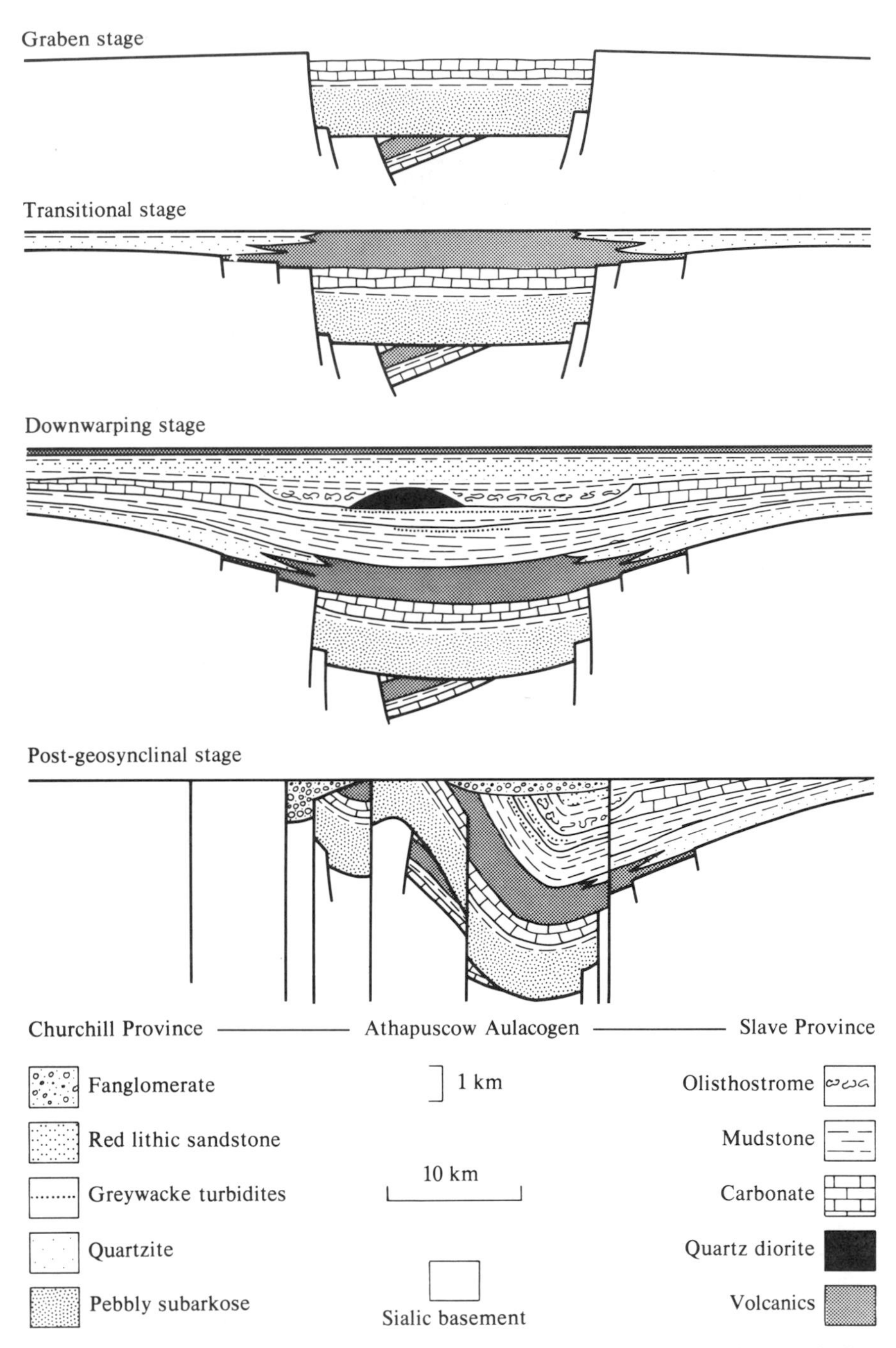

9.2 Schematic transverse cross-sections showing the inferred evolution of the early Proterozoic Athapuscow Aulacogen in northern Canada. After Hoffman *et al.* (1974).

erate of Western Australia (Turner and Walker 1973). Since mixed arenaceous and rudaceous deposits can form in a variety of settings Turner and Walker stress the need to examine also the associated facies and the position in the overall sequence. This, of course, is the essence of facies analysis.

Hunter (1974a, 1974b) has made a comprehensive study of the classic South African Precambrian. The sequence is as shown in Table 9.1.

TABLE 9.1 The Precambrian sequence of the Kaapvaal Craton (see Hunter 1974a, 1974b)

Supergroups			*Thickness (Maximum for each supergroup)*
Waterberg Transvaal Ventersdorp Witwatersrand–Dominion Reef Pongola		Proterozoic	43 km
Swaziland	Moodies Group Fig Tree Group Onverwacht Group	Archaean	21 km

A rigid plate capable of sustaining and preserving ensialic basins, with gently dipping, largely unmetamorphosed deposits, developed at about 3000 m.y. ago, at a time when most other continental areas were still at the greenstone belt stage. This indicates that the Archaean–Proterozoic boundary has no precise time significance across the world.

In the Swaziland Supergroup sediments and volcanics occur in cyclic succession. Petrographic examination of the greywackes of the Fig Tree and Moodies Groups indicates unroofing of a granitic-metamorphic terrain.

The Proterozoic supergroups reflect a general decrease in energy level with time, as demonstrated by the increase in volume of finer siliciclastics and carbonates in successively younger basins, and by estimates of the rate of vertical movement. This latter is achieved by dividing the estimated stratigraphic thickness by the time span, to give kilometres per million years or millimetres per year. There was an apparent migration of the sedimentary basins across the craton from south-east to north-west, with the sediment source being mainly to the north.

Fig. 9.3 illustrates what are interpreted as episodic transgressions and regressions of ? shallow sea over the craton, with a secular change suggestive

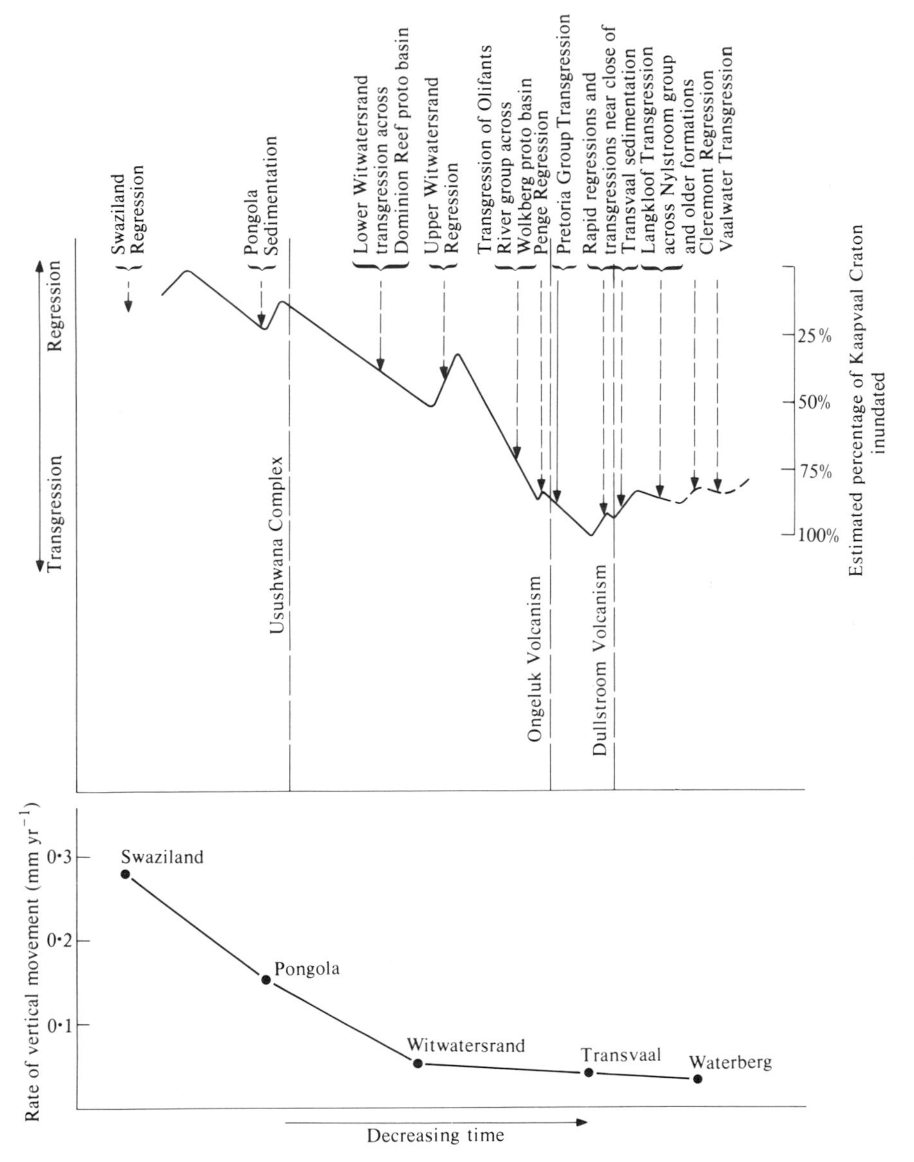

9.3 Plot of estimated percentage of Kaapvaal craton inundated between 3000 and 1800 m.y., and plot of rate of vertical movement with time in the sedimentary basins. Adapted from Hunter (1974b).

of progressive inundation. It would be interesting to know if comparable events can be discerned in other Proterozoic cratons, because this would indicate the reverse of what is demanded by Hargraves' (1976) model.

Evolution of the atmosphere and hydrosphere

It is generally accepted that the atmosphere and hydrosphere were produced by outgassing processes in the primitive Earth. Since volcanic exhalations at the present day lack free oxygen, the initial atmosphere must have been anoxic, and the gas has been generated subsequently either abiotically, by photodissociation, or biotically, through photosynthesis of primitive plants. The common belief is that photosynthesis was the dominant factor, and that the oxygen content of the atmosphere increased at the expense of carbon dioxide progressively through time, but the amount of oxygen at a given time in the early atmosphere is still a matter of considerable controversy. Berkner and Marshall (1965), in a classic paper, postulated that the explosion of Metazoa at the start of the Cambrian was triggered by the achievement of an oxygen content of only 1% of the present level. This extreme view is now discredited and most geologists consider that the atmosphere had reached something like its present oxygen level well before the end of the Precambrian, but there agreement ends. Let us consider the evidence from the stratigraphic record bearing on this problem and the related matter of the character of the hydrosphere.

BANDED IRON FORMATIONS

By far the most voluminous deposits of sedimentary iron ore are represented by the so-called *banded iron formations* (James and Sims 1973), with the Lower Proterozoic examples forming large stratigraphic units ranging up to hundreds of metres in thickness and thousands of kilometres in lateral extent, and world-wide in distribution. They include the deposits of the Lake Superior region and Labrador in North America, Minas Gerais, Brazil, the Ukraine, South Africa and the Hamersley Basin of Western Australia. While the greatest development is in the Lower Proterozoic (2600–1800 m.y.) there are also Archaean examples in greenstone belts ranging back to the oldest known sediments, in the Isua region of West Greenland, dated at 3760 m.y. (Moorbath 1977a). Banded iron formations are very rare and limited in extent in rocks younger than 1800 m.y., and a few Palaeozoic examples have been claimed.

As the name suggests, the most characteristic feature is a fine banding composed of alternating layers of chert and iron minerals, usually in the range of 0.5 to 3 cm per band. These may in turn contain fine laminae of the order of tens of microns in thickness. The iron minerals include haematite, magnetite, siderite and various silicates, such as greenalite, rare chamosite and the products of mild metamorphism stilpnomelane and minnesotaite. Compared with Phanerozoic ironstones the alumina content is very low, and detrital siliciclastics are rare. Although the texture is normally fine grained and homogeneous both ooids and cross bedding occur, and stromatolites have been recorded from the Biwabik Formation of Minnesota.

Because these distinctive rocks virtually disappear from the stratigraphic record after the early Proterozoic there is a natural temptation to assume that they signify an environment very different from anything that was widespread in the last 1800 m.y. The problems in accounting for Phanerozoic ironstones appear modest in comparison with those of the banded iron formations and not surprisingly a wide variety of hypotheses of their origin have been put forward. Particularly troublesome are the fine banding and the conditions controlling the precipitation of so much silica and iron.

The great lateral extent of the Proterozoic deposits suggests to a majority of geologists that they are marine, and indeed it would be difficult to deny a marine origin for the deposits of Archaean greenstone belts. But this only makes the problem of iron mineral precipitation more acute, because of the great rarity of dissolved iron in present-day mildly alkaline sea water. This has led Govett (1966) to suggest a lacustrine origin, because an increase in quantity of dissolved iron has been recorded in the deeper, less oxygenated and more acid waters of some stratified lakes. The banding could therefore possibly signify a seasonal layering. However, bands up to 3 cm thick would imply an unusually high rate of deposition for chemical sediments and if true varves occur they are more likely to be represented by the microns-thick laminae. Eugster and Chou (1973) invoke the evaporative setting of an alkaline playa lake in an anoxic atmosphere, with a sodium silicate gel thought to be the likely precursor of the chert.

Nevertheless it is difficult to conceive of the necessary enormously extensive water bodies as lakes in any normal sense of the word and some form of stratified sea with a different composition from today and a more oxygen-deficient atmosphere seems more plausible (Drever 1974; Holland 1972). Less atmospheric oxygen and correspondingly more carbon dioxide would promote more intense chemical weathering and more acid waters, which should allow more iron to travel to the sea.

In Drever's (1974) model the abundance of chert is related to the evaporation of silica-saturated sea water because there were no organisms to extract the silica. He sees the principal problem in being how to find a mechanism for not precipitating calcium carbonate. Calculations suggest that the hydrogen ions released in shallow water by oxidation of ferrous iron were sufficient to prevent any increase in the ionic activity product of calcite. The depositional environment envisaged is one in which upwelling sea water underwent slight evaporative concentration on marine platforms like the

Great Bahama Bank, with only restricted interchange with the open ocean. This accounts for the absence of siliciclastic detritals.

The alternation of iron-rich and iron-poor layers is related to variations in the intensity of upwelling, which could perhaps be seasonal. There would have been a continuous background precipitation of silica with periods of rapid oxidation of ferrous iron promoting precipitation of iron minerals. The mineralogy of the minerals relates to the amount of organic carbon present. If there were enough to reduce ferric hydroxide to ferrous iron then iron silicates would be precipitated. In deeper anoxic water pyrite would form, while if after reduction of all the iron there were still some reactive carbon then carbon dioxide would be generated and promote the precipitation of siderite. The only important environmental differences required, compared with the present, are a lower atmospheric oxygen and higher carbon dioxide concentration, and an absence of silica-secreting organisms. Holland (1972) considers that only a slightly lower atmospheric oxygen concentration is required.

One problem about these geochemical models is that whereas today there are abundant siliceous organisms that extract most of the silica in sea water—diatoms, radiolaria, silicoflagellates and hexactinellid sponges—none of these was in existence in the late Proterozoic, when the extensive banded iron formations disappear. Yet silica solubility varies little in the likely range of natural waters from pH 4 to pH 9 (Krauskopf 1967), so that a slightly more acid sea would have had little effect. LaBerge (1973) gets out of this difficulty by suggesting that the minute spheroidal structures found in many thin sections of banded ironstones are the remains of primitive organisms that perhaps promoted silica precipitation, but others doubt the presence of any microfossils (Towe 1978), and anyway it is difficult to prove any association with silica precipitation. It is more likely that the chert has preserved any organisms present.

Organic mediators are also required by Cloud (1972, 1973) to account for the great concentration of banded ironstones at around 2000 m.y. ago, the iron oxides of which are regarded as huge oxygen sinks related to the expansion and diversification of microorganisms. The resulting vastly increased biological production of atmospheric oxygen would have acted as an organic poison unless much of it could have been removed by the conversion of ferrous ions in sea water to insoluble ferric oxides. Serious doubt has been thrown on this hypothesis not only by the rarity of conclusively proven microfossils in the ironstones but by the fact that these deposits also occur in much older greenstone belts. They could hardly be as extensive as the Lower Proterozoic ironstones because the greenstone belts are themselves confined in area.

RED BEDS AND EVAPORITES

Red beds should be a good indicator of substantial oxygen in the atmosphere because they signify oxidation of iron during weathering in a subaerial environment while sulphate evaporites should likewise indicate abundant oxygen dissolved in sea water, so that sulphur precipitates as sulphate rather than sulphide. It is important therefore to establish the time of earliest occurrence which should approximate the modern oxygen content of the atmosphere and hydrosphere. Cloud (1972) dated the earliest substantial evaporites as about 800, and the earliest red beds as about 1900 m.y. old. Dimroth and Kimberley (1976), however, record red beds in the early Proterozoic of Labrador that *underlie* the celebrated banded iron formation and are more than 1900 m.y. old. They consist of argillites, sandstones and dolomites together with presumed alluvial haematitic arkoses and conglomerates, containing pebbles of andesite with oxidized weathering crusts.

Similarly, Walker *et al.* (1977) present convincing evidence of widespread and substantial sulphate deposition in Proterozoic rocks of the Australian Northern Territory dated as 1600–1400 m.y. old. This evidence consists of carbonate pseudomorphs after gypsum and anhydrite crystals and chert pseudomorphs after anhydrite nodules; halite casts also occur. Even older extensive evaporite deposition is inferred within a group of excellently preserved Lower Proterozoic sediments of the Great Slave Lake region of northern Canada, that are attributed to alluvial, coastal and shallow marine environments (Badham and Stanworth 1977). Rocks with pseudomorphs after gypsum, anhydrite and halite, and presumed solution breccias, fall within the age range 1900–1700 m.y.

Most remarkable of all, however, is the evidence of evaporite deposition in Archaean greenstone belts. Baritized and silicified evaporites occur in association with silicified carbonates in Western Australia and include chert pseudomorphs after gypsum (Barley *et al.* 1979). Evidence for shallow-water deposition includes desiccation breccias, edgewise conglomerates, climbing ripples and scour-and-fill structures, together with probable stromatolites. Similar shallow-water barite–chert sequences occur in rocks of equivalent age (3500–3400 m.y.) in the Barberton Mountains of South Africa. Such evidence of shallow-water deposition may force us to modify considerably our views on greenstone belt environments.

According to Holland (1972) evaporites are among the most helpful sediments for determining seawater composition and the major chemical constituents should not have departed significantly from present values since the earliest time of substantial evaporite formation.

OTHER EVIDENCE

One of the strongest arguments in favour of an oxygen-deficient atmosphere in the early Proterozoic is the occurrence of what appears to be detrital pyrite and uraninite in several parts of the world, most notably South Africa, in alluvial conglomerates and sandstones of the gold-bearing Dominion Reef and Witwatersrand Systems. It has been generally supposed that neither mineral would be stable in an atmosphere like that of today. As opposed to this interpretation both minerals have been dismissed as diagenetic, having therefore no bearing on the depositional environment (Dimroth and Kimberley 1976). Simpson and Bowles (1977) insist nevertheless that there are definite fine sand-grade detrital grains of pyrite and uraninite associated with quartz in hydraulically equivalent sedimentary assemblages. They also report the existence, however, of detrital grains of uraninite, pyrite and other sulphides together with gold in the Recent alluvium of the River Indus. It seems that relatively small amounts of thorium in the uraninite increase the resistance to attrition and oxidation to a sufficient extent that such material can survive as river detritus under present-day atmospheric conditions. Therefore the case for a reducing atmosphere collapses.

Organic carbon also has a bearing on the question. At the present day only a small fraction of organic carbon survives oxidation but in an oxygen-deficient atmosphere a much higher proportion should have been preserved in the sediments. Yet early Precambrian stromatolites and other sediments show no evidence of this (Dimroth and Kimberley 1976). Conversely it can be argued that the occasional records of early Proterozoic carbonaceous deposits, as in the Witwatersrand System and in the United States, indicate the existence of shallow-water communities of primitive microorganisms despite the presumed lack of an ozone screen to protect them from ultra-violet radiation (Jackson and Moore 1976). Schidlowski *et al.* (1975) found that many Precambrian carbonates have carbon-isotope ratios not greatly different from modern examples, suggesting that organic carbon was approximately as abundant, relative to carbonate carbon, as at the present day. They propose that some 80% of atmospheric and hydrospheric oxygen originated over 3000 m.y. ago.

The earliest occurrence of substantial carbonate deposits should also be relevant, because according to Drever's (1974) model for banded ironstone formation a greater concentration of dissolved ferrous iron in sea water should have inhibited $CaCO_3$ precipitation. The Warrawoona Group of Western Australia is almost the oldest group of sediments yet discovered (3500 m.y.), yet it consists of silicified carbonates interpreted as having been deposited in shallow marine and supratidal environments (Dunlop *et al.* 1978).

In summary, the evidence from a number of fields suggests rather strongly that the composition of the atmosphere and hydrosphere has changed little since the Archaean and probably achieved present values by the late Proterozoic. Formation of the hydrosphere must obviously pre-date the oldest water-laid sediments, dated at 3760 m.y.

Evolution of the biosphere

Living organisms fall into two major categories, the prokaryotes and eukaryotes. Prokaryotes are the more primitive, and characterized by small cells with no clear distinction of nucleus and cytoplasm; they comprise the bacteria and cyanophytes (blue-green algae). All other organisms are eukaryotes, with a clearly defined nucleus. It is natural therefore to suppose that the procaryotes evolved first.

The general presumption is that the earliest organisms were anaerobic heterotrophs dependent for their existence on the free energy available from whatever substances were present. As time passed and resources dwindled there would have been increased selection pressure to utilize alternative energy sources such as light. It has recently been demonstrated that some blue-green algae can photosynthesize in an anoxic environment, but while photosynthesis by green plants releases oxygen to the atmosphere it requires oxygen to get going in the first place, because molecular oxygen 'primes the pump'. Towe (1978) argues that the available biological and biochemical data oppose the general assumption that early Precambrian algae were photosynthesizers supplying free oxygen to the environment, and that photodissociation must be invoked as a primary contributor of oxygen to the early atmosphere.

The days when Precambrian sediments were considered to be abiotic are now long past, the significant breakthroughs being the recognition that stromatolites are organic in origin, and thin-section examination of cherts. Excellently preserved microfossils have been found, such that chert has been called 'The amber of the Precambrian'. This is of course an exaggeration, but it appears that such microfossils, albeit primitive in structure, are by no means rare. Equally exciting has been the discovery of the impressions of highly organized multicellular organisms (Metazoa) in late Precambrian rocks across the world, giving lie to the previously universal belief that there was no evidence for Metazoa before the Cambrian, even though their existence had of necessity been presumed. The most recent of many reviews on Precambrian life are by Schopf (1978) and Ford (1979).

MICROFOSSILS

Spheroidal and filamentous microfossils of organic composition (or, more strictly, nannofossils, because they are often only a few microns in diameter) are now known from all the continents and from almost the whole range of Precambrian time, the oldest being in the Warrawoona Group of Western Australia (3500 m.y.) Rod-shaped and filamentous bodies in the South African Fig Tree Group (3100 m.y.) have been attributed to algae and bacteria. A much more diverse microflora occurs in the Gunflint Formation of the Lake Superior region (2000 m.y.) and in other Proterozoic deposits.

There appears to be a progressive increase in size and diversity through time. The earliest appearance of eukaryotes has been a matter of dispute, with the consensus view favouring a date of 1400 m.y. Kazmierczak (1979) has recently made a case for eukaryotes in the Gunflint Formation. No one has claimed an appearance earlier than 2000 m.y. A variety of microfossils of acritarch type have been described from the late Proterozoic of the Soviet Union.

STROMATOLITES

Stromatolites are widely distributed in the Precambrian and, though it usually cannot be proved, they are generally attributed to the activity of blue-green algae. Whereas they are uncommon in the Archaean, the oldest known being found in the North Pole area of Western Australia and dated as 3400–3500 m.y. (Lowe 1980; Walter *et al.* 1980), they reach a peak of development in Proterozoic carbonates, attaining a size and abundance unmatched in the Phanerozoic. Huge structures in the Lower Proterozoic of northern Canada have areas up to 80 × 45 m and 20 m thickness, with domes exhibiting 2 m of growth relief (Hoffman 1973). Unlike at the present day they appear to have formed extensively in shallow seas. From studies in the Bahamas Garrett (1970) has convincingly argued that the decline of stromatolites can be attributed to the appearance of metazoan grazers. Blue-green algae are diverse and widespread in modern seas but they only form mats that may produce stromatolites in conditions inimical to most organisms, such as intertidal and supratidal flats.

METAZOA

Following the discovery of the famous Ediacara fauna in the late Proterozoic of South Australia (Glaessner and Wade 1966), similar fossils were found in other parts of the world such as England, Namibia and the Soviet Union. The age range is about 600–800 m.y. The commonest fossils are circular to ovate, disc-like impressions, with a variety of concentric, radial and lobate markings, attributed to jellyfish. Other common fossils are large, frond-like impressions plausibly interpreted as pennatulid coelenterates, and probable annelid impressions. A sparse fauna of trace fossils produced by crawlers and burrowers has also been recorded from very late Precambrian strata.

Stanley (1976) has addressed himself to the question of why there was apparently such a long delay between the appearance of the eukaryotic cell and the appearance and diversification of the Metazoa. He believes that the advent of sexuality might have triggered diversification by making possible rapid speciation and hence adaptive radiation. In addition, the near-saturation of the shallow sea floor by blue-green algae, in the absence of cropping by herbivores, might have been an inhibition to evolution. Recent ecological research has demonstrated that the removal of predators from an

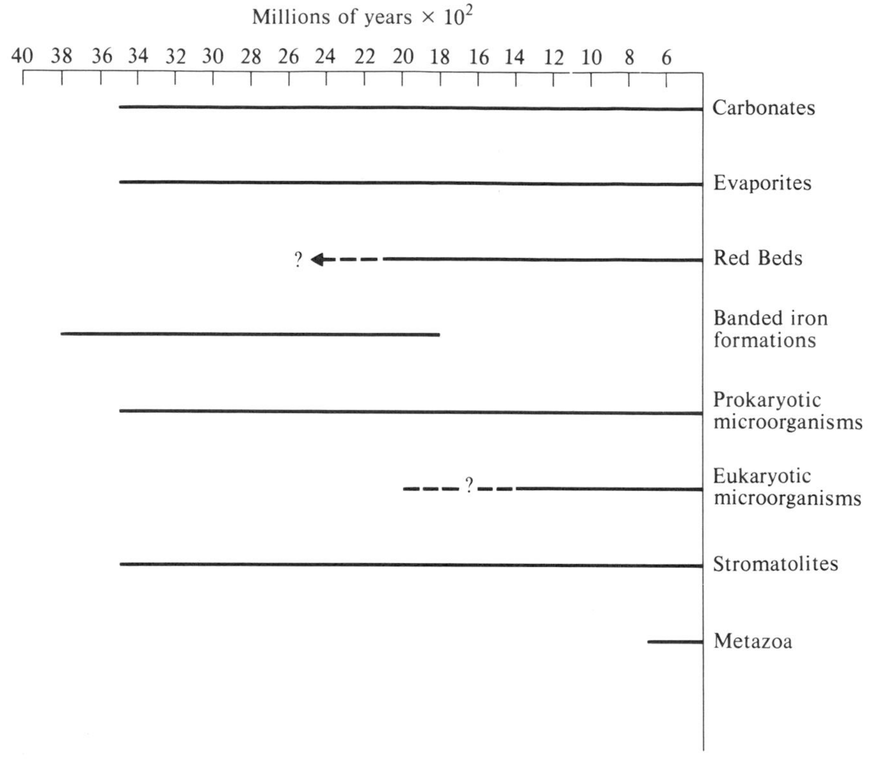

9.4 Distribution through time of Precambrian organisms and environmentally significant sediments.

ecosystem can have the effect of considerably reducing diversity, as one or a few of the prey species of high reproductive potential rapidly expand their populations and come to dominate the biological environment.

The naïve 'physicalist' view that the Cambro-Ordovician metazoan evolutionary explosion was triggered by a rise in atmospheric oxygen concentration to a critical level now seems very dated. By late Precambrian times organisms were sufficiently widespread and abundant to exert a significant control on the environment in a variety of ways (Fig. 9.4). The subsequent evolutionary story will be taken up in the next chapter.

10. Facies and the Phanerozoic fossil record

There is increasing recognition that no attempt at interpreting facies will be adequate unless the ecological information obtainable from the contained fossils is fully taken into account. This desire to obtain the maximum possible information on ancient depositional environments has probably been the primary stimulus to the rapid growth of *palaeoecology*. The subject is not perhaps well named, because there are of necessity fundamental differences in approach from ecology, where environmental parameters can be directly measured and the prime goal is the study of biotic interactions which we can often only guess at in the case of fossils. Nevertheless palaeoecology has grown more biological in outlook in recent years, with the development of interest in fossil communities. It should go without saying that such study must take into account the evidence on environment provided by the lithofacies, but not all the contributors to the beautifully illustrated *The ecology of fossils* (McKerrow 1978) followed this precept.

It is beyond the scope or intention of this book to give a full palaeoecological review, and the reader is referred to other texts, such as the excellent introductory account in chapter 10 of Raup and Stanley (1978). I shall instead concentrate on two topics of outstanding interest, on which facies analysis has some bearing.

That fossils are of immense value both for stratigraphic correlation and environmental interpretation is generally acknowledged, but they also provide the only direct evidence we have of the history of life and the major patterns of evolutionary change. An increasing interest in the more biological aspects of palaeontology is reflected in the coining of the term *palaeobiology*, which has indeed been adopted as the title of a lively new journal. Since it is a truism of Darwinian evolutionists that evolution and extinction have been promoted by environmental change it is natural to seek the best available evidence from both litho- and biofacies of the nature of that change. The last few years have also seen the rise of the subdiscipline of *palaeobiogeography*, stimulated by the widespread acceptance that the continents have moved laterally with respect to each other.

Now within the confines of a single chapter it is impossible to give comprehensive accounts of such major subjects, on which indeed whole books can be written, and I shall deal only with certain important aspects. Phases of mass extinction are a vital part of the study of evolution because the vacation of ecological niches is a necessary condition for subsequent radiation of the survivors. It is also instructive to enquire into the relative importance of plate movements, climate and eustasy in controlling biogeographic change.

One problem that has to be faced at the outset is that facies studies alone are insufficient for a complete analysis of these subjects; the first also involves consideration of biological concepts and the second global tectonics. Nevertheless it is important to establish just where facies interpretation is relevant, if not crucial, to a proper understanding. This will involve frequent referring back to earlier chapters, and the development of a general thesis that hopefully will demonstrate an ultimate connection between diastrophism and the changing biological world.

Diversity changes through time

It is desirable to digress a little in order to say something about organic diversity, and its changes through time.

Ecologists use the term diversity in two different senses, as a sample taxon count and as a measure of relative abundance of the taxa in a given assemblage. There is an increasing tendency to refer to the former as *richness*, restricting *diversity* to a measure of evenness or equitability. Palaeontologists interested in diversity problems have with very few exceptions concerned themselves solely with taxonomic richness, which can be very informative in studying the grosser changes in space and time, and diversity will be used here in this sense.

Sepkoski (1978) presents a graph of the diversity of metazoan orders from the end of the Precambrian (late Vendian) through the whole of the Phanerozoic (Fig. 10.1). He has demonstrated that the metazoan diversification is remarkably consistent with a logistic model which suggests, in its stochastic form, that the number of taxa should rise with time and then fluctuate about some constant equilibrium level. This suggests that the 'explosive' appearance of the early Cambrian fauna was simply one phase of a continuously accelerating diversification and that no special event need be invoked save the earlier initial appearance of the Metazoa. Diversification is probably limited at some point by 'crowding' effects among species. Considerations of diversity-dependent processes suggest that per species rate of extinction should rise as diversity increases.

An equilibrium model following the initial diversification is also favoured by Gould *et al.* (1977). They generated by computer a random model that builds evolutionary trees by allowing lineages to branch and become extinct at equal probabilities, and demonstrate a strong similarity of real and random *clades*, that is, groups with a common evolutionary origin. An overall equilibrium situation is not inconsistent, however, with phases of mass extinction, as brought out for instance by analysing the times of first and last appearance of taxa in the stratigraphic record. Cutbill and Funnell (1967) have attempted a comprehensive analysis of this sort. It proved methodologically desirable to group together different taxonomic rankings but about 60% of the information is at the family level. Their graph for the invertebrates, which have by far the best fossil record (Fig. 10.2), shows

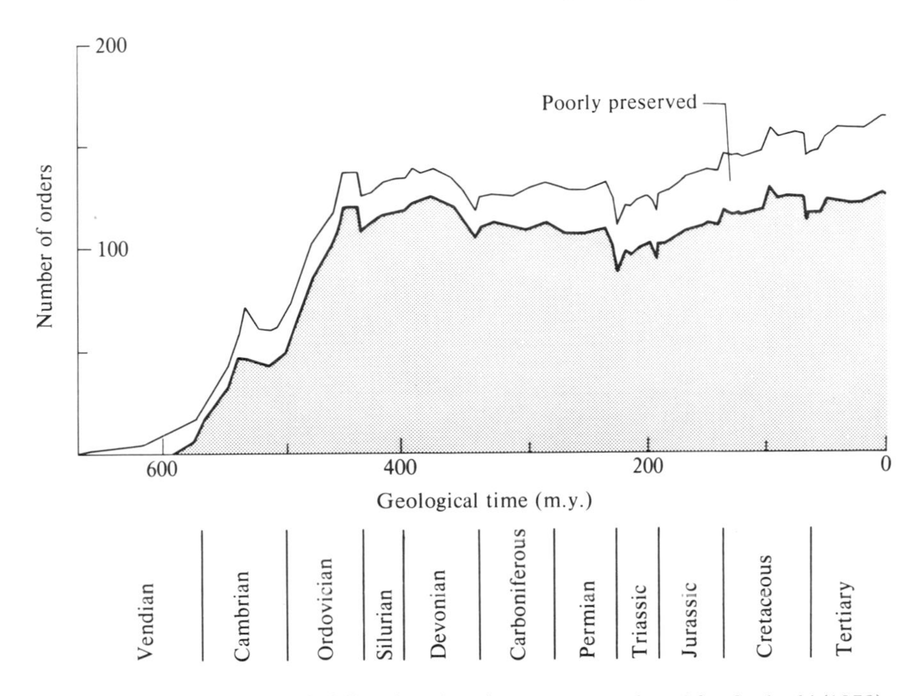

10.1 The Phanerozoic record of diversity of marine metazoan orders. After Sepkoski (1978).

pronounced extinction peaks in the late Cambrian, end Ordovician, late Devonian, late Permian and end Cretaceous, and a greater number of less important ones.

With regard to lower taxonomic levels, a lively controversy has developed about whether marine invertebrate species diversity has increased systematically through the Phanerozoic or whether an equilibrium situation has existed, with the apparent increase through time being an artefact. The leading advocate of the former view is Valentine (1969), who has argued for an order of magnitude increase from the early Palaeozoic to the late Cenozoic, while the steady-state alternative has been propounded principally by Raup.

Raup (1976) estimated the number of species described for each of the geological periods by tabulating new species reported in the *Zoological Record*, and established a strong correlation between apparent species numbers and the present areal distribution of rocks per system. Sheehan (1977) has come up with another interesting finding, having made a study of those palaeontologists expressing interest in fossils of particular systems, using data from the *Directory of Palaeontologists of the World*. It turns out that there is an excellent correlation ($r = 0.94$) between the number of carefully defined 'palaeontologist interest units', and the number of

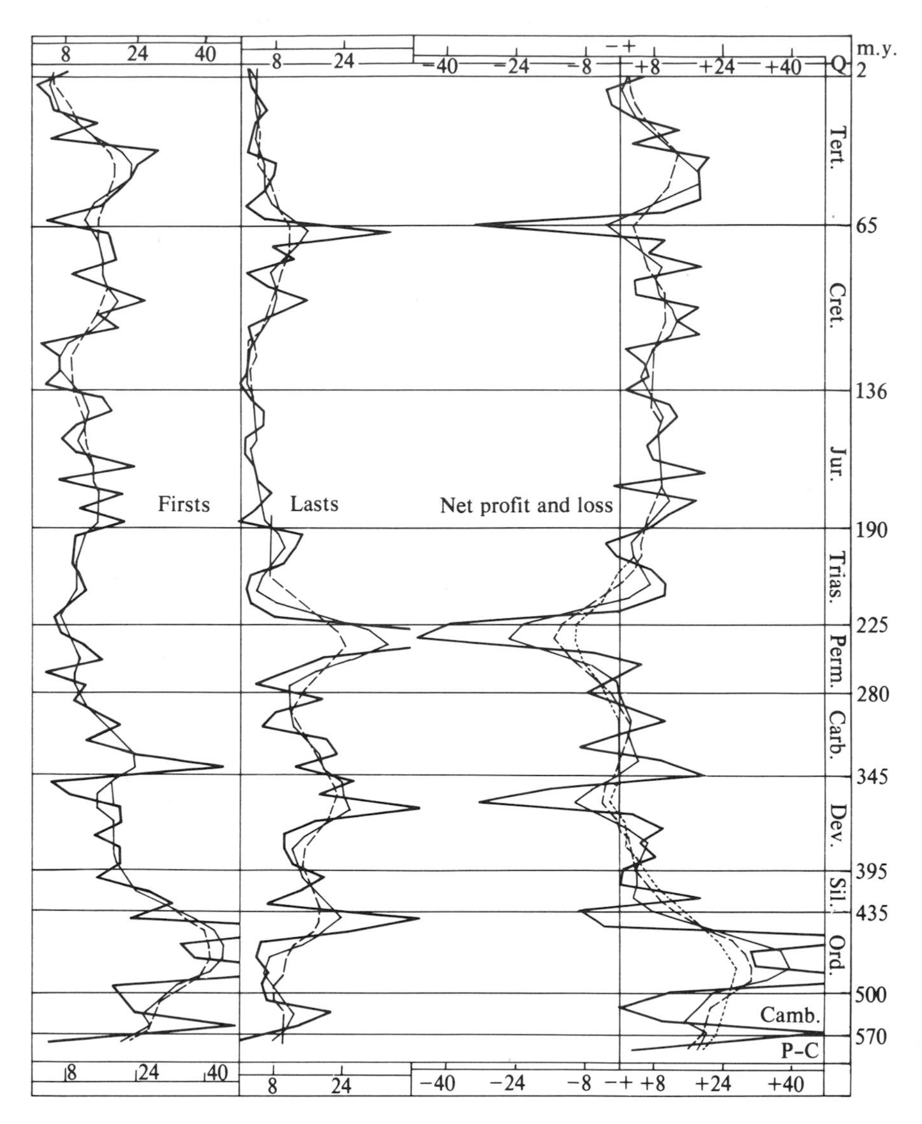

10.2 The invertebrate record of first and last appearances through the Phanerozoic, with net profit and loss. After Cutbill and Funnell (1967).

described species per geological period. Sheehan concludes that the total numbers of described species do not seem to reflect meaningful estimates of the original diversity.

In his reply to Sheehan, Raup acknowledges the good correlation established between the number of species described and the number of interested palaeontologists, as well as outcrop area, but maintains that nothing very positive can be inferred from this. He prefers the view, though he cannot prove it, that geological systems with more available rock have more species and hence more species are described. In other words most palaeontologists are attracted to the most fossiliferous rocks.

Is there any way out of this impasse? Apparently there may be, provided a distinction is drawn between assemblages from different environments, which can be compared from period to period. This has been attempted by Bambach (1977), who analysed data from hundreds of fossil assemblages from three different types of environment, as inferred from the facies. The high stress, marginal marine environment always has the lowest faunal diversity and the open marine environment the highest, while the variable nearshore environment has intermediate values. Bambach's data seem to indicate that within-habitat variation in species numbers is small for long intervals of time, and that the number of species has increased by a factor of about four since the mid-Palaeozoic, with variable nearshore environments showing a less pronounced increase than the open marine, while the high stress environments show no significant change. Thus Valentine seems to be partly vindicated, but substantial problems of interpretation remain, notably the factors controlling within-habitat species diversity. Bambach speculates that changes through time of availability of food resources might be the most significant, but this is clearly an area demanding much further study.

The Cambro-Ordovician radiation

Before turning to the subject of mass extinctions it is desirable to learn something about the nature of the buildup to the 'equilibrium' diversity of the late Ordovician, and any possible environmental control.

Brasier (1979) has produced a detailed analysis of the Cambrian radiation event. A whole range of organisms with preservable hard parts made their first appearance, including calcareous algae, rhizopod protists, archaeocyathids and other sponges, arthropods, molluscs and echinoderms. In addition, acritarchs and soft-bodied organisms became much more diverse, as shown by the record of trace fossils and the excellently preserved fauna of the Burgess Shale of British Columbia (Conway Morris and Whittington 1979). There was also a general tendency towards larger size of the shelled faunas compared with the Vendian and basal Cambrian (Tommotian) faunas. Since it was benthic invertebrates smaller than about 10 mm that tended to develop mineralized shells, there is a strong suggestion that cropping pressures were involved in skeletonization (see chapter 9) because it was precisely these smaller, sessile forms that most risked being eaten.

Most rock sequences through the Upper Vendian to Lower Cambrian tend to exhibit a trace fossil → shelly and tubular fossils → trilobite succession. Thus in an idealized model a typical transgressive sequence on to the early Cambrian platforms would begin with littoral deposits containing a low diversity burrow assemblage (notably *Skolithos, Monocraterion* and *Diplocraterion*) or stromatolites. These would be followed in siliciclastic sequences by sandstones with the above plus *Phycodes* and *Rusophycus* burrows and finally by relatively high diversity argillaceous or calcareous rocks with *Teichichnus* burrows and skeletal fossils. In calcareous sequences a succession might pass from algal biostromes through archaeocyathid reefs into deeper water muds or limestones with hyoliths and trilobites. The Cambrian radiation appears therefore to have an overprint of ecological succession.

The succession of dominant environments through the transgression may have controlled the timing of the development and radiation of different fossil groups. Littoral communities would develop during the initial stages of rising sea level, and rapid transgression would allow them to spread more widely while a contemporary assemblage developed further offshore. In Brasier's model (Fig. 10.3) of episodic transgression it is seen that different communities need not have originated and radiated synchronously. As sea level rose so the littoral deposits tended to be replaced by more offshore calcareous and argillaceous rocks, with a corresponding radiation of trilobites, archaeocyathids, molluscs and other groups. In general the stratigraphic and facies evidence appears to fit the model fairly well.

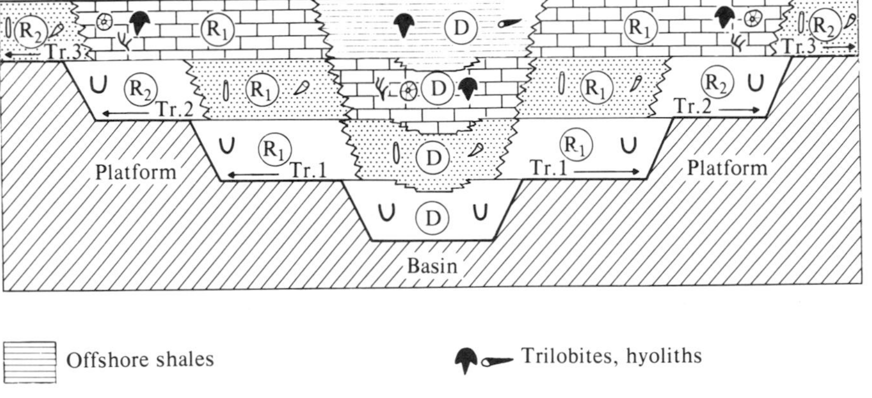

10.3 Model of Cambrian marine transgressions and organic colonization. After Brasier (1979).

House (1967) also invokes the spread of epicontinental seas to account for the further diversity increase in early and mid-Ordovician times. The groups principally affected have calcareous hard parts and include rugose and tabulate corals, stromatoporoids, bryozoans, articulate brachiopods, crinoids and nautiloids. There was also rapid evolutionary diversification among the graptolites.

Episodes of mass extinction

By plotting the percentage of last appearances of animal families through the Phanerozoic Newell (1967b) demonstrated several phases of mass extinction affecting a wide range of unrelated groups, both terrestrial and marine (Table 10.1). Plots of first appearances indicated a series of peaks directly following the extinction phases and were attributed by Newell to episodes of evolutionary diversification following the vacation of ecological niches. Let us examine in more detail some of the extinction events.

The late Devonian event can be tied down more or less precisely to the Frasnian–Famennian boundary. The most significant extinctions occurred in reefal or peri-reefal organisms, with the corals and stromatoporoids being drastically affected. On the other hand conodonts appeared to have suffered comparatively little change (McLaren 1970).

The late Permian extinction phase has been widely regarded as the most important of all. Although Rhodes (1967) played down the sharpness of the change from the Permian to the Triassic, and argued that the diversity decline was less the result of Permian extinction than of lack of Triassic replacement, this view is not borne out by the work of Schopf (1974), who shows that the number of invertebrate families decreased by a factor of two through the Permian. Some groups that became extinct, such as the trilobites and blastoids, had long been in decline, but many others were still flourishing in the late Permian. The sharpest faunal change appears to have taken place across the Permian–Triassic boundary. The best studied of the few sections in the World where marine deposits occurs is in Pakistan. If ammonites are used to determine the boundary, then it can be demonstrated that Permian-type brachiopods and ostracods survived briefly into the Triassic, while a conodont zone straddles the boundary (Kummel and Teichert 1970). At about the Permian–Triassic boundary the families of large herbivorous reptiles, including pareiasaurs, gorgonopsids and hipposaurids, disappeared (Bakker 1977).

Although the late Triassic extinction phase has received comparatively little attention more than a third of all animal families died out according to Newell (1967b). Among the invertebrates the most striking extinction took place among the ammonites. Although they were extremely diverse in the late Triassic (Carnian and Norian) only one family survived into the Jurassic. That the whole group came so close to total extinction is a sobering thought

TABLE 10.1 Major Phanerozoic extinctions

Extinction episode	*Major animal groups strongly affected*	*Percentage of families extinct (after Newell)*
Late Cretaceous	Ammonites* Belemnites Rudistid bivalves* Corals Echinoids Bryozoans Sponges Planktonic foraminifera Dinosaurs* Marine reptiles*	26
Late Triassic	Ammonites Brachiopods Conodonts* Reptiles Fish	35
Late Permian	Ammonites Rugose corals* Trilobites* Blastoids* Inadunate, flexibiliate and camerate crinoids* Productid brachiopods* Fusulinid foraminifera* Bryozoans Reptiles	50
Late Devonian	Corals Stromatoporoids Trilobites Ammonoids Bryozoans Brachiopods Fish	30

* Last appearance of group.

Table 10.1 (*cont.*)

Extinction episode	*Major animal groups strongly affected*	*Percentage of families extinct (after Newell)*
Late Ordovician	Trilobites Brachiopods Crinoids Echinoids	
Late Cambrian	Trilobites Sponges Gastropods	52

for Jurassic and Cretaceous stratigraphers! Even an important group often assumed to be relatively immune to mass extinction events, the bivalves, was severely affected at the generic level, and the megalodontids, a group of large thick-shelled forms favouring reefal or peri-reefal habitats, disappeared completely at the end of the period. Among terrestrial vertebrates, the large herbivorous rhynchosaurs, dicynodonts and aëtosaurian thecodonts became extinct together with the large labyrinthodont amphibians, but the time of extinction seems to have been earlier in the late Triassic, approximately Carnian (Bakker 1977).

The late Cretaceous extinctions have attracted even more attention than those of the late Permian, partly because everyone's favourite prehistoric animals, the dinosaurs, were among the victims. Whereas some groups like the ammonites had been in decline for some time (Hancock 1967; Kennedy 1977) other such as the planktonic foraminifera and dinosaurs were still flourishing in the Maastrichtian, so that interest has been concentrated on the Cretaceous–Tertiary boundary event. One of the best known of the few recorded, fully marine sequences across the boundary occurs in a more or less uniform facies of Maastrichtian and Danian chalk in Denmark, where the boundary is taken at a thin band of smectitic clay, the so-called Fish Clay. Above this the calcareous micro- and nannoplankton undergo an abrupt change, whereas the non-calcareous microplankton represented by dinoflagellates continue from the Cretaceous virtually unchanged. Ammonites and belemnites make their last appearance in the Maastrichtian chalk, but the benthic invertebrate fauna of both the Maastrichtian and Danian is broadly similar, though different at species level.

With regard to terrestrial vertebrates, it is apparent that by no means all groups were severely affected. Thus virtually all families of freshwater aquatic forms, including crocodiles, survived into the Tertiary (Bakker 1977).

Data from the new field of magnetic stratigraphy are now being brought to bear on the problem, because they offer the prospect of extremely precise correlation between stratigraphic sections on land and under the ocean. A biostratigraphically complete sequence of Upper Cretaceous and Palaeogene pelagic limestones at Gubbio, Italy has produced a record of geomagnetic reversals that closely matches the marine magnetic anomaly sequence (Alvarez *et al.* 1977). A detailed analysis of sedimentation rates at Gubbio suggests that the Maastrichtian–Danian faunal overturn might have happened very rapidly, of the order of 10 000 years or less (Kent, D. V. 1977). Palaeontological work on a comparably complete sequence of pelagic deposits at Zumaya, Spain suggests that environmental stresses were first felt by planktonic foraminifera about 10 000 years before the end of the Maastrichtian, following which occurred the major crisis involving extinction of globotruncanid foraminifera and a drastic change in the nannoplankton flora (Percival and Fischer 1977).

Interest naturally focuses on whether or not the dinosaur extinction was coincident in time with that of the planktonic groups. Butler *et al.* (1977) undertook a magnetostratigraphic study of a sequence of terrestrial sediments in New Mexico which they argued exhibits an unbroken transition across the Cretaceous–Tertiary boundary. By correlating with the Gubbio sequence they inferred that the dinosaurs became extinct slightly later than the planktonic foraminifera, perhaps by as much as half a million years. This correlation has been criticized by Alvarez *et al.* (1979), who suggest that an unconformity might be present in the New Mexico sequence. If so, the marine and terrestrial extinction events could indeed have been synchronous. Further information on this issue has been published by Lerbekmo *et al.* (1979) in another magnetostratigraphic study of a terrestrial sequence in Alberta. Their data indicate an approximate synchroneity of dinosaur and foraminiferal extinctions just below anomaly 29, the maximum error being about 100 000 years.

Before proceeding to discuss the possible causes of mass extinctions we need to determine if some types of organism are more susceptible than others to environmental change. MacArthur and Wilson (1967) made a widely adopted ecological distinction between *K*- and *r*- selected species, with *K* being a constant relating to the carrying capacity of the environment and *r* the coefficient of natural increase of populations. *K-selected species* are in general characterized by a high investment in nurturing the comparatively few young, late maturation, greater longevity and trophic specialization (i.e. they consume only a particular type of food). They tend to live in environmentally stable regimes where interaction with other organisms plays an important part. In contrast, *r-selected species* tend to mature rapidly and hence are characteristically small in size, and have a high reproductive potential. Strong fluctuations in population size are the rule, and the organisms are more tolerant of environmental variation and conditions of high physiological stress. They tend to be trophic generalists and dominate the early stages of ecological succession. In other words they are the 'weeds'.

Valuable as this emphasis on population strategies is for evolutionary studies because it touches on some important truths, it is undoubtedly oversimplified and there are many exceptions to such a straightforward distinction. Vermeij (1978) prefers a threefold division into *opportunistic, stress-tolerant* and *biologically competent* species. Opportunistic species correspond more or less to *r*-selected species but stress-tolerant and biologically competent species, while both characterized by long life and low reproductive potential (which are *K*-selected characters) differ in their environmental tolerance and susceptibility to extinction. Whereas physiologically stressful environments, such as shallow marine regimes of fluctuating salinity and temperature, are mostly occupied by stress-tolerant species, the greater susceptibility to extinction of the biologically competent species is enhanced by the destabilizing influence of coevolution, which ensures that selection pressures change continuously. Thus evolution should be more rapid and extinction more frequent (the other side of the coin) for biologically competent species. In the more traditional but limited parlance these would be called the more specialized organisms.

The fossil record appears to bear this out, because tropical reefs are the marine environment most characterized by biotic interactions and stable environments, and reefal and peri-reefal organisms were among the most drastically affected of marine invertebrates. The list is long and includes corals, stromatoporoids, archaeocyathids, rudistid and megalodontid bivalves together with many brachiopods, fusulinids, bryozoans and crinoids. Morover, rapidly evolving groups such as the ammonoids, trilobites and planktonic foraminifera were also relatively vulnerable. As regards terrestrial groups, large organisms, with presumably high longevity but low reproductive potential, were evidently more vulnerable than small.

Causes of mass extinction

If a number of important, physiologically and ecologically unrelated cosmopolitan groups disappeared at about the same time, general rather than particular causes should be sought, which should be of worldwide impact. Thus many of the numerous causes proposed to account for dinosaur extinctions fail to satisfy this criterion. To suggest, for instance, that dinosaurs died out because small mammals ate their eggs both ignores the numerous contemporary extinctions and the fact that, Man apart, animal predators do not usually indulge in overkill, which would disturb the balance of nature. Actually my favourite dinosaur extinction hypothesis relates their demise to the decline of the naked seed plants (gymnosperms). Many of the still extant representatives contain oils with renowned purgative properties. The doleful implication that the dinosaurs died of constipation is unfortunately not supported by the fossil record, which clearly indicates that the angiosperms had replaced the gymnosperms as the dominant land plants well before the end of the Cretaceous.

Only the more widely discussed or plausible causes are considered in the following account.

EXTRATERRESTRIAL AND GEOMAGNETIC REVERSAL EVENTS

Schindewolf (1954) thought that the late Permian extinction event was so extreme that an extraterrestrial event, such as an increase in cosmic radiation flux, must be invoked. With the discovery of evidence of magnetic reversals from the rock record, and the assumption that the weakened geomagnetic field at times of reversal would provide less protection against the biologically harmful effects of cosmic radiation, Uffen (1963) proposed this as a direct cause of mass extinction.

This category of explanation has been strongly criticized by Black (1967) and Waddington (1967). At currently observed cosmic ray and solar particle intensities, the additional dosages produced at sea level during a period of weakening of the geomagnetic field are negligible. Furthermore, the cosmic ray flux is heavily attenuated by water and undergoes a very rapid diminution at shallow depths, and most plankton live some distance below the surface. In addition to these points, it is strange that terrestrial plants, which should surely have been more vulnerable than marine organisms, should have apparently been so little affected, though it needs to be pointed out that Krassilov (1978) recognized an abrupt change in the gymnosperm flora across the Cretaceous–Tertiary boundary. Similarly, the change in Gondwanaland from the Glossopteris to the Dicroidium Flora began in the Lystrosaurus Zone, just above the Permo-Triassic boundary (Bakker 1977).

Nevertheless Hays (1971) observed that, of eight radiolarian species that became extinct in the last 2.5 million years, six disappeared close to the times of geomagnetic reversals as recorded from deep-sea cores, and the correlation statistics appear to render a chance association very unlikely. More generally, Crain (1971) claimed that there was a good correlation between times of increased reversal rate and extinctions through the Phanerozoic (Fig. 10.4). On the basis of a few experiments suggesting that gross behavioural and biochemical abnormalities could occur in organisms in a reduced magnetic field, such as infertility, enzymal changes and reduced feeding activity, he proposed a direct causal link with times of magnetic reversal, independent of any radiation from outer space. However, McElhinny (1971) considers that there is no significant correlation of

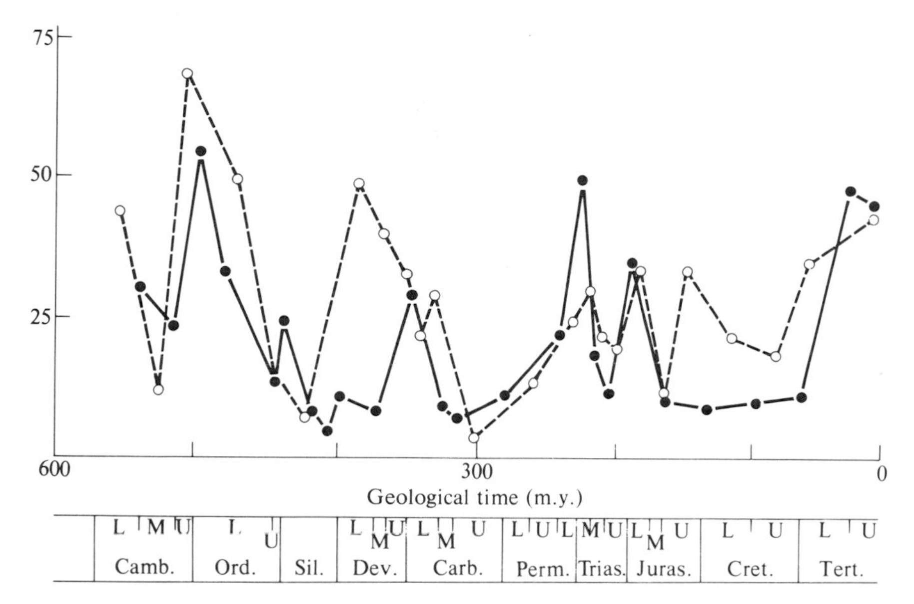

10.4 Comparison of varying magnetic reversal rate (broken line) and biological extinction index (solid line). Vertical scale is in per cent of mixed magnetic measurements and per cent of total existing families extinguished. After Crain (1971).

evolutionary or extinction events with the apparent reversal frequency. Hays (1971) is nevertheless impressed with the fact that both the latest Permian and latest Cretaceous are times of change from a long period of few reversals to a period of frequent reversals, but Fig. 10.4 also shows some striking disparities.

Despite the criticisms that have been put forward Reid *et al.* (1976, 1978) attempt to revive the idea of extraterrestrial events as a cause of extinctions. They suggest increased ultraviolet radiation at the Earth's surface as a result of breakdown of the protective ozone layer or supernova explosions. Quite a different kind of astronomical catastrophe is favoured by McLaren (1970) to account for the late Devonian extinctions, namely the fall of a large meteorite in the Pacific, resulting in vastly increased turbidity and wave disturbance in the ocean system, to which reef organisms would be especially vulnerable.

The trouble with these sorts of hypothesis is that they are difficult if not impossible to prove and take no account of the evidence of the stratigraphic record. Rather than invoking extraterrestrial or geomagnetic events as a kind of *Deus ex machina* it seems methodologically more reasonable to look closely at the record of the rocks and only abandon hope of finding a cause in one or a combination of environmental events of the type most familiar to geologists, if no good correlations emerge.

CLIMATIC AND ATMOSPHERIC CHANGES

Because climatic change has been of paramount importance in the Quaternary it is natural to enquire about extinctions at this time. In fact nothing like mass extinction has occurred. The best-known example of widespread extinctions concerned large mammals in the early Holocene, but this occurred at a time of climatic amelioration and may well be bound up to some extent at least with the rapid expansion of human populations. The most characteristic response of both marine and terrestrial species was to migrate to refuges rather than die out. Further back in the Cenozoic, a good case can be made for climatic control of an important late Eocene extinction phase among deep-sea benthic foraminifera and ostracods, with the development of the psychrosphere (see chapter 7).

It has been a popular belief that dinosaurs died out because of a climatic cooling event at the end of the Cretaceous, with the advent of cold winter seasons that they were ill-equipped to survive. McLean (1978) has recently introduced the novel idea that the extinction was provoked on the contrary by a late Maastrichtian warming event, because physiological studies have demonstrated that reptiles can survive periods of cooling better than periods of warming. He cites one oxygen-isotope study of deep-sea core material in support of his hypothesis but chooses to ignore the wealth of diverse data suggesting that the world's climate did indeed deteriorate across the Cretaceous–Tertiary boundary (see chapter 7). There are, however, at least two major difficulties about the climatic cooling hypothesis. Warmth-loving, stenothermic crocodiles and other freshwater aquatic reptiles were little affected (Bakker 1977), and it is implausible to maintain that the pronounced end-Maastrichtian plankton extinction event, which apparently took place in a geologically very brief period of time, was due to a sudden atmospheric temperature fall, because of the thermal inertia of the ocean system.

Further back in time the evidence for climatic control of mass extinctions is very weak, although it can reasonably be argued that there was an indirect effect in the case of the late Ordovician episode, due to withdrawal of epicontinental seas as a consequence of growth of the Saharan ice sheet. Copper (1977) postulated a climatic cooling event to account for the mass extinctions of reefal and peri-reefal organisms in the late Devonian. In support, he pointed to the widespread Famennian black shale environments of the Western Hemisphere as possible indicators of cold spells and mass mortality, and the survival of purported 'cold climate' infaunal deposit-feeding bivalves and inarticulate brachiopods in the Malvinokaffric Province in the Southern Hemisphere. However, there are better explanations for black shale environments and the survivors of the mass extinction were just the sort of eurytopic, stress-tolerant organisms that have relatively low susceptibility to extinction.

A major problem concerning the end-Cretaceous extinctions is why organisms as different as the marine plankton and the dinosaurs should have been so severely affected. One ingenious solution to this problem was proposed by Tappan (1968). She postulated that the production of marine phytoplankton controlled the oxygen and carbon dioxide levels of the atmosphere through time, and that changes in relative and absolute abundance of these organisms had extensive effects on contemporary biota. Following the major late Cretaceous transgression which preceded the end-Cretaceous regression, the influx of nutrient supply from land dropped as continents were reduced to base level. This and other associated processes led to a marked reduction in marine phytoplankton and a consequential increase of atmospheric carbon dioxide and decrease of oxygen (due to reduced photosynthesis). These atmospheric changes spelled death for the dinosaurs, though estivators and hibernators such as the crocodiles and turtles were able to survive into the Tertiary.

Tappan's argument that widespread transgression should reduce the nutrient supply from the land contradicts other arguments that suggest the reverse (see chapters 5 and 8), while her assumption that a fall in phytoplankton biomass would have significant effects on the well-buffered system of the atmosphere is open to question, to say the least. There is, furthermore, considerable uncertainty about the relative importance of terrestrial and marine plants in contributing oxygen to the atmosphere at the present time.

SALINITY CHANGES

That the Messinian 'salinity crisis' had a drastic effect on the Mediterranean fauna is generally accepted but suggestions that oceanic changes have caused mass extinctions are much more speculative.

Fischer (1964b) developed a hypothesis put forward earlier by Beurlen that the Permo-Triassic extinction phase was a consequence of salinity lowering following the abstraction of salt in the huge Upper Permian evaporite deposits of epicontinental regimes. Data from the Baltic Sea indicate a significant fall in organic diversity below a salinity value of 30‰ with complete loss of stenohaline groups. A withdrawal of about 7×10^{15} tons of salt into evaporite deposits and into refluxed brine in depressions of the deep-sea floor, would effect a lowering of most of the ocean to about 30‰. There is unfortunately no independent test of this interesting idea, because estimates of the evaporite deposits stored in the existing epicontinental basins is insufficient. That huge quantities of salt were stored in deep oceanic brine pools remains entirely speculative. There are other difficulties, notably the fact that the marine extinctions were concentrated at or close to the Permian–Triassic boundary, and that significant contemporary extinctions took place also on the land.

A further shortcoming of the salinity-control hypothesis is that it is unsatisfactory as a *general* explanation for marine mass extinctions, although Berger and Thierstein (1979) have proposed precisely this, on the basis of their intriguing idea of oceanic 'injection events' (Thierstein and Berger 1978). They cite oxygen-isotope data from an exceptionally well-documented Atlantic core, that indicates more or less stable values through the Maastrichtian followed by a sharp change into the basal Danian. This can be interpreted as signifying a sudden increase in water temperature (in which case it comes too late to support McLean's hypothesis) or a salinity decrease. Thierstein and Berger favour the latter interpretation and cite this evidence in support of their hypothesis that the end-Cretaceous marine extinction event was caused by a large-scale invasion of the ocean by low-salinity water from a previously isolated high-latitude basin. However, the existence of extensive high-latitude ice masses at the end of the Cretaceous seems unlikely for a number of reasons (see chapter 7) and it is hard to see how otherwise the requisite large volume of low-salinity water could be created. Thierstein and Berger fail also to provide a convincing explanation of the contemporary dinosaur extinctions. The difficulties are compounded if such speculative injection events are considered a likely cause of other Phanerozoic mass extinctions.

A similar interpretation of the terminal Cretaceous plankton extinctions has been put forward by Gartner and Keany (1978), who argue that a low-salinity surface layer spread over the world ocean during the late Maastrichtian from a previously isolated brackish Arctic Ocean. This has been criticized by Clark and Kitchell (1979) on the grounds that late Maastrichtian microfossils recovered from an Arctic deep-sea core are of normal marine type. Furthermore, Gartner and Keany's data base of nannoplankton from North Sea chalk is apparently invalidated because the core in question has been shown by Shell geologists to contain resedimented slide material (Surlyk 1980).

THE HAUG EFFECT

The synchroneity of maximum orogeny with maximum transgression has been called the Haug Effect, after the great French geologist Emile Haug, who first put forward the idea. Bakker (1977) explains the dinosaur mass extinction in terms of this effect by arguing for exceptionally low topographic diversity at the end of the Cretaceous. This would have led to reduced habitat variety and therefore induced more competition. It is not necessary to postulate an increased extinction rate, merely a reduced immigration rate from the vanishing highlands. Independent checks on this hypothesis are difficult, but it would be surprising if there were few or no late Cretaceous uplands. The thick turbidites of this age in the deep Atlantic appear to suggest otherwise, as does the evidence of pronounced regional regression in the Western United States, associated with the Laramide uplift (Hancock and Kauffman 1979).

ANOXIC EVENTS

By far the most important time of marine invertebrate faunal turnover in the Jurassic was in the early Toarcian, with such important groups as ammonites, bivalves, brachiopods, foraminifera and ostracods all being strongly affected, so that there are hardly any species common to the deposits older and younger than this time. As observed in chapter 8, the early Toarcian is the best candidate for a Jurassic anoxic event of the sort proposed for the Cretaceous. Although it cannot be proved to be worldwide, black shale facies is extremely widespread. The habitable areas of neritic organisms must in consequence have been considerably restricted, which would have provoked extinction by the 'area effect' (see below). It would be interesting to learn if the postulated Cenomanian–Turonian boundary anoxic event is also associated with mass extinction at the species level. Because of the widespread distribution of black shale facies in the Famennian it could be that such an event may prove to be the best available explanation for the late Devonian extinctions.

SEA-LEVEL CHANGES

In an influential paper written many years ago Chamberlin (1909) argued that major faunal changes through time, which provided the basis for biostratigraphic correlation, were under the ultimate control of epeirogenic movements of the continents and ocean basins. Marine regressions corresponded with periods of continental rejuvenation and ocean basin subsidence and led to provincialism and areal restrictions on organism distributions, while transgressions corresponded with periods of reduction of the continents to base level and allowed the spread of new faunas. Moore (1954) suggested further, with reference to Palaeozoic epicontinental seas, that times of regression were more significant for accelerated evolution because of extinctions produced by crowding effects in the more restricted seas. It was Newell (1967b), however, who first proposed a hypothesis explicitly relating faunal extinctions and radiations to eustatic changes of sea level, indicating that a good correlation exists between biotic diversity and the area of habitat available for colonization. Thus a shrinkage of the area of epicontinental sea habitat should have a deleterious effect on neritic organisms and should lead to widespread extinction. Adaptive radiation of the survivors would take place during the expansion of habitat area consequent upon a succeeding transgression.

We have seen that a plausible case can be made relating the rapid Cambro-Ordovician diversity increase to an expansion of epicontinental seas, and reference to the data presented in chapter 6 indicates good general support for Newell's thesis that mass extinction phases correlate with times of significant lowering of sea level. Thus there were major regressions at or near the end of the Cambrian, Ordovician, Permian and Cretaceous. The evidence concerning the late Devonian is less clear, but my own unpublished work confirms an important phase of general regression at the close of the Triassic. The correlation appears to hold also on a smaller scale, because Jurassic ammonite extinctions tend to be concentrated in regressive, and radiations in transgressive intervals (Hallam 1978a). A similar conclusion for Carboniferous goniatite faunas has been reached by Ramsbottom (1979).

The matter can be pursued further by applying the ecologists' well-known species–area relationship. Islands are good ecological laboratories, and it has been shown that a wide range of animals exhibit a simple relationship to island area, expressed in the following equation (MacArthur and Wilson 1967):

$$S = cA^z$$

where S is the number of species, c is a fitted constant which depends on the species analysed, their population density and the nature of the habitat, A is the area and z a parameter generally in the range of 0.2–0.35. c is some sort of measure of environmental 'quality' and should be higher in more stable environments. The work of Preston (1962) suggests that this equation may also hold for habitat 'islands' the size of continents, and Recent mammal and reptile distributions appear to provide support for this (Flessa 1973). This relationship is a vital part of MacArthur and Wilson's widely accepted equilibrium hypothesis. Since a smaller habitat area can only accommodate fewer taxa, reduction in area must lead to lower diversity as the extinction rate increases. Whether the extinction is due to reduced habitat diversity, increased competition, crowding effects or whatever, the basic empirical relationship appears to be well established.

Simberloff (1974) produced a model based on the species–area relationship and, using Schopf's (1974) data, was able to demonstrate an excellent correlation between the changing number of invertebrate families and the estimated area of epicontinental seas through the Permian to early Triassic time interval (Fig. 10.5). A similar exercise on Jurassic bivalve genera also showed a good correlation (Hallam 1977a). Kennedy's (1977) correlation of Mesozoic ammonite radiations and extinctions with estimated areas of continent flooded by sea is likewise impressive (Fig. 10.6).

Sepkoski (1976) undertook a multiple-regression analysis on Raup's data for Phanerozoic invertebrate species diversity which appears to offer good support for the species–area relationship in accord with equilibrium theory, with respect to the area of epicontinental seas, though it should be pointed out that some reservations have since been expressed (Flessa and Sepkoski 1978).

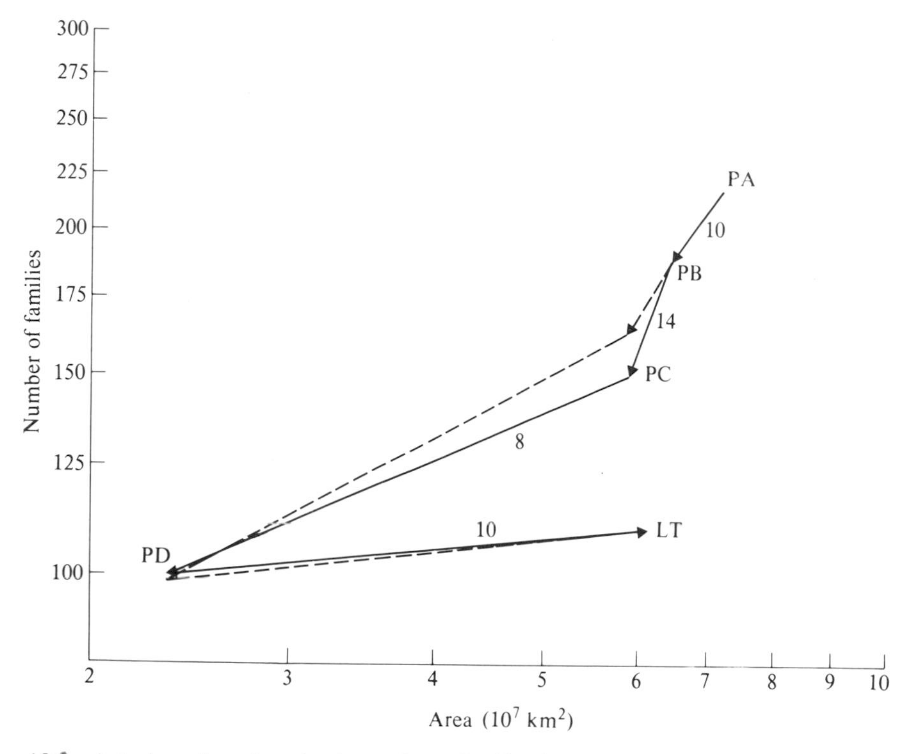

10.5 Actual number of marine invertebrate families (solid line) and number inferred from model based on species–area relationship, for Permian (PA–PD) and Lower Triassic (LT). Numbers on lines denote lengths of intervals in millions of years. After Simberloff (1974).

If anoxic events are indeed associated with transgressions, as argued in chapter 8, then the early Toarcian and perhaps also the late Devonian extinctions are also bound up with sea-level change.

Because of these demonstrated relationships, which take full account of the stratigraphic record and appear to hold over a wide range of time and on a variety of scales, the eustatic control hypothesis appears to be much the most promising way of accounting for alternating extinction and radiation episodes for shallow-marine invertebrates. It may be objected that the striking and rapid glacially-controlled regressions of the Quaternary did not result in widespread extinctions. This is not difficult to explain away if the phenomenon of biological adaptation is taken into account. In the case of the Quaternary sea-level fluctuations, regressions were followed by rapid transgressions after geologically short time intervals, limiting the effect of reduced habitat area and permitting a sufficient number of organisms to survive and expand their populations during the succeeding transgressions.

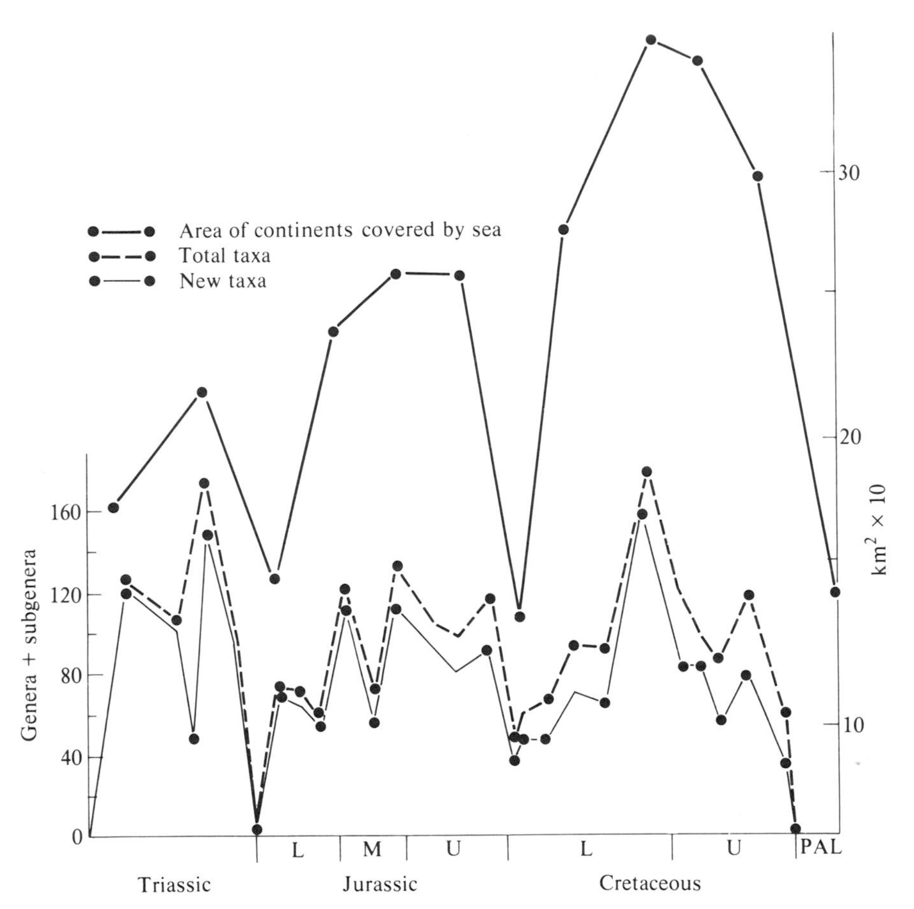

10.6 Ammonite appearances and extinctions in relation to the area of continents covered by sea. After Kennedy (1977).

The situation for the majority of Phanerozoic eustatic changes, not under glacial control, was very different. For long periods of time the epicontinental seas were very extensive and the world climate equable. Under these environmentally stable conditions a progressive 'fine tuning' of biologically competent species to the environment would have taken place, expressed by improved adaptations and in many cases phyletic size increase, generally known as Cope's Rule. These changes would have been promoted especially in reefal and peri-reefal habitats by the phenomenon of co-evolution, because biological interactions were probably considerable. Such organisms would be increasingly susceptible to environmental change, such as a regression-induced reduction of habitat area. Populations would gradually diminish below a critical threshold below which recovery was impossible.

The increased selection pressures would favour the opportunistic and stress-tolerant species, which would provide the primary source for adaptive radiation during the succeeding transgression, during which the evolutionary trend would often tend to be from the *r*-selected to the *K*-selected end of the spectrum. This model is testable in various ways (Hallam 1978b) and the whole subject provides a rich field for future research, on both major and minor extinction events.

A more serious problem concerns the mass extinctions of large land vertebrates, which in the case of the late Permian and late Cretaceous closely coincide in time with the marine extinctions, because the habitat area should have increased during regressive intervals. However, the creation of more extensive landmasses should have led to a greater 'continentality' of climate, with a higher annual range of temperature. The cold winter seasons might have been critical for the larger and more specialized vertebrates, whose smaller population and lower rates of reproduction would have made recovery from adverse environmental conditions more difficult. The feasibility of this proposal requires to be explored much more fully than it has been so far.

Perhaps the most baffling problem concerns the mass extinction of calcareous phyto- and zooplankton species at the end of the Cretaceous, because it is not readily apparent why they should have been seriously affected by regressions of epicontinental seas. Perhaps the coincidence of sea-level fall and general climatic deterioration provided a critical combination of factors, or perhaps there was indeed some unknown factor *X* at work.

An immense amount of work has been devoted to study of the terminal Cretaceous extinction phase, and is the subject of a recent symposium proceedings which contain a wealth of relevant information (Birkelund *et al.* 1979). The discovery that thin clay bands at the Cretaceous–Tertiary boundary at Gubbio and in Denmark are abnormally enriched in iridium has revived interest in a possible extraterrestrial event, such as the impact of a large planetesimal or asteroid (Alvarez *et al.* 1980; Hsü 1980; Napier and Clube 1979; see also Russell 1979). The platinum group metals are depleted in the Earth's crust relative to their cosmic abundance. Therefore concentrations of such elements as iridium in deep-sea sediments are likely to indicate influxes of extraterrestrial material. A continental impact of a large body could conceivably kill off selectively the larger terrestrial vertebrates by blast waves, while the climate might be affected by injection of dust into the stratosphere. An oceanic impact might cause a sharp rise of water temperature, as suggested by the isotopic data cited by Thierstein and Berger (1978). Could it, though, also cause a significant rise of the CCD, to account for the anomalous occurrence of clay bands in the calcareous sequence of Denmark and Gubbio, and a widespread disconformity suggestive of $CaCO_3$ dissolution in the deep ocean? How, furthermore, would an oceanic impact affect terrestrial organisms, or vice versa? Hsü (1980) proposes that large terrestrial organisms were killed off by atmospheric heating caused by a cometary impact, and that the extinction of calcareous marine plankton was a consequence of poisoning by cyanide released by the fallen comet and of a related catastrophic rise of the CCD.

Alternatively, impact of a large Earth-crossing asteroid would inject about 60 times the object's mass into the atmosphere as pulverized rock; a fraction of this dust would stay in the stratosphere for several years and spread across the world; the resulting darkness might suppress photosynthesis, with devastating biological consequences (Alvarez *et al.* 1980).

Nevertheless a much less spectacular explanation has not yet been decisively eliminated, namely that the iridium-rich clay bands merely mark the steady rain of micrometeorite material into the ocean at times when sedimentation rate was for some reason or other greatly reduced. Such material has, after all, long been known to be concentrated in abyssal red clay. At the present time virtually nothing is known of the amount of iridium present in other parts of the stratigraphic record, and many more analyses need to be undertaken before it can be safely claimed that the recorded occurrences are highly anomalous.

A further serious problem for the catastrophists concerns the recent challenge by the Italian geologist F. C. Wezel to the view that the classic Gubbio section records an undisturbed transition from the Cretaceous to the Tertiary. According to Wezel's detailed researches many of the presumed pelagic deposits are resedimented turbidites, with reworked foraminifera (Surlyk 1980).

It looks as though the end-Cretaceous extinctions provide a further example of what has been called the 'Vogt-Holden effect' (Birkelund *et al.* 1979). 'New data, regardless of reliable source or high quality, have scarcely ever ruled out any past theory, but have fueled the promulgation of newer and even more outlandish proposals.'

Those not so inclined to be as cynical or pessimistic might console themselves with Oscar Wilde's slightly less bleak dictum: truth is never pure and rarely simple.

Faunal provinces

With the wealth of information that has been published recently on palaeobiogeographic distributions and our greatly improved knowledge of past environments we should be in good position to draw reasonable conclusions about the controls on faunal provinciality. The two most widely accepted types of control involve climate and plate movements.

CLIMATE AND SEA-LEVEL CHANGES

Since climate has manifestly been of paramount importance in controlling faunal provinciality in the Quaternary it is not surprising that many palaeontologists have considered it to be a dominant influence in earlier times. In the equable Mesozoic, however, climatic effects must have played a lesser part, correspondingly more difficult to disentangle from other factors. The Jurassic is a good test case both because of the abundant data available and because of the insignificance of plate movements involving the continents, the only event of importance being the creation of a narrow ocean in the central part of the Atlantic towards the end of the period.

The most obvious large-scale effect of climate is to promote latitudinal differentiation of organisms, though factors such as seasonality and diurnal illumination may be at least as important as simple temperature change. As outlined in chapter 7 tropical and temperate belts can be distinguished among both the terrestrial plants and marine invertebrates, but the more interesting question concerns the origin and differentiation of the marine *Tethyan* and *Boreal* Realms or Superprovinces. These appeared fitfully in the early Jurassic but became well established in the mid-Jurassic and persisted into the early Cretaceous. Detailed documentation of the faunas is presented in Hallam (1975) and Casey and Rawson (1973).

With few exceptions, differentiation of a distinctive boreal fauna confined to the northern parts of Eurasia and North America took place only among the ammonites and belemnites. Otherwise the Boreal Realm is broadly characterized by reduced abundance and diversity of such characteristic Tethyan groups as lituolid foraminifera, hermatypic corals, hydrozoans and rudistid bivalves. On the other hand, most bivalve groups are at least as diverse in the Boreal Realm. The change from one realm to the other was gradational and fluctuated geographically with time.

The most popular interpretation has been that the boreal fauna evolved in waters that were too cool for the Tethyan faunas to invade, but this raises a problem. Contemporary terrestrial floras indicate that temperature gradients poleward from the tropics were much more modest than today, yet Jurassic ammonite faunas can change markedly within a degree of latitude, with few or even, in extreme cases, no genera common to the two realms, despite the fact that marine conditions are more equable than terrestrial.

Europe is the region best suited to a detailed analysis of this subject, and some years ago I undertook a study to investigate any possible relationship with facies (Hallam 1969). There indeed appears to be a relationship throughout most of the Jurassic, with the Tethyan Realm as defined by ammonites being generally characterized by calcareous and the Boreal Realm by siliciclastic facies (Fig. 10.7). (Elsewhere in the world Tethyan faunas may occur in siliciclastic facies.) Taking into account also the reduced faunal diversity to the north, the interpretation I favoured involved primarily salinity control, with the boreal faunas in Europe occupying a shallow sea

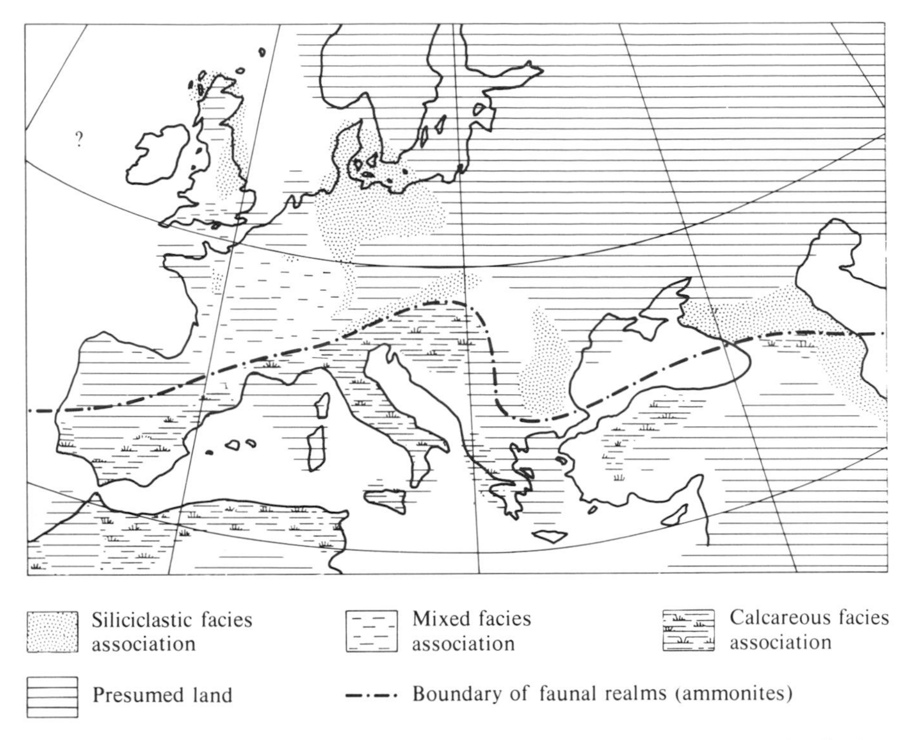

10.7 Palaeogeographic sketch map for the Pliensbachian of Europe, showing the distribution of facies associations and faunal realms. After Hallam (1969).

of slightly reduced salinity compared with the more oceanic Tethys, because of freshwater influx from the land in a comparatively restricted environment. This would also explain the dominance of siliciclastic deposits.

This hypothesis also has shortcomings, notably the presence in the boreal sea of organisms generally regarded as stenotopic marine, such as brachiopods, corals, echinoderms and, of course, ammonites, even though they might have been of reduced diversity. While such groups no doubt contained species that could tolerate slight reductions of salinity, it was difficult to conceive how such a sea could have persisted in a fairly constant state through such a long period of time. The more obviously brackish 'Rhaetic' lagoon was, after all, quite rapidly replaced in the early Jurassic by more normal sea following a eustatic rise (see chapter 5). Thus I abandoned the salinity-control hypothesis in favour of one invoking Sanders' (1968) concept of environmental stability, with the Tethyan faunas being more stenotopic and unable to tolerate a less stable or predictable environment in the shallow epicontinental boreal sea, with fluctuating salinity, temperature and perhaps oxygen content (Hallam 1975). Though there may well be some truth in this, it is still unsatisfactory as a comprehensive explanation, especially as Sanders' stability-time hypothesis has recently been under attack (e.g. Abele and Walters 1979). What appears to be lacking is an additional palaeogeographic component. It can hardly be a coincidence that the times when the Tethyan–Boreal ammonite differentiation was most acute (in the Bathonian and Tithonian/Volgian, so that precise correlation has so far proved impossible) correspond with phases of marine regression. A major episode of shallowing and withdrawal of the sea from extensive part of the North Atlantic region took place early in the mid-Jurassic, and the end-Jurassic regression was apparently world-wide. Conversely, times of transgression coincide with phases of faunal spread and radiation, as noted earlier.

The matter can be explored further with reference to one of the most abundant and diverse groups of Jurassic invertebrates, the bivalves. An analysis of genera across the world indicates a clear inverse correlation between the incidence of endemism and the area of continents covered by sea, which presumably reflects the relative height of sea level (Fig. 10.8).

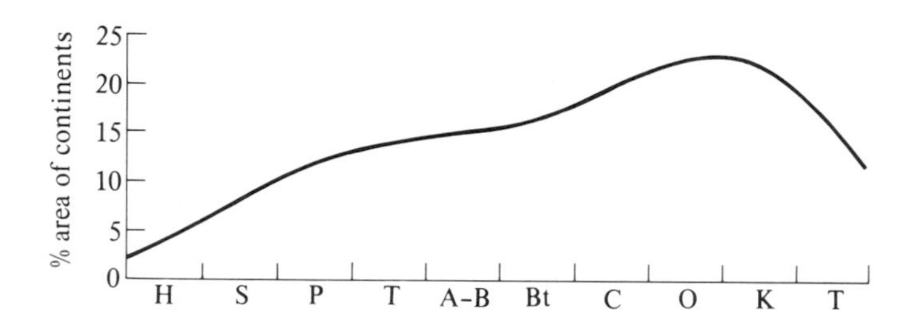

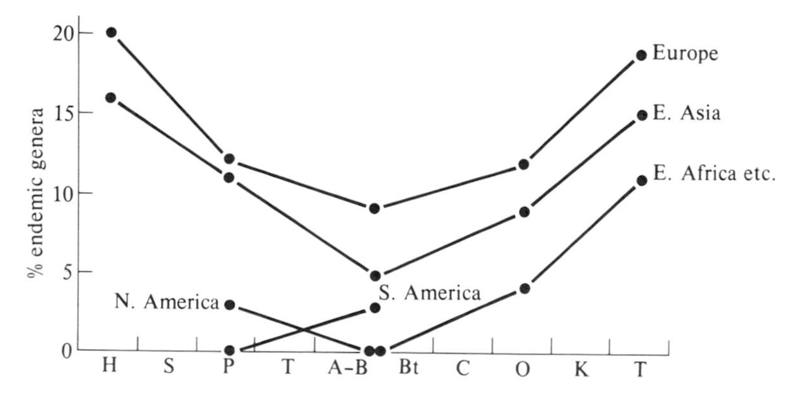

10.8 Temporal changes in endemism of Jurassic bivalve genera for different regions (lower graphs) and percentage of continents covered by sea through Jurassic time. For full stage names see appendix. After Hallam (1977a).

This is just the sort of relationship it is reasonable to expect. At times of low sea level and restriction of seas faunal migrations between continental shelf areas would be rendered more difficult and with less gene flow there would be more local speciation among the less dispersible organisms which occupied shallower water habitats.

It was argued at some length in chapter 5 that epicontinental seas were environments distinct in a number of ways from the open ocean system, and in which consequently one might well expect endemic faunas to evolve, with the degree of endemism exhibiting some relationship to the degree of restriction. In the Upper Cretaceous of the United States Western Interior there is an endemic fauna of ammonites, associated with an invertebrate fauna of reduced diversity compared with the Gulf Coast region (Kennedy and Cobban 1976). Analogy offers itself with the more extensive Jurassic–Lower Cretaceous boreal endemism, yet no one has to my knowledge suggested a simple climatic control. Other examples of epicontinental sea endemism among trilobites are mentioned below. Arkell (1956) proposed a significant palaeogeographic component to the Tethyan–Boreal endemism, a view also adopted by Fürsich and Sykes (1977) in their study of Oxfordian mollusc distributions in north-west Europe. There also may well have been an overprint of other factors such as climate, but I suspect that they were of subordinate importance. Perhaps more significant are biotic interactions. Once a boreal fauna had become established it might have competitively resisted Tethyan invasions. Thus Tethyan ammonites launched a mass invasion of northern Europe in the Toarcian only after the Pliensbachian endemic boreal family, the Amalthidae, had become extinct.

The proposed relationship between relative height of sea level and endemism needs to be tested for other fossil groups and systems. It may be thought to provide an alternative interpretation to the climatic one favoured by Cocks and McKerrow (1973) to account for an increase in brachiopod endemism from the Silurian to the early Devonian, because this seems to correlate broadly with the spread of epicontinental seas (see chapter 6). Indeed, Johnson and Boucot (1973) argue that the reduction of brachiopod endemism from the early to the late Devonian was the result of increasingly free communications as seaways became less restricted.

Further Palaeozoic examples evidently include Cambrian trilobites and Permian fusulinids. As epicontinental seas spread progressively through Cambrian time, endemism among the shallow-water trilobites should have decreased, while the late Permian drop in sea level should correlate with an increase in endemism among the fusulinid foraminifera, which appear to have favoured shallow-water habitats. The changing incidence of endemism through time in fact corresponds to the predictions in both cases (A. R. Palmer and T. Ozawa, personal communications).

PLATE MOVEMENTS

In attempting to establish the relationship between changing continental positions and faunal distributions it is important to disentangle the influence of facies, because properly defined faunal provinces should as far as possible be facies-independent. This is well illustrated by Lower Ordovician trilobites. Whittington and Hughes (1972) by means of a statistical analysis distinguished the following four provinces, which were used as a guide to reconstructing continental positions:

Selenopeltis	(Great Britain, France, Morocco, Czechoslovakia, together with China in the analysis of families)
Asaphid	(Poland, Sweden, Estonia)
Bathyurid	(Newfoundland, north-east USSR, USA, Spitsbergen, Kazakhstan)
Asaphopsis	(South America, Australia)

Fortey (1975) showed, however, that representatives of all provinces but the Selenopeltis were present in Spitsbergen, which was assigned by Whittington and Hughes exclusively to the Bathyurid Province. He was able to demonstrate that the Spitzbergen faunas could be grouped into three communities related to distinctive facies interpreted in terms of depth of sea (Fig. 10.9). Only the shallowest water, illaenid–cheirurid community exhibits high endemicity, and is therefore the one best suited to studies of provinciality. An analogous facies-related distribution of Cambrian trilobites was also recognized by Palmer (1973) with the relatively pandemic agnostids being the least useful for palaeobiogeographic analysis.

Such patterns can be recognized for other groups and other systems. Thus the well-known, presumed depth-related Silurian brachiopod communities

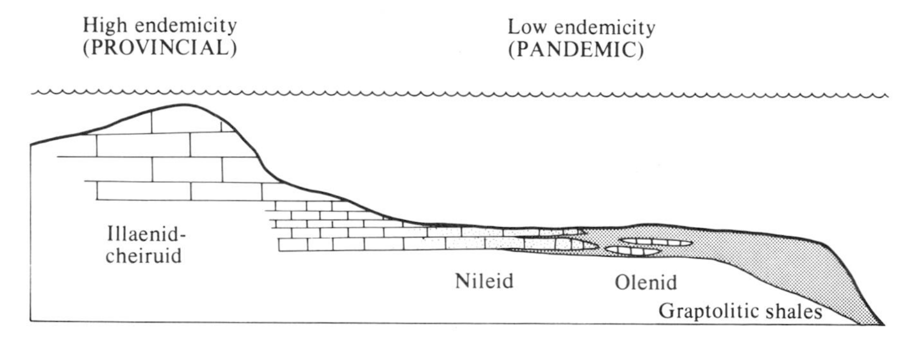

10.9 Palaeogeographic model showing disposition of community types across an early Ordovician epicontinental sea. Adapted from Fortey (1975).

(Cocks and McKerrow 1978) spring readily to mind. Similarly it has been possible to show that some Jurassic bivalve groups, notably certain pterioids, had relatively wide distributions and tended to inhabit deeper waters, while others such as the hippuritoids and trigonioids have the highest proportion of endemic genera (Hallam 1977a).

By taking into account what is known about the dispersion potential of living invertebrate larvae I have proposed the distinction of two patterns of changing faunal distribution with time, related to plate movements (Hallam 1973). *Convergence* refers to the degree of resemblance of faunas in different regions increasing from an earlier to a later time (Fig. 10.10) and *divergence* refers to the reverse phenomenon. Table 10.2 lists a number of examples, which also include non-marine vertebrates.

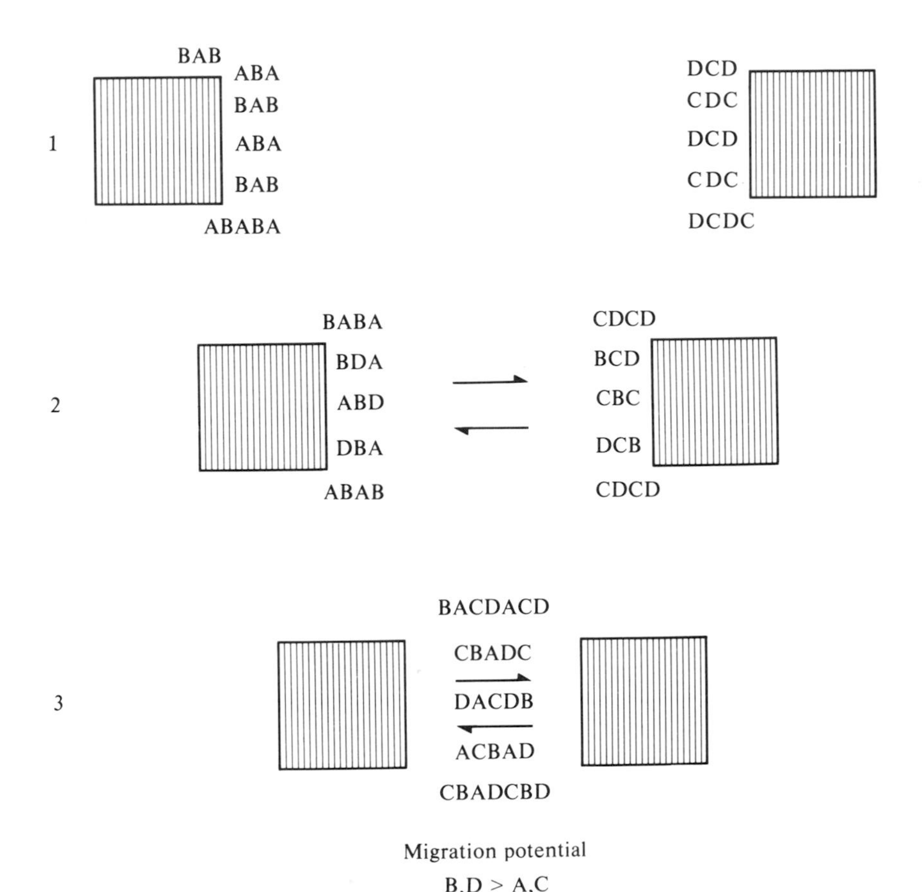

10.10 Schematic representation of convergence of faunas of different continents, signified by vertical lines. As the continents approach each other as a result of seafloor spreading, the faunas (symbolized by A, B, C and D) progressively merge. After Hallam (1974).

TABLE 10.2 Correlation of plate tectonic events and changes in faunal distributional patterns*

Plate tectonic event	*Convergence*	*Divergence*
Closure of Proto-Atlantic (Ordovician, Silurian)	Trilobites, graptolites, corals, brachiopods, conodonts, anaspids and thelodonts of the two continents flanking the Proto-Atlantic	
Closure of Urals Seaway	Post-Permian continental vertebrates of Eurasia	
Opening of Atlantic (Cretaceous, Tertiary)		Cretaceous bivalves and benthic foraminifera of Caribbean and Mediterranean. Upper Cretaceous ammonites of USA and W. Europe–N. Africa. Post-Lower Eocene mammals of North America and Europe. Tertiary mammals of Africa and South America
Opening of Indian Ocean (Cretaceous)		Bivalves of East African and Indian shelves
Closure of Tethys (late Cretaceous) (mid-Tertiary)	?Ammonites of Eurasia and Africa–Arabia. Mammals of Eurasia and Africa	Molluscs, foraminifera etc. of Indian Ocean and Mediterranean–Atlantic

* References cited in Hallam (1974).

The early Palaeozoic closure of the Iapetus Ocean in the North Atlantic region is a particularly good example of convergence which affected many groups, and is independent of the influence of facies. Full documentation is provided by Ziegler *et al.* (1977). McKerrow and Cocks (1976) have sought to establish the relative rates of migration of different faunal groups across the ocean, and have attempted estimates, on the basis of pelagic larval migration and subduction rate data, of the changing width of the ocean with time. According to their interpretation the pelagic animals (graptolites) crossed first, followed later by animals (trilobites, brachiopods) with pelagic larval stages, but animals without a pelagic larval stage (benthic ostracods) were not able to cross until the ocean had closed at one point, though not necessarily everywhere along its length. Finally, faunas limited to fresh water or brackish water (like many Devonian fish) did not cross until there were non-marine connections between the continents on either side of the closing ocean (Fig. 10.11).

A third biogeographic pattern I have termed *complementarity*. Complementarity in the distributional changes of contiguous marine and terres-

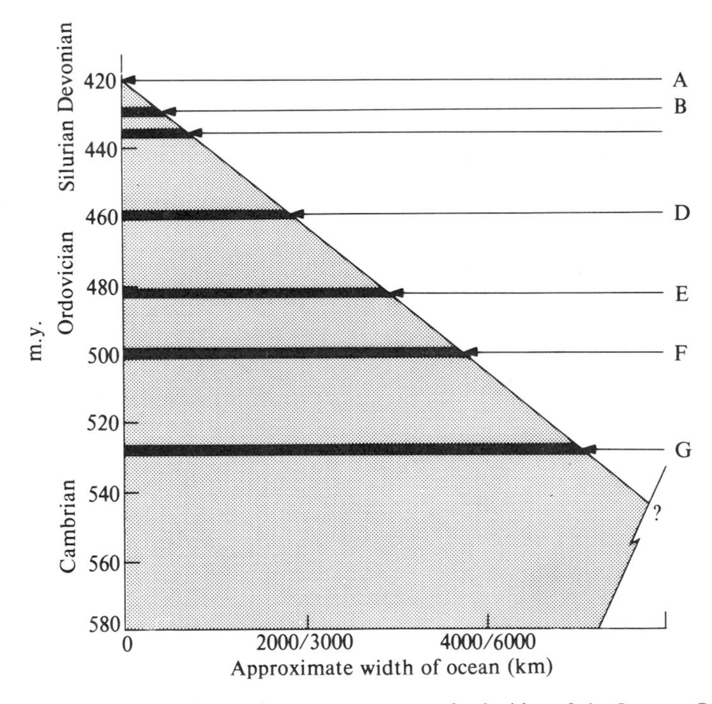

10.11 The times at which faunas became common to both sides of the Iapetus Ocean. A, Closure of ocean (Norway); B, freshwater fish; C, benthic ostracods; D, trilobite and brachiopod species; E, trilobite and brachiopod genera; F, *Didymograptus bifidus*; G, *Dictyonema*. After McKerrow and Cocks (1976).

trial animals is recognizable when one group exhibits convergence and the other divergence. This happens, for instance, when a land connection is created between two hitherto isolated areas of continent, so allowing convergence of the terrestrial faunas to take place, while severing of a once continuous land mass gives rise to divergence as a result of genetic isolation. Just the converse is true for the faunas of the seas which envelop the land masses (Fig. 10.12).

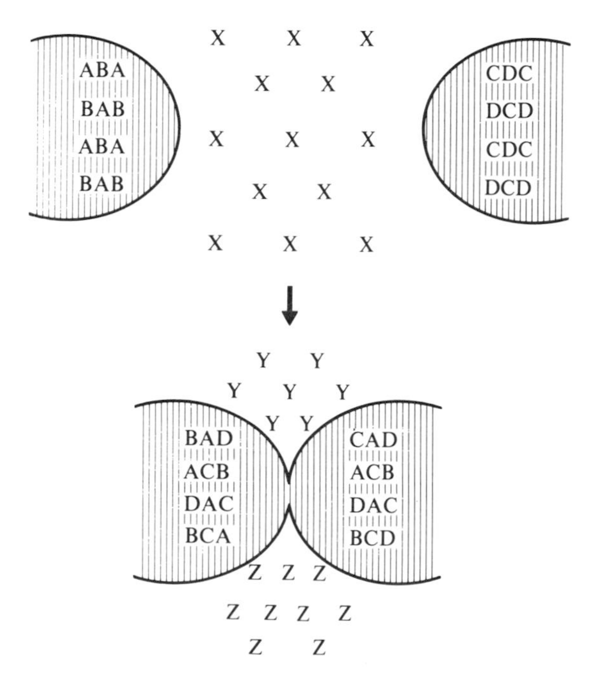

10.12 Complementarity in the distribution of terrestrial and marine faunas as two continents approach each other and eventually become sutured together. The terrestrial faunas exhibit convergence because of the creation of a corridor, while the marine organisms diverge from common ancestors X to faunas Y and Z. After Hallam (1974).

A classic example concerns the creation of the Panama Isthmus in the late Pliocene, which appears to be associated with a young phase of spreading between the Cocos and Nazca plates east of the East Pacific Rise (Atwater 1973). This led to pronounced cross migration of mammals between the previously isolated North and South American continents. Webb's (1976) palaeoecological study of this celebrated interchange reveals two major phases. The first, in the early Pliocene, was limited to a few broadly adapted herbivores and carnivores which might have been able to cross narrow water gaps. The second and more important phase occurred in the late

Pliocene–early Pleistocene and involved a diversity of adaptive types including freshwater herbivores, a sure indication of land connection.

The existence on the Pacific and Caribbean sides of the isthmus of numerous closely related *cognate* or *geminate* species indicates recent derivation from common ancestors which lived at a time when no land barrier existed (Vermeij 1978). Keigwin (1978) has established from his studies of planktonic foraminifera that a significant divergence of faunas took place in the time interval 3.5–3.1 million years ago, which must correspond to the time of formation of the isthmus.

A second example relates to the closure of Tethys in the Old World, a phenomenon apparently bound up with collision of Africa–Arabia and Eurasia. In the early Tertiary, invertebrate faunas exhibited a high degree of similarity throughout the length of the Tethys, but in the late Oligocene or early Miocene this general unity was abruptly ended. Indian Ocean benthic foraminifera and molluscs differ strikingly from those of the Mediterranean region, implying closure of the seaway in the Middle East by creation of a land corridor between Africa and Asia (Adams 1967; Hallam 1967c). At about the same time, significant cross migration of African and Eurasian mammals commenced across the Middle East corridor (Coryndon and Savage 1973).

This poses an intriguing problem. Dewey *et al.* (1973), in their comprehensive plate tectonic analysis of the Tethyan region, maintain that a continuous sialic connection from Arabia to Asia was not established until as late as Pliocene times, as the northern margin of Arabia collided with a trench-arc complex on the southern margin of Iran, and the Zagros Crush Zone was initiated. Yet there is unequivocal evidence along a zone extending from Oman through southern Iran and Turkey to the eastern Mediterranean of ophiolite obduction in the Maastrichtian, after which sedimentation patterns changed drastically; this must surely relate to a major collision event (see chapter 4). There is presumed to have been another subduction zone through Mesozoic times further north, along the line of the Caucasus and Pontide ranges of northern Turkey, so that a marine strait could have existed until quite late in the Tertiary, but no convincing story of continental collision has yet been put forward for this region.

It looks therefore as though the late Oligocene complementarity phenomenon relates directly to the worldwide episode of regression at that time. Previous to that time but after the late Cretaceous collision event a shallow marine strait on continental crust must have persisted along the line of the Tethys, allowing free migration of invertebrates. So once again sea-level change appears to rear its (I hope not ugly) head.

There are also difficulties concerning the Dewey *et al.* reconstruction for the Mediterranean region, which presents a very complex kinematic jigsaw puzzle which is as yet far from being adequately resolved. Dewey *et al.* argue for Miocene collision of North Africa with the Iberian Peninsula along the line of the Betic Cordillera, and the conversion of several oceanic straits into collisional sutures in the Alpine–Appenine region at this time. On the other hand, Laubscher and Bernoulli (1977) criticize the Dewey scheme in many points of detail and indicate the ubiquity of evidence of compressive tectonic movements from the Apennines and Alps to Turkey and beyond, from mid-Cretaceous times. Milnes's (1978) detailed analysis of the Alps suggests that the final stages of continental underthrusting, leading to locking together of the separate landmasses, had been completed by the early Oligocene.

A comparable disparity between plate tectonic reconstructions and local geological and palaeontological evidence is evident in the case of India. According to Smith and Briden's (1977) maps collision of India with Asia along the line of the Himalayas did not take place until the early Miocene, but Powell and Conaghan (1973) infer from the dating of ophiolite obduction and associated sedimentation that it must have occurred shortly before the mid-Eocene. This interpretation is supported by Molnar and Tapponier (1975), who point out a sharp reduction in seafloor spreading rate in the plate carrying India subsequent to the late Eocene. Continued spreading at a reduced rate after Eocene collision produced further convergence of India against continental crust, which has caused extensive deformation by strike-slip faulting over a large area of the Asian landmass. Further confirmation comes from recent palaeomagnetic work (Klootwijk and Peirce 1979) and from palaeobiogeography. Sahni and Kumar (1974) establish that the earliest Indian mammals are of mid-Eocene age and constitute quite a rich fauna exhibiting strong affinities to the Mongolian fauna; a free land connection is evidently required.

Yet a further example concerns the western end of Tethys, between North and South America. This region has been analysed by Ladd (1976) on the basis of the implications of Atlantic spreading data and consideration of Caribbean tectonics. According to Ladd, an appreciable separation of the two American continents by seaway was in existence in the late Cretaceous and earliest Tertiary, with convergent motion between them not commencing until the late Eocene to Miocene interval. Yet Rage (1978) insists, because of land vertebrate distributions, that a continental connection must have existed shortly before the end of the Cretaceous.

Have the standard plate tectonic reconstructions therefore grossly overestimated the width of the Mesozoic Tethyan Ocean? The problem disappears in the novel reconstruction of Owen (1976), but he argues for an Earth that has been rapidly expanding since the Palaeozoic, an interpretation justifiably viewed with considerable scepticism by most Earth scientists. But that is another story.

Appendix

List of the more widely used pre-Quaternary stratigraphic stage terms. Worldwide correlation of the Ordovician and Silurian is usually specified at the level of series, while there are no generally applied stage or series terms for the Cambrian.

SYSTEM	STAGE
Pliocene	Astian Zanclian
Miocene	Messinian Tortonian Langhian Burdigalian Aquitainian
Oligocene	Chattian Rupelian Lattorfian
Eocene	Bartonian Lutetian Ypresian
Palaeocene	Thanetian Danian
Cretaceous	Maastrichtian Campanian Santonian Coniacian Turonian Cenomanian Albian Aptian Barremian Hauterivian Valanginian Berriasian

SYSTEM		STAGE
Jurassic		Tithonian
		Kimmeridgian
		Oxfordian
		Callovian
		Bathonian
		Bajocian
		Aalenian
		Toarcian
		Pliensbachian
		Sinemurian
		Hettangian
Triassic		Norian
		Carnian
		Ladinian
		Anisian
		Scythian
Permian		Tatarian
		Kazanian
		Artinskian
		Sakmarian
Carboniferous		Stephanian
		Westphalian
		Namurian
	Dinanian	Visean
		Tournaisian
Devonian		Famennian
		Frasnian
		Givetian
		Emsian
		Siegenian
		Gedinnian
Silurian		Pridoli
		Ludlow
		Wenlock
		Llandovery

SYSTEM	STAGE
Ordovician	Ashgill Caradoc Llandeilo Llanvirn Arenig Tremadoc
Cambrian	

Bibliography

Abele, L. G. and Walters, K. (1979). Marine benthic diversity: a critique and alternative explanation. *J. Biogeog.* **6,** 115–126.

Adams, C. G. (1967). Tertiary foraminifera in the Tethyan, American and Indo-Pacific provinces. In: Adams, C. G. and Ager, D. V. (eds), *Aspects of Tethyan biogeography*, Publ. System. Ass. no. 7, 195–218.

Adams, C. G., Benson, R. H., Kidd, R. B., Ryan, W. B. F. and Wright, R. C. (1977). The Messinian salinity crisis and evidence of late Miocene eustatic changes in the world ocean. *Nature* **269,** 383–386.

Ager, D. V. (1973). *The nature of the stratigraphic record.* Macmillan, London.

Allen, J. R. L. (1964). Studies in fluviatile sedimentation: six cyclothems from the Lower Old Red Sandstone, Anglo-Welsh Basin. *Sedimentology* **3,** 163–198.

Allen, J. R. L. (1970). *Physical processes of sedimentation.* Allen and Unwin, London.

Allen, J. R. L. (1974). Studies in fluviatile sedimentation: implications of pedogenic carbonate units, Lower Old Red Sandstone, Anglo-Welsh outcrop. *Geol. J.* **9,** 181–208.

Allen, P. (1975). Wealden of the Weald: a new model. *Proc. geol. Ass. Lond.* **86,** 389–436.

Alvarez, L. W., Alvarez, W., Asaro, F. and Michel, H. V. (1980). Extraterrestrial cause for the Cretaceous–Tertiary extinction: experiment and theory. *Science* **208,** 1095–1108.

Alvarez, W. *et al.* (1977). Type section for the Late Cretaceous-Paleocene geomagnetic reversal time scale. *Bull. geol. Soc. Am.* **88,** 367–389.

Alvarez, W. *et al.* (1979). Comments and replies on biostratigraphy and magneto-stratigraphy of Paleocene terrestrial deposits, San Juan Basin, New Mexico. *Geology* **7,** 66–71.

Anderson, R. Y. and Kirkland, D. W. (1966). Intrabasin varve correlation. *Bull. geol. Soc. Am.* **77,** 241–256.

Arkell, W. J. (1933). *The Jurassic System in Great Britain.* Oxford Univ. Press.

Arkell, W. J. (1956). *Jurassic geology of the world.* Oliver and Boyd, Edinburgh.

Atwater, T. (1973). Studies of sea floor spreading and plate movements in the Pacific Basin. In: Tarling, D. H. and Runcorn, S. K. (eds), *Implications of continental drift to the earth sciences.* Academic Press, London and New York, 213–217.

Aubouin, J. (1965). *Geosynclines.* Elsevier, Amsterdam.

Badham, J. P. N. and Stanworth, C. W. (1977). Evaporites from the lower Proterozoic of the East Arm, Great Slave Lake. *Nature* **268,** 516–518.

Bakker, R. T. (1977). Tetrapod mass extinctions—a model of the regulation of speciation rates and immigration by cycles of topographic diversity. In: A. Hallam (ed), *Patterns of evolution*. Elsevier, Amsterdam, 439–468.

Bambach, R. K. (1977). Species richness in marine benthic habitats through the Phanerozoic. *Paleobiology* **3,** 152–167.

Barley, M. E., Dunlop, J. S. R., Glover, J. E. and Groves, D. I. (1979). Sedimentary evidence for an Archaean shallow-water volcanic-sedimentary facies, eastern Pilbara Block, Western Australia. *Earth planet. Sci. Lett.* **43,** 74–84.

Barnard, P. D. W. (1973). Mesozoic floras. In: Hughes, N. F. (ed), *Organisms and continents through time.* Spec. Pap. Palaeont. no. 12, 175–188.

Bathurst, R. G. C. (1975). *Carbonate sediments and their diagenesis*, 2nd edn. Elsevier, Amsterdam.

Benson, R. A. (1975). The origin of the psychrosphere as recorded in changes in deep-sea ostracode assemblages. *Lethaia* **8,** 69–83.

Berger, W. H. and Thierstein, H. R. (1979). On Phanerozoic mass extinctions. *Naturwissenschaften* **66,** 46–47.

Berger, W. H. and Winterer, E. L. (1974). Plate stratigraphy and the fluctuating carbonate line. In: Hsü, K. J. and Jenkyns, H. C. (eds), *Pelagic sediments: on land and under the sea.* Spec. Publ. int. Ass. Sediment. no. 1, 11–48.

Berggren, W. A. and Hollister, C. D. (1977). Plate tectonics and paleocirculation—commotion in the ocean. *Tectonophysics* **38,** 11–48.

Berkner, L. V. and Marshall, L. C. (1965). On the origin and rise of oxygen concentration in the earth's atmosphere. *J. atmos. Sci.* **22,** 225–261.

Bernoulli, D. (1972). North Atlantic and Mediterranean Mesozoic facies; a comparison. *Init. Rep. Deep Sea Drilling Project* **9,** 801–871.

Bernoulli, D. and Jenkyns, H. C. (1974). Alpine, Mediterranean and Central Atlantic Mesozoic facies in relation to the early evolution of the Tethys. In: Dott, R. H. and Shaver, R. H. (eds), *Modern and ancient geosynclinal sedimentation.* Spec. Publ. Soc. econ. Paleont. Miner. no. 19, 129–160.

Bernoulli, D. and Wagner, C. W. (1971). Subaerial diagenesis and fossil caliche in the Calcare Massicio Formation (Lower Jurassic, Central Apennines, Italy). *N. Jb. Geol. Paläont. Abh.* **138,** 135–149.

Beuf, S., Biju-Duval, B., Charpal, O., Rognon, P., Gabriel, O. and Bennacef, A. (1971). *Les grès du paléozoique inférieur au Sahara.* Editions Technip, Paris.

Bird, J. M. and Dewey, J. F. (1970). Lithospheric plate-continental margin tectonics and the evolution of the Appalachian orogen. *Bull. geol. Soc. Am.* **81,** 1031–1060.

Birkelund, T., Bromley, R. G. and Christensen, W. K. (eds) (1979). *Cretaceous—Tertiary boundary events.* Univ. Copenhagen, 2 vols.

Black, D. I. (1967). Cosmic ray effects and faunal extinctions at geomagnetic field reversals. *Earth planet. Sci. Lett.* **3,** 225–236.

Blatt, H., Middleton, G. V. and Murray, R. C. (1979). *Origin of sedimentary rocks.* 2nd edn. Prentice Hall, New Jersey.

Bond, G. (1978). Speculations on real sea-level changes, and vertical motions of continents at selected times in the Cretaceous and Tertiary periods. *Geology* **6,** 247–250.

Borchert, H. and Muir, R. O. (1964). *Salt deposits: the origin, metamorphism and deformation of evaporites.* Van Nostrand, London.

Bott, M. H. P. (ed) (1976). Sedimentary basins of continental margins and cratons. *Tectonophysics* **36,** 1–314.

Boucot, A. J. (1974). Silurian and Devonian biogeography. In: Ross, C. A. (ed),

Paleogeographic provinces and provinciality. Spec. Publ. Soc. econ. Paleont. Miner. no. 21, 165–176.

Bradley, W. H. (1929). The varves and climate of the Green River epoch. *Prof. Pap. U.S. geol. Surv.* no. 158, 87–110.

Bradley, W. H. (1964). Geology of Green River Formation and associated Eocene rocks in southwestern Wyoming and adjacent parts of Colorado and Utah. *Prof. Pap. U.S. geol. Surv.* no. 496–A.

Bradshaw, M. J. (1978). *A facies analysis of the Bathonian of eastern England.* Unpubl. DPhil. thesis, Univ. Oxford.

Brasier, M. D. (1979). The Cambrian radiation event. In: House, M. R. (ed), *The origin of major invertebrate groups*. Academic Press, London and New York, 103–159.

Bridges, P. H. (1976). Lower Silurian transgressive barrier islands. *Sedimentology* **23,** 347–362.

Bromley, R. G. (1967). Marine phosphorites as depth indicators. *Mar. Geol.* **5,** 503–509.

Brookfield, M. E. (1971). An alternative to the 'clastic trap' interpretation of oolitic ironstone facies. *Geol. Mag.* **108,** 137–143.

Buchardt, B. (1978). Oxygen isotope palaeotemperatures from the Tertiary period in the North Sea area. *Nature* **275,** 121–123.

Bull, W. B. (1972). Recognition of alluvial fan deposits in the stratigraphic record. In: Rigby, J. K. and Hamblin, W. K. (eds), *Recognition of ancient sedimentary environments*. (Spec. Publ. Soc. econ. Miner. Paleont. no. 16, 63–83.

Burke, K. and Dewey, J. F. (1973). Plume-generated triple junctions: key indicators in applying plate tectonics to old rocks. *J. Geol.* **81,** 406–433.

Butler, R. F., Lindsay, E. H., Jacobs, L. L. and Johnson, N. M. (1977). Magnetostratigraphy of the Cretaceous–Tertiary boundary in the San Juan Basin, New Mexico. *Nature* **267,** 318–323.

Calvert, S. E. (1966). Origin of diatom-rich, varved sediments from the Gulf of California. *J. Geol.* **74,** 546–565.

Calvert, S. E. (1974). Deposition and diagenesis of silica in marine sediments. In: Hsü, K. J. and Jenkyns, H. C. (eds), *Pelagic sediments: on land and under the sea.* Spec. Publ. int. Ass. Sedimentol. no. 1, 273–299.

Carroll, D. (1958). Role of clay minerals in the transportation of iron. *Geochim. Cosmochim. Acta* **14,** 1–27.

Casey, R. and Rawson, P. F. (eds) (1973). *The boreal Lower Cretaceous.* Seel House Press, Liverpool.

Chaloner, W. S. and Creber, G. T. (1973). Growth rings in fossil woods as evidence of past climates. In: Tarling, D. H. and Runcorn, S. K. (eds), *Implications of continental drift to the Earth Sciences.* Academic Press, London and New York, 425–437.

Chaloner, W. G. and Lacey, W. S. (1973). The distribution of late Palaeozoic floras. In: Hughes, N. F. (ed), *Organisms and continents through time.* Spec. Pap. Palaeont. no. 12, 271–289.

Chamberlin, T. C. (1909). Diastrophism as the ultimate basis of correlation. *J. Geol.* **17,** 689–693.

Chaney, R. W. (1944). Summary and conclusions. In: Chaney, R. W. (ed), *Pliocene floras of California and Oregon.* Publ. Carnegie Inst. Washington no. 553, 353–383.

Chang, K. H. (1975). Concepts and terms of unconformity–bounded units as formal stratigraphic units of distinct category. *Bull. geol. Soc. Am.* **86,** 1544–1552.

Churkin, M. (1974). Paleozoic marginal ocean-basin-volcanic arc systems in the Cordilleran Foldbelt. In: Dott, R. H. and Shaver, R. H. (eds), *Modern and ancient geosynclinal sedimentation*, Spec. Publ. Soc. econ. Paleont. Miner. no. 19, 174–192.

Ciesielski, P. F. and Weaver, F. M. (1974). Early Pliocene temperature changes in the Antarctic seas. *Geology* **2,** 511–515.

Cita, M. B. (1973). Mediterranean evaporite: paleontological arguments for a deep-basin desiccation model. In: Drooger, C. W. (ed.), *Messinian events in the Mediterranean.* North-Holland, Amsterdam, 206–228.

Clark, D. L. and Kitchell, J. A. (1979). Comment on 'The terminal Cretaceous event: a geologic problem with an oceanographic solution'. *Geology* **7,** 228–229.

Clemmensen, L. (1978). Alternating aeolian, sabkha and shallow-lake deposits from the Middle Triassic Gipsdalen Formation, Scoresby Land, East Greenland. *Palaeogeog., Palaeoclimatol., Palaeoecol.* **24,** 111–135.

Clemmey, H. (1978). A Proterozoic lacustrine interlude from the Zambian Copperbelt. In: Matter, A. and Tucker, M. (eds), *Modern and ancient lake sediments.* Spec. Publ. int. Ass. Sedimentol. no. 2, 259–278.

Cloud, P. (1972). A working model of the primitive earth. *Am. J. Sci.* **272,** 537–548.

Cloud, P. E. (1973). Paleoecological significance of the banded iron-formation. *Econ. Geol.* **68,** 1135–1145.

Coates, A. G. and Oliver, W. A. (1973). Coloniality in zoantharian corals. In: Boardman, R. S., Cheetham, A. H. and Oliver, W. A. (eds), *Animal colonies: development and function through time.* Dowden, Hutchinson and Ross, Strandsburg, Pa, 3–27.

Cocks, L. R. M. and McKerrow, W. S. (1973). Brachiopod distributions and faunal provinces in the Silurian and Lower Devonian. In: Hughes, N. R. (ed), *Organisms and continents through time.* Spec. Pap. Palaeont. no. 12, 291–304.

Cocks, L. R. M. and McKerrow, W. S. (1978). Silurian. In: McKerrow, W. S. (ed), *The ecology of fossils.* Duckworth, London, 93–124.

Coleman, J. M. (1966). *Recent coastal sedimentation: central Louisiana Coast.* Louisiana State Univ. Press. Coastal Studies Series no. 17.

Coleman, J. M. and Wright, L. D. (1975). Modern river deltas: variability of processes and sand bodies. In: Broussard, M. L. (ed), *Deltas, models for exploration.* Houston geol. Soc., Houston, 99–149.

Collinson, J. D. (1972). The Røde Ø Conglomerate of Inner Scoresby Sund and the Carboniferous (?) and Permian rocks west of the Schuchert Flod. *Medd. Grønland* **192,** 1–48.

Collinson, J. D. (1978a). Alluvial sediments. In: Reading, H. G. (ed), *Sedimentary environments and facies.* Blackwell, Oxford, 15–60.

Collinson, J. D. (1978b). Deserts. In: Reading, H. G. (ed), *Sedimentary environments and facies.* Blackwell, Oxford, 80–96.

Collinson, J. D. (1978c). Lakes. In: Reading, H. G. (ed), *Sedimentary environments and facies.* Blackwell, Oxford, 80–96.

Conant, L. C. and Swanson, V. E. (1961). Chattanooga Shale and related rocks of Central Tennessee and nearby areas. *Prof. Pap. U.S. geol. Surv.* no. 357.

Conway Morris, S. and Whittington, H. B. (1979). The Animals of the Burgess Shale. *Sci. Am.* **241** (1), 122–133.

Cook, T. D. and Bally, A. W. (1977). *Stratigraphic atlas of North and Central America.* Princeton Univ. Press.

Coope, G. R. (1977). Fossil coleopteran assemblages as sensitive indicators of climatic changes during the Devensian (Last) cold stage. *Phil. Trans. R. Soc.* B**280,** 313–340.

Cooper, M. R. (1977). Eustacy during the Cretaceous: its implications and importance. *Palaeogeog., Palaeoclimatol., Palaeoecol.* **22,** 1–60.

Copper, P. (1977). Paleolatitudes in the Devonian of Brazil and the Frasnian–Famennian mass extinction event. *Palaeogeog., Palaeoclimatol., Palaeoecol.* **21,** 165–208.

Corliss, B. H. (1979). Response of deep-sea benthonic Foraminifera to development of the psychrosphere near the Eocene/Oligocene boundary. *Nature* **282,** 63–65.

Coryndon, S. C. and Savage, R. J. G. (1973). The origin and affinities of African mammal faunas. In: Hughes, N. F. (ed), *Organisms and continents through time.* Spec. Pap. Palaeont. no. 12, 121–135.

Crain, I. K. (1971). Possible direct causal relation between geomagnetic reversals and biological extinctions. *Bull. geol. Soc. Am.* **82,** 2603–2606.

Crowell, J. C. (1974). Origin of late Cenozoic basins in southern California. In: Dickinson, W. R. (ed), *Tectonics and sedimentation.* Spec. Publ. Soc. econ. Paleont. Mineral. no. 22, 190–204.

Crowell, J. C. (1978). Gondwanan glaciation, cyclothems, continental positioning, and climate change. *Am. J. Sci.* **278,** 1345–1372.

Crowell, J. C. and Frakes, L. A. (1975). The late Palaeozoic glaciation. In: Campbell, K. S. W. (ed), *Gondwana geology*. Aust. Nat. Univ. Press, Canberra.

Cummins, W. A. (1962). The greywacke problem. *Liverpool Manchester Geol. J.* **3,** 51–72.

Curtis, C. D. and Spears, D. A. (1968). The formation of sedimentary iron minerals. *Econ. Geol.* **63,** 257–270.

Cutbill, J. L. and Funnell, B. M. (1967). Computer analysis of the fossil record. In: Harland, W. B. *et al.* (eds), *The fossil record.* Geol. Soc. Lond., 791–820.

Daley, B. (1972). Some problems concerning the early Tertiary climate of southern Britain. *Palaeogeog., Palaeoclimatol., Palaeoecol.* **11,** 177–190.

Davies, D. K., Ethridge, F. G. and Berg, R. R. (1971). Recognition of barrier environments. *Bull. Am. Ass. petrol. Geol.* **55,** 550–565.

Davies, G. R. and Ludlam, S. D. (1973). Origin of laminated and graded sediments, Middle Devonian of western Canada. *Bull. geol. Soc. Am.* **84,** 3527–3546.

Davies, T. A., Hay, W. W., Southam, J. R. and Worsley, T. R. (1977). Estimates of Cenozoic oceanic sedimentation rates. *Science* **197,** 53–55.

Degens, E. T. and Stoffers, P. (1976). Stratified waters as a key to the past. *Nature* **263,** 22–27.

Degens, E. T. and Stoffers, P. (1980). Environmental events recorded in Quaternary sediments of the Black Sea. *J. geol. Soc. Lond.* **137,** 131–138.

De Raaf, J. F. M., Boersma, J. R. and Van Gelder, A. (1977). Wave-generated structures and sequences from a shallow marine succession, Lower Carboniferous, County Cork, Ireland. *Sedimentology* **24,** 451–483.

Dewey, J. F., Pitman, W. C., Ryan, W. B. F. and Bonnin, J. (1973). Plate tectonics and the evolution of the Alpine system. *Bull. geol. Soc. Am.* **84,** 3134–3180.

Dickinson, W. R. (ed) (1974). *Tectonics and sedimentation.* Spec. Publ. Soc. econ. Paleont. Miner. no. 22.

Dimroth, E. and Kimberley, M. E. (1976). Precambrian atmospheric oxygen: evidence in the sedimentary distributions of carbon, sulfur, uranium and iron. *Can. J. Earth Sci.* **13,** 1161–1185.

Donovan, D. T. and Jones, E. J. W. (1979). Causes of world-wide changes in sea level. *J. geol. Soc. Lond.* **136,** 187–192.

Dörjes, J. and Hertweck, G. (1975). Recent biocoenoses and ichnocoenoses in shallow-water marine environments. In: Frey, R. W. (ed), *The study of trace fossils.* Springer Verlag, New York, 459–491.

Dorman, F. H. and Gill, E. D. (1959). Oxygen isotope paleotemperature determinations of Australian Cainozoic fossils. *Science* **130,** 1576.

Dott, R. H. (1974). The geosynclinal concept. In: Dott, R. H. and Shaver, R. H. (eds), *Modern and ancient geosynclinal sedimentation.* Spec. Publ. Soc. econ. Paleont. Miner. no. 19, 1–13.

Dott, R. H. and Shaver, R. H. (eds) (1974). *Modern and ancient geosynclinal sedimentation.* Spec. Publ. Soc. econ. Paleont. Miner. no. 19.

Douglas, R. G. and Savin, S. M. (1975). Oxygen and carbon isotope analyses of Cretaceous and Tertiary microfossils from Shatsky Rise and other sites in the North Pacific Ocean. *Init. Rep. Deep Sea Drilling Project* **32,** 509–520.

Drever, J. I. (1974). Geochemical model for the origin of Precambrian banded iron formations. *Bull. geol. Soc. Am.* **85,** 1099–1106.

Drewry, G. E., Ramsay, A. T. S. and Smith, A. G. (1974). Climatically controlled sediments, the geomagnetic field and trade wind belts in Phanerozoic time. *J. Geol.* **82,** 531–553.

Drooger, C. W. (ed) (1973). *Messinian events in the Mediterranean.* North-Holland, Amsterdam.

Du Dresnay, R. (1977). Le milieu récifal fossile du Jurassique inférieur (Lias) dans le domaine des Chaines atlasiques du Maroc. *Mém. Bull. Rech. Géol. Min.* no. 89, 296–312.

Duff, P. McL. D., Hallam, A. and Walton, E. K. (1967). *Cyclic sedimentation.* Elsevier, Amsterdam.

Dunbar, K. O. and Rodgers, J. (1957). *Principles of stratigraphy.* Wiley, New York.

Dunham, R. J. (1969). Vadose pisolite in the Capitan Reef (Permian), New Mexico and Texas. In: Friedman, G. M. (ed), *Depositional environments in carbonate rocks.* Spec. Publ. Soc. econ. Paleont. Miner. no. 14, 182–191.

Dunham, R. J. (1972). Guide for study and discussion for individual reinterpretation of the sedimentation and diagenesis of the Permian Capitan geologic reef and associated rocks, New Mexico and Texas. In: *Permian Basin section.* Publ. Soc. econ. Paleont. Miner. no. 72–14.

Dunlop, J. S. R., Muir, M. D., Milne, V. A. and Groves, D. I. (1978). A new microfossil assemblage from the Archaean of Western Australia. *Nature* **274,** 676–678.

Durham, J. W. (1950). Cenozoic marine climates of the Pacific coast. *Bull. geol. Soc. Am.* **61,** 1243–1264.

Dzulynski, S. and Walton, E. K. (1965). *Sedimentary features of flysch and greywackes.* Elsevier, Amsterdam.

Edwards, M. B. (1975). Glacial retreat sedimentation in the Smalfjord Formation, Late Precambrian, North Norway. *Sedimentology* **22,** 75–94.

Edwards, M. B. (1978). Glacial environments. In: Reading, H. G. (ed), *Sedimentary environments and facies.* Blackwell, Oxford, 416–438.

Elliott, T. (1978a). Deltas. In: Reading, H. G. (ed), *Sedimentary environments and facies.* Blackwell, Oxford, 97–142.

Elliott, T. (1978b). Clastic shorelines. In: Reading, H. G. (ed), *Sedimentary environments and facies*, Blackwell, Oxford, 143–177.

Emiliani, C. (1955). Pleistocene temperatures. *J. Geol.* **63,** 538–578.

Epstein, S., Buchsbaum, H. A., Lowenstam, H. A. and Urey, H. C. (1953). Revised carbonate-water isotopic temperature scale. *Bull. geol. Soc. Am.* **64,** 1315–1326.

Eugster, H. P. and Chou, I. -M. (1973). The depositional environments of Precambrian banded iron-formations. *Econ. Geol.* **68,** 1114–1168.

Eugster, H. P. and Surdam, R. C. (1973). Depositional environment of the Green River Formation of Wyoming: a preliminary report. *Bull. geol. Soc. Am.* **84,** 1115–1120.

Fairbridge, R. W. (1961). Eustatic changes of sea level. In: Ahrens, L. H. *et al.* (eds), *Physics and chemistry of the Earth.* Pergamon, London, 99–185.

Ferm, J. C. (1970). Alleghany deltaic deposits. In: Morgan, J. P. and Shaver, R. H. (eds), *Deltaic sedimentation, modern and ancient.* Spec. Publ. Soc. econ. Paleont. Miner. no. 15, 246–255.

Fischer, A. G. (1964a). The Lofer cyclothems of the Alpine Triassic. *Bull. geol. Surv. Kansas* **169,** 107–149.

Fischer, A. G. (1964b). Brackish oceans as the cause of the Permo–Triassic marine faunal crisis. In: Nairn, A. E. M. (ed), *Problems in paleoclimatology.* Interscience, London and New York, 566–574.

Fischer, A. G. and Arthur, M. A. (1977). Secular variations in the pelagic realm. In: Cook, H. E. and Enos, P. (eds), *Deep-water carbonate environments*, Spec. Publ. Soc. econ. Paleont. Miner. no. 25, 19–50.

Fisher, W. L. (1964). Sedimentary patterns in Eocene cyclic deposits, northern Gulf region. *Bull. geol. Surv. Kansas* **169,** 151–170.

Fisher, W. L., Brown, L. F., Scott, A. J. and McGowen, J. H. (1969). *Delta systems in the exploration for oil and gas.* Bur. econ. Geol. Univ. Texas, Austin.

Fleming, C. A. (1962). New Zealand biogeography; a paleontologist's approach. *Tuatara* **10,** 53–108.

Flessa, K. W. (1973). Biogeographic models for extinction, diversification and evolutionary rate. *Abstr. geol. Soc. Am.* 623.

Flessa, K. W. and Sepkowski, J. J. (1978). On the relationship between Phanerozoic diversity and changes in habitable area. *Paleobiology* **4,** 359–366.

Flint, R. F. (1971). *Glacial and Quaternary geology.* Wiley, New York.
Ford, T. D. (1979). Precambrian fossils and the origin of the Phanerozoic phyla. In: House, M. R. (ed), *The origin of major invertebrate groups.* Spec. Vol. System. Ass. **12,** 7–21.
Forney, G. G. (1975). Permo–Triassic sea level change. *J. Geol.* **83,** 773–779.
Fortey, R. A. (1975). Early Ordovician trilobite communities. *Fossils and Strata* **4,** 339–360.
Frakes, L. A. (1979). *Climates throughout geologic time.* Elsevier, Amsterdam.
Francheteau, J. *et al.* (1979). Massive deep-sea sulphide ore deposits discovered on the East Pacific Rise. *Nature* **277,** 523–528.
Frey, R. W. (ed) (1975). *The study of trace fossils.* Springer Verlag, New York.
Friedman, G. M. (1961). Distinction between dune, beach, and river sands from their textural characteristics. *J. sedim. Petrol.* **31,** 514–529.
Friedman, G. M. (1967). Dynamic processes and statistical parameters compared for size frequency distributions of beach and river sands. *J. sedim. Petrol.* **37,** 327–354.
Friedman, G. R. (1979). Differences in size distributions of populations of particles among sands of various origins. *Sedimentology* **26,** 3–32.
Funnell, B. M. (1967). Foraminifera and Radiolaria as depth indicators in the marine environment. *Mar. Geol.* **5,** 333–347.
Funnell, B. M. (1978). Productivity control of chalk sedimentation. *Abstr. 10th Int. Congr. Sedimentol.*, Jerusalem, 228.
Fürsich, F. T. and Sykes, R. M. (1977). Palaeobiogeography of the European Boreal Realm during Oxfordian (Upper Jurassic) times: a quantitative approach. *N. Jb. Geol. Paläont. Abh.* **155,** 137–161.
Galloway, W. E. (1975). Process framework for describing the morphologic and stratigraphic evolution of the deltaic depositional systems. In: Broussard, M. L. (ed), *Deltas, models for exploration.* Houston geol. Soc., Houston, 87–98.
Garrels, R. M. (1960). *Mineral equilibria at low temperature and pressure.* Harper, New York.
Garrels, R. M. and Mackenzie, F. T. (1971). *Evolution of sedimentary rocks.* Norton, New York.
Garrett, P. (1970). Phanerozoic stromatolites: noncompetitive ecological restriction by grazing and burrowing animals. *Science* **169,** 171–173.
Garrett, P., Smith, D. L., Wilson, A. O. and Patriguin, D. (1971). Physiography, ecology and sediments of two Bermudan patch reefs. *J. Geol.* **79,** 647–668.
Garrison, R. E. and Fischer, A. G. (1969). Deep-water limestones and radiolarites of the Alpine Jurassic. In: Friedman, G. M. (ed), *Depositional environments in carbonate rocks, a symposium.* Spec. Publ. Soc. econ. Paleont. Miner. no. 14, 20–56.
Gartner, S. and Keany, J. (1978). The terminal Cretaceous event: a geologic problem with an oceanographic solution. *Geology* **6,** 708–712.
Gazdzicki, A. (1974). Rhaetian microfacies, stratigraphy and facial development in the Tatra Mountains. *Acta Geol. Pol.* **24,** 17–96.
Gebelein, C. D. (1969). Distribution, morphology and accretion rate of recent subtidal algal stromatolites. *J. sedim. Petrol.* **39,** 49–69.

George, T. N. (1978). Eustasy and tectonics: sedimentary rhythms and stratigraphical units in British Dinantian correlation. *Proc. Yorks. geol. Soc.* **42,** 229–262.

Gignoux, M. (1955). *Stratigraphic geology.* Freeman, San Francisco.

Gill, E. D. (1961). The climates of Gondwanaland in Kainozoic time. In: Nairn, A. E. M. (ed), *Descriptive palaeoclimatology.* Interscience, New York, 332–353.

Ginsburg, R. N. (ed) (1975). *Tidal deposits.* Springer Verlag, Berlin and New York.

Glaessner, M. F. and Wade, M. (1966). The late Precambrian fossils from Ediacara, South Australia. *Palaeontology* **9,** 599–628.

Glennie, K. W. (1970). *Desert sedimentary environments.* Elsevier, Amsterdam.

Glennie, K. W. (1972). Permian Rotliegendes of northwest Europe interpreted in light of modern desert sedimentation studies. *Bull. Am. Ass. petrol. Geol.* **56,** 1048–1071.

Glennie, K. W., Boeuf, M. S. A., Hughes Clark, M. W., Moody-Stuart, M., Pilaar, W. F. H. and Reinhardt, B. M. (1973). Late Cretaceous nappes in Oman Mountains and their geologic evolution. *Bull. Am. Ass. petrol. Geol.* **57,** 5–27.

Goldbery, R. (1979). Sedimentology of the Lower Jurassic flint clay-bearing Mish hor formation, Makhtesh Ramon, Israel. *Sedimentology* **26,** 229–251.

Gordon, W. A. (1975). Distribution by latitude of Phanerozoic evaporite deposits. *J. Geol.* **53,** 671–684.

Goudie, A. (1973). *Duricrusts in tropical and subtropical landscapes.* Clarendon Press, Oxford.

Gould, S. J. (1965). Is uniformitarianism necessary? *Am. J. Sci.* **263,** 223–228.

Gould, S. J., Raup, D. M., Sepkowski, J. J., Schopf, T. J. M. and Simberloff, D. S. (1977). The shape of evolution: a comparison of real and random clades. *Paleobiology* **3,** 23–40.

Govett, G. J. S. (1966). Origin of banded iron formations. *Bull. geol. Soc. Am.* **77,** 1191–1212.

Grabau, A. W. (1936). Oscillation or pulsation? *Rep. 16th Int. geol. Congr.* **1,** 539–552.

Greensmith, J. T. (1978). *Petrology of the sedimentary rocks*, 6th edn. Allen and Unwin, London.

Gressly, A. (1838). Observations géologiques sur le Jura Soleurois. *N. Denkschr. allg. schweiz. Ges. ges. Naturw.* **2,** 1–112.

Griffin, J. J., Windom, H. and Goldberg, E. D. (1968). Distribution of clay minerals in the world ocean. *Deep-Sea Res.* **15,** 433–461.

Grim, R. E. (1968). *Clay mineralogy*, 2nd edn. McGraw-Hill, New York.

Hall, J. (1859). Description and figures of the organic remains of the lower Helderberg Group and the Oriskany Sandstone. *Natural history of New York, Palaeontology.* Geol. Surv., Albany, NY.

Hallam, A. (1960). A sedimentary and faunal study of the Blue Lias of Dorset and Glamorgan. *Phil. Trans. R. Soc.* B**243,** 1–44.

Hallam, A. (1963). Major epeirogenic and eustatic changes since the Cretaceous and their possible relationship to crustal structure. *Am. J. Sci.* **261,** 397–423.

Hallam, A. (ed) (1967a). Depth indicators in marine sedimentary environments. *Mar. Geol.* **5,** 329–567.

Hallam, A. (1967b). The depth significance of shales with bituminous laminae. *Mar. Geol.* **5,** 481–493.

Hallam, A (1967c). The bearing of certain palaeozoogeographic data on continental drift. *Palaeogeog., Palaeoclimatol., Palaeoecol.* **3,** 201–241.

Hallam, A. (1969). Faunal realms and facies in the Jurassic. *Palaeontology* **12,** 1–18.

Hallam, A. (1973). Distributional patterns in contemporary terrestrial and marine animals. In: Hughes, N. F. (ed), *Organisms and continents through time.* Spec. Pap. Palaeont. no. 12, 93–105.

Hallam, A. (1974). Changing patterns of provinciality and diversity of fossil animals in relation to plate tectonics. *J. Biogeog.* **1,** 213–225.

Hallam, A. (1975). *Jurassic environments.* Cambridge Univ. Press.

Hallam, A. (1976). Geology and plate tectonics interpretation of the sediments of the Mesozoic radiolarite-ophiolite complex in the Neyriz region, southern Iran. *Bull. geol. Soc. Am.* **87,** 47–52.

Hallam, A. (1977a). Jurassic bivalve biogeography. *Paleobiology* **3,** 58–73.

Hallam, A. (1977b). Secular changes in marine inundation of U.S.S.R and North America through the Phanerozoic. *Nature* **269,** 762–772.

Hallam, A. (1978a). Eustatic cycles in the Jurassic. *Palaeogeog., Palaeoclimatol., Palaeoecol.* **23,** 1–32.

Hallam, A. (1978b). How rare is phyletic gradualism? Evidence from Jurassic bivalves. *Paleobiology* **4,** 16–25.

Hallam, A. and Bradshaw, M. J. (1979). Bituminous shales and oolitic ironstones as indicators of transgressions and regressions. *J. geol. Soc. Lond.* **136,** 157–164.

Hallam, A. and Sellwood, B. W. (1976). Middle Mesozoic sedimentation in relation to tectonics in the British area. *J. Geol.* **84,** 302–321.

Hamilton, W. (1977). Subduction in the Indonesian region. In: *Island arcs, deep Sea Trenches and Back-Arc Basins.* Maurice Ewing Series **1,** 15–31. Am. Geophys. Un.

Hancock, J. M. (1967). Some Cretaceous–Tertiary marine faunal changes. In: W. B. Harland *et al.* (eds), *The fossil record.* Geol. Soc. Lond., 91–103.

Hancock, J. M. (1975). The petrology of the Chalk. *Proc. geol. Ass. Lond.* **86,** 499–535.

Hancock, J. M. and Kauffman, E. G. (1979). The great transgressions of the Late Cretaceous. *J. geol. Soc. Lond.* **136,** 175–186.

Haq, B. U., Premoli-Silva, I. and Lohman, G. P. (1977). Calcareous plankton paleobiogeographic evidence for major climatic fluctuations in the early Cenozoic Atlantic Ocean. *J. geophys. Res.* **82,** 3861–3876.

Harder, H. (1970). Boron content of sediments as a tool in facies analysis. *Sedim. geol.* **4,** 153–175.

Hargraves, R. B. (1976). Precambrian geologic history. *Science* **193,** 363–371.

Harland, W. B., Herod, K. N. and Krinsley, D. H. (1966). The definition and identification of tills and tillites. *Earth-Sci. Rev.* **2,** 225–256.

Haxby, W. F., Turcotte, D. L., Bird, J. M. (1976). Thermal and mechanical evolution of the Michigan Basin. *Tectonophysics* **36,** 57–75.

Hays, J. D. (1971). Faunal extinctions and reversals of the earth's magnetic field. *Bull. geol. Soc. Am.* **82,** 2433–2447.

Hays, J. D. and Pitman, W. C. (1973). Lithospheric plate motion, sea level changes and climatic and ecological consequences. *Nature* **246,** 16–22.

Hays, J. D., Imbrie, J. and Shackleton, N. J. (1976). Variations in the Earth's orbit: pacemaker of the ice ages. *Science* **194,** 1121–1132.

Heckel, P. H. (1972). Recognition of ancient shallow marine environments. In: Rigby, J. K. and Hamblin, W. H. (eds), *Recognition of ancient sedimentary environments.* Spec. Publ. Soc. econ. Paleont. Miner. no. 16, 226–286.

Heckel, P. H. (1974). Carbonate buildups in the geologic record: a review. In: Laporte, L. (ed), *Reefs in time and space*, Spec. Publ. Soc. econ. Paleont. Miner. no. 18, 90–154.

Heckel, P. H. (1977). Origin of phosphatic black shale facies in Pennsylvanian cyclothems of mid-continent North America. *Bull. Am. Ass. petrol. Geol.* **61,** 1045–1068.

Heezen, B. C. and Hollister, C. D. (1971). *The face of the deep.* Oxford Univ. Press, New York.

Heezen, B. C. and McGregor, I. D. (1973). The evolution of the Pacific. *Sci. Amer.* **229**(5), 102–112.

Hoffman, H. J. (1973). Stromatolites: characteristics and utility. *Earth-Sci. Rev.* **9,** 339–373.

Hoffman, P., Dewey, J. F. and Burke, K. (1974). Aulacogens and their genetic relation to geosynclines, with a Proterozoic example from Great Slave Lake, Canada. In: Dott, R. H. and Shaver, R. H. (eds), *Modern and ancient geosynclinal sedimentation*. Spec. Publ. Soc. econ. Paleont. Miner no. 19, 38–55.

Holland, H. D. (1972). The geologic history of sea water—an attempt to solve the problem. *Geochim. Cosmochim. Acta* **36,** 637–651.

House, M. R. (1967). Fluctuations in the evolution of Palaeozoic invertebrates. In: Harland, W. B. *et al. The fossil record.* Geol. Soc. Lond., 41–54.

House, M. R. (1974). Facies and time in Devonian tropical areas. *Proc. Yorks. geol. Soc.* **40,** 233–288.

Howard, J. D., Frey, R. W., Reineck, H. E., Gadow, S., Wunderlich, F., Dörjes, J. and Hertweck, G. (1972). Georgia coastal region, Sapelo Island, USA: sedimentology and biology. *Senckenberg. marit.* **4,** 3–223.

Hsü, K. J. (1971). Franciscan melanges as a model for eugeosynclinal sedimentation and underthrusting tectonics. *J. geophys. Res.* **76,** 1162–1170.

Hsü, K. J. (1972a). Origin of saline giants: a critical review after the discovery of the Mediterranean evaporite. *Earth-Sci. Rev.* **8,** 371–396.

Hsü, K. J. (1972b). When the Mediterranean dried up. *Sci. Am.* **277**(6), 27–36.

Hsü, K. J. (1974). Melanges and their distinction from olistostromes. In: Dott, R. H. and Shaver, R. H. (eds), *Modern and ancient geosynclinal sedimentation.* Spec. Publ. Soc. econ. Paleont. Miner. no. 19, 321–333.

Hsü, K. J. (1980). Terrestrial catastrophe caused by cometary impact at the end of the Cretaceous. *Nature* **285,** 201–203.

Hsü, K. J. and Jenkyns, H. C. (eds) (1974). *Pelagic sediments: on land and under the sea*. Spec. Publ. int. Ass. Sediment. no. 1.

Hsü, K. J., Cita, M. B. and Ryan, W. B. F. (1972). The origin of the Mediterranean evaporites. *Init. Rep. Deep Sea Drilling Project* **13**, 1203–1231.

Hsü, K. J. *et al.* (1977). History of the Mediterranean salinity crisis. *Nature* **267,** 399–403.

Huber, N. K. and Garrels, R. M. (1953). Relation of pH and oxidation potential to sedimentary iron mineral formation. *Econ. Geol.* **48,** 337–357.

Hubert, J. F. (1978). Paleosol caliche in the New Haven Arkose, Newark Group, Connecticut. *Palaeogeog., Palaeoclimatol., Palaeoecol.* **24,** 151–168.

Hubert, J. F., Butera, J. G. and Rice, R. F. (1972). Sedimentology of Upper Cretaceous Cody-Parkman delta, southwestern Powder River Basin, Wyoming. *Bull. geol. Soc. Am.* **83,** 1649–1670.

Hudson, J. D. (1962). The stratigraphy of the Great Estuarine Series (Middle Jurassic) of the Inner Hebrides. *Trans. Edinb. geol. Soc.* **19,** 139–165.

Hudson, J. D. (1963). The recognition of salinity-controlled mollusc assemblages in the Great Estuarine Series of the Inner Hebrides. *Palaeontology* **6,** 318–326.

Hudson, J. D. (1966). Hugh Miller's reptile bed and the Mytilus Shales, Middle Jurassic, Isle of Eigg, Scotland. *Scott. J. Geol.* **2,** 265–281.

Hudson, J. D. (1970). Algal limestones with pseudomorphs after gypsum from the Middle Jurassic of Scotland. *Lethaia* **3,** 11–40.

Hudson, J. D. (1977a). Stable isotopes and limestone lithification. *J. geol. Soc. Lond.* **133,** 637–660.

Hudson, J. D. (1977b). Oxygen isotope studies on Cenozoic temperatures, oceans and ice accumulation. *Scott. J. Geol.* **13,** 313–325.

Hughes Clarke, M. W. and Keij, A. J. (1973). Organisms as producers of carbonate sediment and indicators of environment in the southern Persian Gulf. In: Purser, B. H. (ed.), *The Persian Gulf: Holocene carbonate sedimentation and diagenesis in a shallow epicontinental sea.* Springer Verlag, Berlin and New York, 33–56.

Hunt, B. G. (1979). The effects of past variations of the Earth's rotation rate on climate. *Nature* **281,** 188–191.

Hunter, D. R. (1974a). Crustal development in the Kaapvaal Craton, I. The Archaean. *Precamb. Res.* **1,** 259–294.

Hunter, D. R. (1974b). Crustal development in the Kaapvaal Craton, II. The Proterozoic. *Precamb. Res.* **1,** 295–326.

Hutchinson, R. W. and Engels, G. G. (1970). Tectonic significance of regional geology and evaporite lithofacies in northeastern Ethiopia. *Phil. Trans. R. Soc.* A**267,** 313–329.

Imbrie, J. and Kipp, N. G. (1971). A new micropaleontological method of quantitative paleoclimatology: application to a late Pleistocene Caribbean core. In: Turekian, K. (ed), *The late Cenozoic glacial ages.* Yale Univ. Press, New Haven, 71–182.

Imbrie, J., Van Donk, J. and Kipp, N. G. (1973). Paleoclimatic investigation of a Late Pleistocene Caribbean deep-sea core: comparison of isotopic and faunal methods. *J. Quat. Res.* **3,** 10–38.

Irwin, M. L. (1965). General theory of epeiric clear water sedimentation. *Bull. Am. Ass. petrol. Geol.* **49,** 445–459.

Jackson, T. A. and Moore, C. B. (1976). Secular variations in kerogen structure and carbon, nitrogen and phosphorus concentrations in pre-Phanerozoic and Phanerozoic sedimentary rocks. *Chem. Geol.* **18,** 107–136.

James, H. L. (1966). Chemistry of the iron-rich sedimentary rocks. *Prof. Pap. U.S. geol. Surv.* no. 440-W.

James, H. L. and Sims, P. K. (eds) (1973). Precambrian iron formations of the world. *Econ. Geol.* **68,** 913–1179.

Jeans, C. V. (1978). The origin of the Triassic clay assemblages of Europe with special reference to the Keuper Marl and Rhaetic of parts of England. *Phil. Trans. R. Soc.* A**289,** 549–639.

Jeans, C. V., Merriman, R. J. and Mitchell, J. G. (1977). Origin of Middle Jurassic and Lower Cretaceous fuller's earths in England. *Clay Miner.* **12,** 11–44.

Jenkyns, H. C. (1974). Origin of red nodular limestones (Ammonitico Rosso, Knollenkalke) in the Mediterranean Jurassic: a digenetic model. In: Hsü, K. J. and Jenkyns, H. C. (eds), *Pelagic sediments: on land and under the sea.* Spec. Publ. int. Ass. Sedimentol. no. 1, 249–271.

Jenkyns, H. C. (1978). Pelagic environments. In: Reading, H. G. (ed), *Sedimentary environments and facies.* Blackwell, Oxford, 314–371.

Jenkyns, H. C. (1980). Cretaceous anoxic events: from continents to oceans. *J. geol. Soc. Lond.* **137,** 171–188.

Johnson, H. D. (1978). Shallow siliciclastic seas. In: Reading, H. G. (ed), *Sedimentary environments and facies.* Blackwell, Oxford, 207–258.

Johnson, J. G. and Boucot, A. J. (1973). Devonian brachiopods. In: Hallam, A. (ed), *Atlas of palaeobiogeography.* Elsevier, Amsterdam, 89–96.

Johnson, N. M., Opdyke, N. D. and Lindsay, E. H. (1975). Magnetic polarity stratigraphy of Pliocene–Pleistocene terrestrial deposits and vertebrate faunas. *Bull. geol. Soc. Am.* **86,** 5–12.

Jones, D. L., Silberling, N. J. and Hillhouse, J. (1977). Wrangellia—a displaced terrane in northwestern North America. *Can. J. Earth Sci.* **14,** 2565–2577.

Jørgensen, C. B. (1955). Quantitative aspects of filter feeding in invertebrates. *Biol. Rev.* **30,** 391–454.

Kauffman, E. G. (1978). Benthic environments and paleoecology of the Posidonienschiefer (Toarcian). *N. Jb. Geol. Paläont. Abh.* **157,** 1836.

Kay, M. (1951). North American geosynclines. *Mem. geol. Soc. Am.* no. 48.

Kazmierczak, J. (1979). The eukaryotic nature of *Eosphaera*-like ferriferous structures from the Precambrian Gunflint Iron Formation, Canada: a comparative study. *Precamb. Res.* **9,** 1–22.

Keigwin, L. D. (1978). Pliocene closing of the Isthmus of Panama, based on biostratigraphic evidence from nearby Pacific Ocean and Caribbean Sea cores. *Geology* **6,** 630–634.

Keith, M. L., Anderson, G. M. and Eichler, R. (1964). Carbon and oxygen isotopic composition of mollusk shells from marine and fresh-water environments. *Geochim. Cosmochim. Acta* **28,** 1757–1786.

Kemp, E. M. (1978). Tertiary climatic evolution and vegetation history in the southeast Indian Ocean region. *Palaeogeog., Palaeoclimatol., Palaeoecol.* **24,** 169–208.

Kendall, C. G. St C. (1969). An environmental re-interpretation of the Permian evaporite/carbonate shelf sediments of the Guadalupe Mountains. *Bull. geol. Soc. Am.* **80,** 2503–2526.

Kennedy, W. J. (1977). Ammonite evolution. In: Hallam, A. (ed), *Patterns of evolution.* Elsevier, Amsterdam, 251–304.

Kennedy, W. J. and Cobban, W. A. (1976). *Aspects of ammonite biology, biostratigraphy and biogeography.* Spec. Pap. Palaeontol. no. 17.

Kennedy, W. J. and Garrison, R. E. (1975a). Morphology and genesis of nodular phosphates in the Cenomanian Glauconitic Marl of southeast England. *Lethaia* **8,** 339–360.

Kennedy, W. J. and Garrison, R. E. (1975b). Morphology and genesis of nodular chalks and hardgrounds in the Upper Cretaceous of southern England. *Sedimentology* **22,** 311–386.

Kennedy, W. J. and Juignet, P. (1974). Carbonate banks and slump beds in the Upper Cretaceous (Upper Turonian–Santonian) of Haute Normandie, France. *Sedimentology* **21,** 1–42.

Kennett, J. P. (1977). Cenozoic evolution of Antarctic glaciation, the circum-Antarctic ocean, and their impact on global paleoceanography. *J. geophys. Res.* **82,** 3843–3860.

Kent, D. V. (1977). An estimate of the duration of the faunal change at the Cretaceous–Tertiary boundary. *Geology* **5,** 769–771.

Kent, P. E. (1977). The Mesozoic development of aseismic continental margins. *J. geol. Soc. Lond.* **134,** 1–18.

Keulegen, G. H. and Krumbein, W. H. (1949). Stable configuration of bottom slope in a shallow sea and its bearing on geological processes. *Trans. Am. geophys. Un.* **30,** 555–861.

Kimberley, M. M.(1974). Origin of iron ore by diagenetic replacement of calcareous oolite. *Nature* **250,** 319–320.

Kimberley, M. M. (1979). Origin of oolitic iron minerals. *J. sedim. Petrol.* **49,** 110–132.

Kitchell, J. A., Kitchell, J. F., Johnson, G. L. and Hunkins, K. L. (1978). Abyssal traces and megafauna: comparison of productivity, diversity and density in the Arctic and Antarctic. *Paleobiology* **4,** 171–180.

Klein, G. de V. (1971). A sedimentary model for determining a paleotidal range. *Bull. geol. Soc. Am.* **82,** 2585–2592.

Klinkhammer, G., Bender, M. and Weiss, R. F. (1977). Hydrothermal manganese in the Galapagos Rift. *Nature* **269,** 319–320.

Klootwijk, D. T. and Peirce, J. W. (1979). India's and Australia's pole path since the late Mesozoic and the India–Asia collision. *Nature* **282,** 605–607.

Knauth, L. P. and Epstein, S. (1976). Hydrogen and oxygen isotope ratios in nodular and bedded cherts. *Geochim. Cosmochim. Acta* **40,** 1095–1108.

Knoll, A. H. (1978). Did emerging continents trigger metazoan evolution? *Nature* **276,** 701–703.

Knox, R. W. O'B (1971). An alternative to the 'clastic trap' hypothesis. *Geol. Mag.* **108,** 544–545.

Krassilov, V. A. (1975). Climatic changes in eastern Asia as indicated by fossil floras. II. Late Cretaceous and Danian. *Palaeogeog., Palaeoclimatol., Palaeoecol.* **17,** 157–172.

Krassilov, V. A. (1978). Late Cretaceous gymnosperms from Sakhalin, U.S.S.R. and the terminal Cretaceous event. *Palaeontology* **21,** 893–906.

Krauskopf, K. B. (1967). *Introduction to geochemistry*. McGraw-Hill, New York.

Krinsley, D. H. and Donahue, J. (1968). Environmental interpretation of sand grain surface textures by electron microscopy. *Bull. geol. Soc. Am.* **79,** 743–748.

Krinsley, D. and Doornkamp, J. C. (1973). *Atlas of quartz sand surface textures.* Cambridge Univ. Press.

Kuenan, Ph.H and Migliorini, C. I. (1950). Turbidity currents as a cause of graded bedding. *J. Geol.* **58,** 91–127.

Kukla, G. J. (1977). Pleistocene land-sea correlations. 1. Europe. *Earth-Sci. Rev.* **13,** 307–374.

Kummel, B. and Teichert, C. (1970). Stratigraphy and paleontology of the Permian–Triassic boundary beds, Salt Range and Trans-Indus ranges, West Pakistan. *Spec. Publ. Univ. Kansas Dept Geol.* no. 4.

LaBerge, G. L. (1973). Possible biological origin of Precambrian iron formations. *Econ. Geol.* **68,** 1098–1109.

Ladd, J. W. (1976). Relative motion of South America with respect to North America and Caribbean tectonics. *Bull. geol. Soc. Am.* **87,** 969–976.

Laporte, L. F. (1971). Paleozoic carbonate facies of the Central Appalachian shelf. *J. sedim. Petrol.* **41,** 724–740.

Laporte, L. F. (ed) (1974). *Reefs in time and space.* Spec. Publ. Soc. econ. Paleont. Miner. no. 18.

Laubscher, H. and Bernoulli, D. (1977). Mediterranean and Tethys. In: Nairn, A. E. M., Kanes, W. H. and Stehli, F. G. (eds), *The ocean basins and margins.* Plenum, New York, vol. 4a, 1–28.

Leeder, M. R. and Nami, M. (1979). Sedimentary models for the non-marine Scalby Formation (Middle Jurassic) and evidence for late Bajocian/Bathonian uplift of the Yorkshire Basin. *Proc. Yorks. geol. Soc.* **42,** 461–482.

Lees, A. (1975). Possible influences of salinity and temperature on modern shelf carbonate sedimentation. *Mar. Geol.* **19,** 159–198.

Lees, A. and Buller, A. T. (1972). Modern temperature-water and warm-water shelf carbonate sediments contrasted. *Mar. Geol.* **13,** M67–73.

Leggett, J. K. (1980). British Lower Palaeozoic black shales and their palaeo-oceanographic significance. *J. geol. Soc. Lond.* **137,** 139–156.

Leggett, J. K., McKerrow, W. S. and Eales, M. H. (1979). The Southern Uplands of Scotland: a Lower Palaeozoic accretionary prism. *J. geol. Soc. Lond.* **136,** 755–770.

Lerbekmo, J. F., Evans, M. E. and Baudsgaard, H. (1979). Magnetostratigraphy, biostratigraphy and geochronology of Cretaceous–Tertiary boundary sediments, Red Deer Valley. *Nature* **279,** 26–30.

Lisitzin, E. (1974). *Sea level changes.* Elsevier, Amsterdam, Oceanog. Ser. 8.

Logan, B. W., Rezak, R. and Ginsburg, R. N. (1964). Classification and environmental significance of stromatolites. *J. Geol.* **72,** 68–83.

Lonsdale, P. (1977). Clustering of suspension-feeding macrobenthos near abyssal hydrothermal vents at oceanic spreading centers. *Deep-Sea Res.* **24,** 857–863.

Lowe, D. R. (1980). Stromatolites 3,400-Myr old from the Archaean of Western Australia. *Nature* **284,** 441–443.

Luyendyk, B. P., Forsyth, D. and Phillips, J. D. (1972). Experimental approach to the paleocirculation of the oceanic surface waters. *Bull. geol. Soc. Am.* **83,** 2649–2666.

MacArthur, R. H. and Wilson, E. O. (1967). *The theory of island biogeography.* Princeton Univ. Press.

McCabe, P. J. (1977). Deep distributary channels and giant bedforms in the Upper Carboniferous of the Central Pennines, northern England. *Sedimentology* **24,** 271–290.

McClure, H. A. (1978). Early Palaeozoic glaciation in Arabia. *Palaeogeog., Palaeoclimatol., Palaeoecol.* **25,** 315–326.

McElhinny, M. W. (1971). Geomagnetic reversals during the Phanerozoic. *Science* **172,** 157–159.

McElhinny, M. W. (1973). *Palaeomagnetism and plate tectonics.* Cambridge Univ. Press.

McIntyre, A., *et al*. (1976). The surface of the ice-age earth. *Science* **191,** 1131–1137.

McKee, E. D. (1966). Structures of dunes at White Sands National Monument, New Mexico (and comparison with structures of dunes from other selected areas). *Sedimentology* **7,** 1–69.

McKenzie, D. P. (1978). Some remarks on the development of sedimentary basins. *Earth planet. Sci. Lett.* **40,** 25–32.

McKenzie, J. A., Jenkyns, H. C. and Bennet, G. G. (1980). Stable isotope study of the cyclic diatomite-claystones from the Tripoli Formation, Sicily: a prelude to the Messinian salinity crisis. *Palaeogeog., Palaeoclimatol., Palaeoecol.* **29,** 125–141.

McKerrow, W. S. (ed) (1978). *The ecology of fossils.* Duckworth, London.

McKerrow, W. S. (1979). Ordovician and Silurian changes in sea level. *J. geol. Soc. Lond.* **136,** 137–145.

McKerrow, W. S. and Cocks, L. R. M. (1976). Progressive faunal migration across the Iapetus Ocean. *Nature* **263,** 304–306.

McLaren, D. J. (1970). Presidential address: time, life and boundaries. *J. Paleont.* **44,** 801–815.

McLean, D. M. (1978). A terminal Mesozoic 'greenhouse': lessons from the past. *Science* **201,** 401–406.

Matter, A. and Tucker, M. E. (eds) (1978). *Modern and ancient lake sediments.* Spec. Publ. int. Ass. Sedimentol. no. 2.

Matthews, S. C. and Cowie, J. W. (1979). Early Cambrian transgression. *J. geol. Soc. Lond.* **136,** 133–135.

Maxwell, W. G. H. (1968). *Atlas of the Great Barrier Reef.* Elsevier, Amsterdam.

Menzies, R. J., George, R. Y. and Rowe, G. T. (1973). *Abyssal environments and ecology of the world oceans.* Wiley, New York.

Meyerhoff, A. A. (1970). Continental drift, II. High-latitude evaporite deposits and geologic history of Arctic and North Atlantic Oceans. *J. Geol.* **78,** 406–444.

Middleton, G. V. (1973). Johannes Walther's law of correlation of facies. *Bull. geol. Soc. Am.* **84,** 979–988.

Middleton, G. V. and Hampton, M. A. (1976). Subaqueous sediment transport and deposition by sediment gravity flows. In: Stanley, D. J. and Swift, D. J. P. (eds), *Marine sediment transport and environmental management.* Wiley, New York, 197–218.

Millot, G. (1970). *Geology of clays.* Chapman and Hall, London.

Milnes, A. G. (1978). Structural zones and continental collision, Central Alps. *Tectonophysics* **47,** 369–392.

Mitchell, A. H. G. and McKerrow, W. S. (1975). Analogous evolution of the Burma orogen and the Scottish Caledonides. *Bull. geol. Soc. Am.* **86,** 305–315.

Mitchell, A. H. G. and Reading, H. G. (1978). Sedimentation and tectonics. In: Reading, H. G. (ed), *Sedimentary environments and facies.* Blackwell, Oxford, 439–476.

Mohr, R. E. (1975). Measured periodicities of the Biwabik (Precambrian) stromatolites and their geophysical significance. In: Rosenberg, G. D. and Runcorn, S. K. (eds), *Growth rhythms and the history of the Earth's rotation.* Wiley, New York, 43–55.

Molnar, P. and Tapponier, P. (1975). Cenozoic tectonics of Asia: effects of a continental collision. *Science* **189,** 419–426.

Monty, C. (1967). Distribution and structure of recent stromatolitic algal mats, eastern Andros Island, Bahamas. *Bull. Ann. Soc. Geol. Belg.* **88,** B269–276.

Moorbath, S. (1977a). Ages, isotopes and evolution of Precambrian continental crust. *Chem. Geol.* **20,** 151–187.

Moorbath, S. (1977b). The oldest rocks and the growth of continents. *Sci. Am.* **236**(3), 92–103.

Moore, H. B. (1931). Muds of the Clyde Sea area. III. Chemical and physical conditions; rate and nature of sedimentation; and fauna. *J. mar. Biol. Ass. U.K.* **17,** 325–366.

Moore, J. C. *et al.* (1979). Progressive accretion in the Middle America Trench, Southern Mexico. *Nature* **281,** 638–642.

Moore, R. C. (1954). Evolution of late Paleozoic invertebrates in response to major oscillations of shallow seas. *Bull. Mus. comp. Zool. Harv.* **122,** 259–286.

Moore, T. C., Van Andel, Tj. H., Sancetta, C. and Pisias, N. (1978). Cenozoic hiatuses in pelagic sediments. *Micropaleontology* **24,** 113–138.

Mörner, N.-A. (1976). Eustasy and geoid changes. *J. Geol.* **84,** 123–151.

Morris, K. A. (1980). A comparison of major sequences of organic-rich mud deposition in the British Jurassic. *J. geol. Soc. Lond.* **137,** 157–170.

Mutti, E. (1974). Examples of ancient deep-sea fan deposits from circum-Mediterranean geosynclines. In: Dott, R. H. and Shaver, R. H. (eds), *Modern and ancient geosynclinal sedimentation.* Spec. Publ. Soc. econ. Paleont. Miner. no. 19, 92–105.

Napier, W. M. and Clube, S. V. M. (1979). A theory of terrestrial catastrophism. *Nature* **282,** 455–459.

Nelson, C. H. and Nilson, T. (1974). Submarine fans and channels. In: Dott, R. H. and Shaver, R. H. (eds), *Modern and ancient geosynclinal sedimentation.* Spec. Publ. Soc. econ. Paleont. Miner. no. 19, 54–76.

Neumann, A. C., Kofoed, J. W. and Keller, G. H. (1977). Lithoherms in the Straits of Florida. *Geology* **5,** 4–10.

Newell, N. D. (1967a). Paraconformities. *Spec. Publ. Univ. Kansas Dept Geol.* no. 2, 349–367.

Newell, N. D. (1967b). Revolutions in the history of life. *Spec. Pap. geol. Soc. Am.* no. 89, 63–91.

Newell, N. D., Rigby, J. K., Fischer, A. G., Whiteman, A. J., Hickox, J. E., and Bradley, J. S. (1953). *The Permian reef complex of the Guadaloupe Mountains region, Texas and New Mexico.* Freeman, San Francisco.

Nicolas, J. and Bildgen, P. (1979). Relations between the location of the karst bauxites in the Northern Hemisphere, the global tectonics and the climate variations during geological time. *Palaeogeog., Palaeoclimatol., Palaeogeog.* **28,** 205–239.

Nio, S.-D. (1976). Marine transgressions as a factor in the formation of sand wave complexes. *Geol. Mijnb.* **55,** 18–40.

Norris, R. J., Carter, R. M. and Turnbull, I. M. (1978). Cainozoic sedimentation in basin adjacent to a major continental transform boundary in southern New Zealand. *J. geol. Soc. Lond.* **135,** 191–205.

Osgood, R. G. (1970). Trace fossils of the Cincinatti area. *Paleontogr. Am.* **6**(41), 281–444.

Owen, H. G. (1976). Continental displacement and expansion of the earth during the Mesozoic and Cenozoic. *Phil. Trans. R. Soc.* A**281,** 223–291.

Palmer, A. R. (1973). Cambrian Trilobites. In: Hallam, A. (ed), *Atlas of palaeobiogeography.* Elsevier, Amsterdam, 3–11.

Palmer, T. J. (1979). The Hampen Marly and White Limestone formations: Florida-type carbonate lagoons in the Jurassic of central England. *Palaeontology* **22,** 189–228.

Pannella, G. (1975). Palaeontological clocks and the history of the Earth's rotation. In: Rosenberg, G. D. and Runcorn, S. K. (eds), *Growth rhythms and the history of the Earth's rotation.* Wiley, New York, 253–283.

Payton, C. E. (ed) (1977). Stratigraphic interpretation of seismic data. *Mem. Am. Ass. petrol. Geol.* no. 26.

Percival, S. F. and Fischer, A. G. 1977. Changes in calcareous nannoplankton in the Cretaceous–Tertiary biotic crisis at Zumaya, Spain. *Evol. Theory* **2,** 1–35.

Picard, M. D. and High, L. R. (1972a). Criteria for recognising lacustrine rocks. In: Rigby, J. K. and Hamblin, W. K. (eds), *Recognition of ancient sedimentary environments.* Spec. Publ. Soc. econ. Paleont. Miner. no. 16, 108–145.

Picard, M. D. and High, L. R. (1972b). Paleoenvironmental reconstructions in an area of rapid facies change. Parachute Creek Member of Green River Formation (Eocene), Uinta Basin, Utah. *Bull. geol. Soc. Am.* **83,** 2689–2708.

Pitman, W. C. (1978). Relationship between eustasy and stratigraphic sequences of passive margins. *Bull. geol. Soc. Am.* **89,** 1389–1403.

Potter, P. E. and Pettijohn, F. J. (1977). *Paleocurrents and basin analysis*, 2nd edn. Springer Verlag, Berlin and New York.

Potter, P. E., Shimp, N. F. and Witters, J. (1963). Trace elements in marine and fresh-water argillaceous sediments. *Geochim. Cosmochim. Acta* **27,** 669–694.

Poulton, T. P. and Callomon, J. H. (1977). A new species of trigoniid bivalve from the boreal Bathonian (Jurassic) of central East Greenland. *Bull. geol. Soc. Denmark* **26,** 115–159.

Powell, C. McA. and Conaghan, P. J. (1973). Plate tectonics and the Himalayas. *Earth planet. Sci. Lett.* **20,** 1–12.

Preston, F. W. (1962). The canonical distribution of commonness and rarity. *Ecology* **43,** 185–215 and 410–432.

Pryor, W. A. (1971). Petrology of the Permian Yellow Sands of northeastern England and their North Sea basin equivalents. *Sedim. Geol.* **6,** 221–254.

Purser, B. H. (ed) (1973). *The Persian Gulf: Holocene carbonate sedimentation and diagenesis in a shallow epicontinental sea.* Springer Verlag, Berlin and New York.

Rage, J.-C. (1978). Une connection continentale entre Amerique du Nord et Amerique du Sud au Crétacé supérieur? L'exemple des Vertébrés continentaux. *C. r. somm. Soc. géol. Fr.* fasc. 6, 281–285.

Ramsay, A. T. S. (1973). A history of organic siliceous sediments in oceans. In: Hughes, N. F. (ed), *Organisms and continents through time.* Spec. Pap. Palaeontol. no. 12, 199–234.

Ramsbottom, W. H. C. (1979). Rates of transgression and regression in the Carboniferous of NW Europe. *J. geol. Soc. Lond.* **136,** 147–153.

Raup, D. M. (1976). Species diversity in the Phanerozoic: an interpretation. *Paleobiology* **2,** 289–297.

Raup, D. M. and Stanley, S. M. (1978). *Principles of paleontology,* 2nd edn. Freeman, San Francisco.

Reading, H. G. (1975). Strike-slip fault systems; an ancient example from the Cantabrians. *9th Int. sediment. Congr.*, Nice 1975, Theme 4 (2), 289–292.

Reading, H. G. (ed) (1978). *Sedimentary environments and facies.* Blackwell, Oxford.

Reading, H. G. and Walker, R. G. (1966). Sedimentation of Eocambrian tillites and associated sediments in Finmark, northern Norway. *Palaeogeog., Palaeoclimatol., Palaeoecol.* **2,** 177–212.

Reid, E. M. and Chandler, M. E. J. (1933). *The flora of the London Clay.* Brit. Mus. (nat. Hist.), London.

Reid, G. C., Isaksen, I. S. A., Holzer, T. E. and Crutzen, P. J. (1976). Influence of ancient solar-proton events on the evolution of life. *Nature* **259,** 177–179.

Reid, G. C., McAfee, J. R. and Crutzen, P. J. (1978). Effects of intense stratospheric ionisation events. *Nature* **275,** 489–492.

Reineck, H.-E. and Singh, I. B. (1973). *Depositional sedimentary environments.* Springer Verlag, Berlin and New York.

Rhoads, D. C. and Young, D. K. (1970). The influence of deposit-feeding organisms on sediment stability and community trophic structure. *J. mar. Res.* **28,** 150–178.

Rhodes, F. H. T. (1967). Permo-Triassic extinction. In: Harland, W. B. *et al.* (eds), *The fossil record*, Geol. Soc. Lond., 57–76.

Richter-Bernburg, G. (1960). Zeitmessung geologischer Vorgange nach Warven. Korrelation im Zechstein. *Geol. Rdsch.* **49,** 132–148.

Robertson, A. H. F. and Hudson, J. D. (1974). Pelagic sediments in the Cretaceous and Tertiary history of the Troodos Massif, Cyprus. In: Hsü, K. J. and Jenkyns, H. C. (eds), *Pelagic sediments: on land and under the sea.* Spec. Publ. int. Ass. Sediment. no. 1, 403–406.

Robinson, P. C. (1971). A problem of faunal replacement on Permo–Triassic continents. *Palaeontology* **14,** 131–153.

Rouchy, J.-M. (1976). Sur la genèse de deux principaux types de gypse (finement lité et en chevrons) du Miocène terminale de Sicile et d'Espagne meridionale. *Rev. Géog. phys. Géol. dynam.* **18,** 347–364.

Ruddiman, W. F. and McIntyre, A. (1973). Time-transgressive deglacial retreat of polar waters from the North Atlantic. *J. Quat. Res.* **3,** 117–130.

Ruddiman, W. F., Sancetta, C. D. and McIntyre, A. (1977). Glacial/interglacial response rate of subpolar North Atlantic waters to climatic change: the record in oceanic sediments. *Phil. Trans. R. Soc.* B**280,** 119–141.

Rupke, N. A. (1978). Deep clastic seas. In: Reading, H. G. (ed), *Sedimentary environments and facies.* Blackwell, Oxford, 272–415.

Rusnak, G. A. (1960). Sediments of Laguna Madre, Texas. In: Shepard, F. P., Phleger, F. B. and Van Andel, T. H. (eds), *Recent sediments, northwest Gulf of Mexico.* Am. Ass. Petrol. Geol., Tulsa, 153–196.

Russell, D. A. (1979). The enigma of the extinction of the dinosaurs. *Ann. Rev. Earth planet. Sci.* **7,** 163–182.

Ryan, W. B. F. and Cita, M. B. (1977). Ignorance concerning episodes of ocean-wide stagnation. *Mar. Geol.* **23,** 197–215.

Ryther, J. H. (1963). Geographic variations in productivity. In: Hill, M. N. (ed), *The sea.* Interscience, New York, **2,** 347–380.

Sahni, A. and Kumar, V. (1974). Palaeogene palaeobiogeography of the Indian subcontinent. *Palaeogeog., Palaeoclimatol., Palaeoecol.* **15,** 209–226.

Saito, T. and Van Donk, J. (1974). Oxygen and carbon isotope measurements of Late Cretaceous and Early Tertiary foraminifera. *Micropaleontology* **20,** 152–177.

Salvan, H. (1960). Les phosphate de chaux sédimentaires du Maroc. *Notes maroc. Soc. géog. Maroc.* no. 14, 7–20.

Sanders, H. L. (1968). Marine benthic diversity: a comparative study. *Am. Nat.* **102,** 243–282.

Savin, S. M., Douglas, R. G. and Stehli, F. G. (1975). Tertiary marine paleotemperatures. *Bull. geol. Soc. Am.* **86,** 1499–1510.

Schäfer, W. (1972). *Ecology and palaeoecology of marine environments.* Oliver and Boyd, Edinburgh.

Schermerhorn, L. J. G. (1976). Reply to discussions by G. M. Young and G. E. Williams. *Am. J. Sci.* **276,** 375–384.

Schidlowski, M., Eichmann, R. E. and Junge, C. E. (1975). Precambrian sedimentary carbonates: carbon and oxygen isotope geochemistry and implications for the terrestrial oxygen budget. *Precamb. Res.* **2,** 1–69.

Schindewolf, O. H. (1954). Über die möglichen Ursachen der grossen erdgeschichtlichen Faunenschnitte. *N. Jb. Geol. Paläont. Abh.* **10,** 457–465.

Schlanger, S. O. and Jenkyns, H. C. (1976). Cretaceous oceanic anoxic events: causes and consequences. *Geol. Mijnb.* **55,** 179–184.

Schluger, P. R. and Robertson, H. E. (1975). Mineralogy and chemistry of the Patapsco Formation, Maryland, related to the ground-water geochemistry and flow system: a contribution to the origin of red beds. *Bull. geol. Soc. Am.* **86,** 153–158.

Schmalz, R. F. (1969). Deep-water evaporite deposition: a genetic model. *Bull. Am. Ass. petrol. Geol.* **53,** 798–823.

Schneider, S. H. and Thompson, S. L. (1979). Ice ages and orbital variations: some simple theory and modelling. *Quat. Res.* **12,** 188–203.

Scholl, D. W. and Marlow, M. S. (1974). Sedimentary sequence in modern Pacific trenches and the deformed circum-Pacific eugeosyncline. In: Dott, R. H. and Shaver, R. H. (eds), *Modern and ancient geosynclinal sedimentation.* Spec. Publ. Soc. econ. Paleont. Miner. no. 19, 193–211.

Schopf, J. W. (1978). The evolution of the earliest cells. *Sci. Am.* **239**(3), 84–102.

Schopf, T. J. M. (1974). Permo-Triassic extinctions: relation to sea floor spreading. *J. Geol.* **82,** 129–143.

Schuchert, C. (1955). *Atlas of paleogeographic maps of North America.* Wiley, New York.

Schwarzacher, W. (1975). *Sedimentation models and quantitative stratigraphy.* Elsevier, Amsterdam.

Sclater, J. G., Hellinger, S. and Tapscott, C. (1977). The paleobathymetry of the Atlantic Ocean from the Jurassic to the present. *J. Geol.* **85,** 509–552.

Scoffin, T. P. (1972). Fossilisation of Bermudan patch reefs. *Science* **178,** 1280–1282.

Scrutton, C. T. (1978). Periodic growth features in fossil organisms and the length of the day and month. In: Broche, P. and Sundermann, J. (eds), *Tidal friction and the Earth's rotation.* Springer Verlag, Berlin, 154–196.

Seibold, E. (1958). Jahreslagen in Sedimenten der Mittleren Adria. *Geol. Rdsch.* **47,** 100–117.

Seilacher, A. (1967). Bathymetry of trace fossils. *Mar. Geol.* **5,** 413–428.

Selley, R. C. (1965). Diagnostic characters of fluviatile sediments of the Torridonian Formation (Precambrian of northwest Scotland). *J. sedim. Petrol.* **35,** 366–380.

Selley, R. C. (1976a). *An introduction to sedimentology.* Academic Press, London.

Selley, R. C. (1976b). The habitat of North Sea oil. *Proc. geol. Ass. Lond.* **87,** 359–388.

Selley, R. C. (1978). *Ancient sedimentary environments*, 2nd edn. Chapman and Hall, London.

Sellwood, B. W. (1978). Shallow-water carbonate environments. In: Reading, H. G. (ed), *Sedimentary environments and facies.* Blackwell, Oxford, 259–313.

Sellwood, B. W. and Hallam, A. (1974). Bathonian volcanicity and North Sea rifting. *Nature* **252,** 27–28.

Sepkoski, J. J. (1976). Species diversity in the Phanerozoic: species–area effects. *Paleobiology* **2,** 298–303.

Sepkoski, J. J. (1978). A kinetic model of Phanerozoic taxonomic diversity. I. Analysis of marine orders. *Paleobiology* **4,** 223–251.

Shackleton, N. J. (1967). Oxygen isotope analyses and Pleistocene temperatures reassessed. *Nature* **215,** 15–17.

Shackleton, N. J. and Kennett, J. P. (1975). Paleotemperature history of the Cenozoic and the initiation of Antarctic glaciation: oxygen and carbon isotope analysis in DSDP sites 277, 279 and 281. *Init. Rep. Deep Sea Drilling Project* **29,** 743–755.

Shackleton, N. J. and Opdyke, N. D. (1977). Oxygen isotope and palaeomagnetic evidence for early Northern Hemisphere glaciation. *Nature* **270,** 216–219.

Shaw, A. B. (1964). *Time in stratigraphy.* McGraw-Hill, New York.

Shearman, D. J. (1966). Origin of marine evaporites by diagenesis. *Trans. Inst. Min. Metall.*, sect. B, **75,** 208–215.

Shearman, D. J. (1970). Recent halite rock, Baja California, Mexico. *Trans. Inst. Min. Metall.*, sect. B, **79,** 155–162.

Shearman, D. J. and Fuller, J. G. C. M. (1969). Anhydrite diagenesis, calcitisation, and organic laminites, Winnipegosis Formation, Middle Devonian, Saskatchewan. *Bull. Can. petrol. Geol.* **17,** 496–525.

Sheehan, P. (1977). Species diversity in the Phanerozoic—a reflection of labor by systematists? *Paleobiology* **3,** 325–328.

Sheldon, R. P. (1963). Physical stratigraphy and mineral resources of Permian rocks in western Wyoming. *Prof. Pap. U.S. geol. Surv.* no. 313-B.

Sheldon, R. P. (1964a). Paleolatitudinal and paleogeographic distribution of phosphorite. *Prof. Pap. U.S. geol. Surv.* no. 501–C, 106–113.

Sheldon, R. P. (1964b). Exploration for phosphorite in Turkey—a case history. *Econ. Geol.* **59,** 1159–1175.

Shelton, J. W. (1965). Trend and genesis of lowermost sandstone unit of Eagle Sandstone at Billings, Montana. *Bull. Am. Ass. petrol. Geol.* **49,** 1385–1397.

Shinn, E. A. (1968). Practical significance of birdseye structures in carbonate rocks. *J. sedim. Petrol.* **38,** 215–223.

Shinn, E. A. (1969). Submarine lithification of Holocene carbonate sediments in the Persian Gulf. *Sedimentology* **12,** 109–144.

Simberloff, D. (1974). Permo-Triassic extinctions: effects of an area on biotic equilibrium. *J. Geol.* **82,** 267–274.

Simpson, P. R. and Bowles, J. F. W. (1977). Uranium mineralisation of the Witwatersrand and Dominion Reef Systems. *Phil. Trans. R. Soc.* A**286,** 527–548.

Sleep, N. H. (1976). Platform subsidence mechanisms and eustatic sea level changes. *Tectonophysics* **36,** 45–56.

Sloss, L. L. and Speed, R. C. (1974). Relationship of cratonic and continental margin tectonic episode. In: Dickinson, W. R. (ed), *Tectonics and sedimentation.* Spec. Publ. Soc. econ. Paleont. Miner. no. 22, 89–119.

Smith, A. G. and Briden, J. C. (1977). *Mesozoic and Cenozoic paleocontinental maps.* Cambridge Univ. Press.

Smith, A. G., Woodcock, N. H. and Naylor, M. A. (1979). The structural evolution of a Mesozoic continental margin, Othris Mountains, Greece. *J. geol. Soc. Lond.* **136,** 589–603.

Spencer, A. M. (1971). Late Precambrian glaciation in Scotland. *Mem. geol. Soc. Lond.* no. 6.

Spooner, E. T. C. (1976). The strontium isotopic composition of sea water, and seawater-oceanic crust interaction. *Earth planet. Sci. Lett.* **31,** 167–174.

Stanley, K. O., Jordan, W. M. and Dott, R. H. (1971). New hypothesis of early Jurassic paleogeography and sediment dispersal for western United States. *Bull. Am. Ass. petrol. Geol.* **55,** 10–19.

Stanley, S. M. (1976). Ideas on the timing of metazoan evolution. *Paleobiology* **2,** 209–219.

Steel, R. J. (1974). New Red Sandstone floodplain and piedmont sedimentation in the Hebridean Province. *J. sedim. Petrol.* **44,** 336–357.

Steel, R. J. and Wilson, A. C. (1975). Sedimentation and tectonism (?Permo-Triassic) on the margin of the North Minch Basin, Lewis. *J. geol. Soc. Lond.* **131,** 183–202.

Stehli, F. G. (1970). A test of the Earth's magnetic field during Permian time. *J. geophys. Res.* **75,** 3325–3342.

Stevens, G. R. (1971). The relationship of isotopic temperatures and faunal realms to Jurassic–Cretaceous paleogeography, particularly in the S.W. Pacific. *J. R. Soc. N.Z.* **1,** 145–158.

Stewart, F. H. (1954). Permian evaporites and associated rocks in Texas and New Mexico compared with those of northern England. *Proc. Yorks. geol. Soc.* **29,** 185–235.

Stewart, F. H. (1963). Marine evaporites. *Prof. Pap. U.S. geol. Surv.* no. 440–Y.

Suess, E. (1906). *The face of the Earth*, vol. 2. Clarendon Press, Oxford.

Surlyk, F. (1978). Submarine fan sedimentation along fault scarps on tilted fault blocks (Jurassic–Cretaceous boundary, East Greenland). *Bull. Grønl. geol. Undersøg. Bull.* no. 128.

Surlyk, F. (1980). The Cretaceous–Tertiary boundary event. *Nature* **285,** 187–188.

Swett, K., Klein, G. de V. and Smit, D. E. (1971). A Cambrian tidal sand body—the Eriboll Sandstone of Northwest Scotland: an ancient-recent analog. *J. Geol.* **79,** 400–415.

Swift, D. J. P. (1969). Inner shelf sedimentation: processes and products. In: Stanley, D. J. (ed), *The new concepts of continental margin sedimentation: application to the geological record.* Am. Geol. Inst., Washington, DS-4-1–DS-4-46.

Tan, F. C. and Hudson, J. D. (1974). Isotopic studies of the palaeoecology and diagenesis of the Great Estuarine Series (Jurassic) of Scotland. *Scott. J. Geol.* **10,** 91–128.

Tanner, W. F. (1965). Upper Jurassic paleogeography of the Four Corners Region. *J. sedim. Petrol.* **35,** 564–574.

Tappan, H. (1968). Primary production, isotopes, extinctions and the atmosphere. *Palaeogeog., Palaeoclimatol., Palaeoecol.* **4,** 187–210.

Taylor, J. H. (1949). Petrology of the Northampton Sand Ironstone Formation. *Mem. geol. Surv. G.B.*

Teichert, C. (1958). Concepts of facies. *Bull. Am. Ass. petrol. Geol.* **42,** 2718–2744.

Thiede, J. and Van Andel, T. H. (1977). The palaeoenvironment of anaerobic sediments in the late Mesozoic South Atlantic Ocean. *Earth planet. Sci. Lett.* **33,** 301–309.

Thierstein, H. R. and Berger, W. H. (1978). Injection events in ocean history. *Nature* **276,** 461–466.

Thompson, R. W. (1975). Tidal flat sediments of the Colorado River delta. In: Ginsburg, R. N. (ed), *Tidal deposits.* Springer Verlag, Berlin and New York, 57–65.

Till, R. (1978). Arid shorelines and evaporites. In: Reading, H. G. (ed), *Sedimentary environments and facies.* Blackwell, Oxford, 178–206.

Towe, K. M. (1978). Early Precambrian oxygen: a case against photosynthesis. *Nature* **274,** 657–661.

Trümpy, R. (1960). Paleotectonic evolution of the Central and Western Alps. *Bull. geol. Soc. Am.* **71,** 843–908.

Turcotte, D. L. and Burke, K. (1978). Global sea-level changes and the thermal structure of the earth. *Earth planet. Sci. Lett.* **41,** 341–346.

Turner, G. C. and Walker, R. G. (1973). Sedimentology, stratigraphy and crustal evolution of the Archaean greenstone belt near Sioux Lookout, Ontario. *Can. J. Earth Sci.* **10,** 817–845.

Tyson, R. V., Wilson, R. C. L. and Downie, C. (1979). A stratified water column environmental model for the type Kimmeridge Clay. *Nature* **277,** 377–380.

Uffen, R. J. (1963). Influence of the Earth's core on the origin and evolution of life. *Nature* **198,** 143–144.

Vail, P. R. *et al.* (1977). Seismic stratigraphy and global changes of sea level. In: Payton, C. E. (ed), *Stratigraphic interpretation of seismic data.* Mem. Am. Ass. petrol. Geol. no. 26, 49–212.

Vakhrameev, V. A. (1964). Jurassic and early Cretaceous floras of Eurasia and the paleofloristic provinces of this period. *Tr. geol. Inst. Moscow* **102,** 1–263 (in Russian).

Valentine, J. W. (1969). Patterns of taxonomic and ecological structure of the shelf benthos during Phanerozoic time. *Palaeontology* **12,** 684–709.

Valentine, J. W. (1973). *Evolutionary paleoecology of the marine biosphere.* Prentice-Hall, New Jersey.

Van Andel, Tj. H. (1975). Mesozoic/Cenozoic calcite compensation depth and the global distribution of calcareous sediments. *Earth planet. Sci. Lett.* **26,** 187–194.

Van Andel, Tj. H., Heath, G. R. and Moore, T. C. (1975). Cenozoic history and paleoceanography of the central equatorial Pacific. *Mem geol. Soc. Am.* no. 143.

Van Andel, Tj. H., Thiede, J., Sclater, J. G. and Hay, W. W. (1977). Depositional history of the South Atlantic Ocean during the last 125 million years. *J. Geol.* **85,** 651–698.

Van Houten, F. B. (1964). Cyclic lacustrine sedimentation, Upper Triassic Lockatong Formation, Central New Jersey and adjacent Pennsylvania. *Bull. geol. Surv. Kansas* **169,** 495–531.

Van Houten, F. B. (1965). Composition of Triassic Lockatong and associated formations of Newwark Group, Central New Jersey and adjacent Pennsylvania. *Am. J. Sci.* **263,** 825–863.

Van Houten, F. B. (1973). Origin of red beds: a review. *Ann. Rev. Earth planet. Sci.* **1,** 39–61.

Van Houten, F. B. (1974). Northern Alpine molasse and similar Cenozoic sequences of southern Europe. In: Dott, R. H. and Shaver, R. H. (eds), *Modern and ancient geosynclinal sedimentation.* Spec. Publ. Soc. econ. Paleont. Miner. no. 19, 260–273.

Vermeij, G. J. (1978). *Biogeography and adaptation: patterns of marine life.* Harvard Univ. Press.

Vinogradov, A. P. (ed) (1967–69). *Atlas of the lithological-palaeogeographic maps of the U.S.S.R.* Ministry of Geology, Moscow, 4 vols.

Visher, G. S. (1969). Grain size distributions and depositional processes. *J. sedim. Petrol.* **39,** 1074–1106.

Von Brunn, V. and Hobday, D. K. (1976). Early Precambrian tidal sedimentation in the Pongola Supergroups of South Africa. *J. sedim. Petrol.* **46,** 670–679.

Waddington, C. J. (1967). Paleomagnetic field reversals and cosmic radiation. *Science* **158,** 913–915.

Walker, R. G. (1967). Turbidite sedimentary structures and their relationship to proximal and distal depositional environments. *J. sedim. Petrol.*, **37,** 25–43.

Walker, R. G. (ed) (1979). *Facies models.* Geoscience Canada, Reprint Series 1. Geol. Soc. Canada.

Walker, R. G. and Harms, J. C. (1972). Eolian origin of Flagstone beds, Lyons Sandstone (Permian), type area, Boulder County, Colorado. *Mountain Geol.* **9,** 279–288.

Walker, R. G. and Pettijohn, F. J. (1971). Archaean sedimentation: analysis of the Minnitaki basin, north-western Ontario. *Bull. geol. Soc. Am.* **82,** 2099–2130.

Walker, R. N., Muir, M. D., Diver, W. L., Williams, N. and Wilkins, N. (1977). Evidence of major sulphate evaporite deposits in the Proterozoic McArthur Group, Northern Territory, Australia. *Nature* **265,** 526–529.

Walker, T. R. (1967). Formation of red beds in ancient and modern deserts. *Bull. geol. Soc. Am.* **78,** 353–368.

Walter, M. R. (ed) (1976). *Stromatolites.* Elsevier, Amsterdam.

Walter, M. R., Buick, R. and Dunlop, J. S. R. (1980). Stromatolites 3,400–3,500 Myr old from the North Pole area, Western Australia. *Nature* **284,** 443–445.

Walther, J. (1894). *Einleitung in die Geologie als historische Wissenschaft.* Fischer Verlag, Jena, 3 vols.

Waterhouse, J. B. and Bonham-Carter, G. (1972). Permian paleolatitudes judged from brachiopod diversities. *Rep. 24th Int. geol. Congr.*, sect. 7, 350–361.

Webb, S. D. (1976). Mammalian faunal dynamics of the great American interchange. *Paleobiology* **2,** 220–234.

West, I. M. (1975). Evaporites and associated sediments of the basal Purbeck formation (Upper Jurassic) of Dorset. *Proc. geol. Ass. Lond.* **86,** 205–225.

West, R. G. (1979). *Pleistocene geology and biology*. 2nd edn, Longmans, London.

Whittington, H. B. and Hughes, C. P. (1972). Ordovician geography and faunal provinces deduced from trilobite distribution. *Phil. Trans. R. Soc.* B**263,** 235–278.

Will, H.-J. (1969). Untersuchungen zur Stratigraphie und Genese des Oberkeupers in Nordwestdeutschland. *Beih. Geol. Jb.* no. 54.

Williams, G. E. (1969). Characteristics and origin of a Precambrian pediment. *J. Geol.* **77,** 183–207.

Wilson, J. L. (1975). *Carbonate facies in geologic history.* Springer Verlag, Berlin and New York.

Windley, B. F. (1977a). *The evolving continents.* Wiley, New York.

Windley, B. F. (1977b). Timing of continental growth and emergence. *Nature* **270,** 426–428.

Wise, D. U. (1974). Continental margins, freeboard and the volumes of continents and oceans through time. In: Burke, K. and Drake, C. L. (eds), *The geology of continental margins.* Springer Verlag, Berlin and New York, 45–58.

Wolfe, J. A. (1978). A paleobotanical interpretation of Tertiary climates in the northern hemisphere. *Amer. Sci.* **66,** 694–703.

Zankl, H. (1971). Upper Triassic carbonate facies in the northern Limestone Alps. In: Müller, G. (ed), *Sedimentology of parts of Central Europe.* Guidebook VIII Int. sedim. Congr. Kramer, Frankfurt, 147–185.

Zankl, H. and Schroeder, J. H. (1972). Interaction of genetic processes in Holocene reefs off North Eleuthera Island, Bahamas. *Geol. Rdsch.* **61,** 520–541.

Ziegler, A. M., Scotese, C. R., Johnson, M. E., McKerrow, W. S. and Bambach, R. K. (1977). Paleozoic biogeography of continents bordering the Iapetus (pre-Caledonian) and Rheic (pre-Hercynian) oceans. *Spec. Publ. Milwaukee Public Mus., Biol. Geol.* no. 2, 1–22.

Ziegler, P. A. (1977). Geology and hydrocarbon provinces of the North Sea. *Geojournal* **1,** 7–32.

Name index

Italic figures refer to pages with relevant illustrations

Subject index

Italic figures refer to pages with relevant illustrations